Agricultural Intensification, Environmental Conservation, Conflict and Co-Existence at Lake Naivasha, Kenya

Africa-Europe Group for Interdisciplinary Studies

VOLUME 34

The titles published in this series are listed at *brill.com/agis*

Agricultural Intensification, Environmental Conservation, Conflict and Co-Existence at Lake Naivasha, Kenya

Edited by

Gerda Kuiper
Eric M. Kioko
Michael Bollig

BRILL

LEIDEN | BOSTON

Cover illustration: Greenhouses at Lake Naivasha, 2016, photo by Bernard Calas

Library of Congress Cataloging-in-Publication Data

Names: Kuiper, Gerda, editor. | Kioko, Eric, editor. | Bollig, Michael, editor.
Title: Agricultural intensification, environmental conservation, conflict and co-existence at Lake Naivasha, Kenya : contribution to a dialectical anthropology / edited by Gerda Kuiper, Eric Kioko, Michael Bollig.
Description: Leiden ; Boston : Brill, [2024] | Series: Africa-europe group for interdisciplinary studies, 1574-6925 ; vol. 34 | Includes index.
Identifiers: LCCN 2024020212 (print) | LCCN 2024020213 (ebook) | ISBN 9789004695412 (paperback) | ISBN 9789004695429 (ebook)
Subjects: LCSH: Sustainable agriculture–Kenya–Naivasha, Lake, Region. | Lake ecology–Kenya–Naivasha, Lake. | Naivasha, Lake, Region (Kenya)–Social conditions–21st century. | Naivasha, Lake, Region (Kenya)–Economic conditions–21st century. | Naivasha, Lake (Kenya)
Classification: LCC S473.K42 L3553 2024 (print) | LCC S473.K42 (ebook) | DDC 306.3/4909676271–dc23/eng/20240517
LC record available at https://lccn.loc.gov/2024020212
LC ebook record available at https://lccn.loc.gov/2024020213

Typeface for the Latin, Greek, and Cyrillic scripts: "Brill". See and download: brill.com/brill-typeface.

ISSN 1574-6925
ISBN 978-90-04-69541-2 (paperback)
ISBN 978-90-04-69542-9 (e-book)
DOI 10.1163/9789004695429

This book is printed on acid-free paper and produced in a sustainable manner.

Contents

PART 3
Agricultural Intensification in and around Naivasha

PART 4
Resource Conflicts and Co-Existence in and around Naivasha

PART 5
The Future of the Lake Naivasha Basin

Acknowledgements

This volume is the outcome of a workshop that brought a varied group of researchers together, all of whom have conducted extensive research in and about the Lake Naivasha area. Due to the on-going COVID-19 pandemic, the workshop was held online on October 5 and 6, 2020. It was hosted by the Global South Studies Centre (GSSC) of the University of Cologne. The workshop was initiated by and for researchers who had worked in three large research projects in the years before: *Resilience, Collapse and Reorganization in Social Ecological Systems of African Savannahs*, a Research Unit (FOR1501) funded by the German Research Foundation (DFG) between 2011 and 2018; the EU-Funded Marie Curie Innovative Training Network *Resilience in East African Landscapes* (REAL), which ran from 2013 to 2017; and the ongoing DFG-funded Collaborative Research Centre TRR 228 *Future Rural Africa: Future-Making and Social-Ecological Transformation*, which started in 2018. In addition, a good number of researchers with no affiliation to these projects were also invited to the workshop.

We first and foremost thank the contributors to this volume, who have been extremely dedicated and patient throughout the publication process. We in addition would like to thank the following workshop participants, who did not submit a chapter but provided valuable input during the workshop through presentations or moderation: David Anderson, Alexander Follmann, Susan Jepkemoi, Anna Lisa Ramella, Dorothea Schulz, and Nicole Wagner.

We are grateful to Marie Thomalla Arellano, Lena Fried, Angelika Mikusz, Youssef El Mouzakki, and Lisa Walterscheid for their hard work during various steps of the publication process, as they assisted us with copy-editing, formatting, and background research. We therewith also acknowledge the contribution of the DAAD Global Climate and Environment Center *Future African Savannas*, a project which four of the five mentioned research assistants were or are part of.

We are thankful to P.X. Amphlett for the thorough copy-editing of the book chapters and to M. Feinen for the creation of several maps for the volume. We acknowledge the financial support received through the CRC *Future Rural Africa* for these measures.

We furthermore would like to thank the anonymous peer reviewer who took on the difficult task of reviewing this eclectic volume. We also thank Brill's editor Franca de Kort, and Gregor Dobler and Manuel Ramos, the editors of the series this volume is part of, for enabling the publication of this volume.

Finally, we wish to extend our deepest appreciation to the people of Lake Naivasha and its environs, including research participants, members of Naivasha's diverse communities, and officials from state and non-state institutions, who have provided their hospitality to and shared their insights and knowledge with the many researchers who contributed to this volume. Without their interest in our work, this volume would not have come into existence.

Figures, Tables and Maps

Figures

Tables

Maps

Abbreviations

AR	Annual Report
BMU	Beach Management Unit
CSR	Corporate Social Responsibility
DFG	German Research Foundation
DFI	Development Financial Institution
EMCA	Environment Management and Coordination Act
EPRA	Energy and Petroleum Regulatory Authority
ESIA	Environmental and Social Impact Assessment
FGD	Focus-group Discussion
GCC	Global Commodity Chains
GDC	Geothermal Development Company
GOGA	Greater Olkaria Geothermal Area
GPN	Global Production Networks
HCD	Horticultural Crops Directorate
ICD	Inland Container Depot
ITCZ	Intertropical Convergence Zone
KFC	Kenya Flower Council
KFS	Kenya Forest Service
KMFRI	Kenya Marine and Fisheries Research Institute
KNA	Kenya National Archives
KSh	Kenyan Shilling
KWS	Kenya Wildlife Services
LANAWRUA	Lake Naivasha Water Resource Users Association
LNRA	Lake Naivasha Riparian Association
LNROA	Lake Naivasha Riparian Owners Association
MoE	Ministry of Energy
NEMA	National Environmental Management Agency
NK	Nature Kenya
NMK	National Museums of Kenya
NP	National Park
NT	National Treasury
PAP	Project-affected Person
PES	Payment for Ecosystem Services
RAP	Resettlement Action Plan
SEZ	Special Economic Zone
SEZA	Special Economic Zones Authority
SGR	Standard Gauge Railway

SPV	Special Purpose Vehicle
WAP	Water Allocation Plan
WRMA	Water Resources Management Authority
WRUA	Water Resource Users Association

Notes on Contributors

Anne Achieng
is a passionate anthropologist with a profound interest in the study of women and the diverse ways in which people lead their lives. With an ongoing MA in anthropology from the University of Nairobi, she has dedicated her career to unravelling the intricate connections between gender, culture, and society. Driven by a deep curiosity, she has conducted fieldwork in a remote community in Kenya, exploring the roles, challenges, and contributions of women in various circumstances. Her aim is to foster cross-cultural understanding and empower marginalised voices to drive positive social change.

Christine Adongo
holds an MSc in environmental sciences from Kenyatta University, Nairobi. Her research focuses on the anthropology of energy, green growth and climate change in Kenyan peripheries. She has been a research associate at the Institute of Ethnology and Social Anthropology at the *Ecole des Hautes Études en Sciences Sociales* (EHESS), Paris, under the Marie Curie ITN *Resilience in East African Landscapes* (*REAL*) project. She is currently finishing up her PhD in social anthropology, where she is exploring the anthropology of renewable energy interventions, environmental change and local adaptation in Naivasha.

Michael Bollig
is professor of social and cultural anthropology at the University of Cologne. His key interest is the environmental anthropology of Sub-Saharan Africa. His current research projects focus on the social-ecological dynamics of large-scale conservation projects, the commodification of nature and the political ecology of pastoralism. He holds the ERC advanced grant *REWILDING* and is the current speaker of the Collaborative Research Centre *Future Rural Africa: Future-Making and Social-Ecological Transformation*. He is also the author of *Risk Management in a Hazardous Environment* (2006), co-author of *African Landscapes* (2009) with O. Bubenzer, and co-editor of the volumes *Resilience and Collapse in African Savannahs* (2017) with D. Anderson and *African Futures* (2022) with C. Greiner and S. van Wolputte.

Bernard Calas
is a professor in geography at Bordeaux-Montaigne University, Researcher at the *Les Afriques dans le monde* (*UMR 5115 LAM*) research centre. He is specialised in economic and political geography with a focus on African emerging

metropolis and African integration in the global economy. Most of his research was carried in East Africa.

Johannes Dittmann
is a post-doctoral researcher at the University of Bonn (Department of Geography) and associated member of the Collaborative Research Centre *Future Rural Africa: Future-Making and Social-Ecological Transformation*. His key interests lie in development geography and political ecology of Sub-Saharan Africa. As part of his master thesis, he did qualitative social research on the politics fishery at Lake Naivasha in collaboration with Antony F. Ogolla. In his PhD he focused on the politics of transboundary conservation in northeastern Namibia. As part of his post-doctoral project, he is concerned with the relativity of failure and success in development and adaptation to climate change.

Selina Emmanuel
holds an Honours Bachelor of social sciences in international development and globalisation from the University of Ottawa and is currently completing her graduate studies in geography of environmental risks and human security at the United Nations University in Bonn, Germany. Nurturing her passion for equitable and climate-resilient human and natural systems, she is exploring urban food systems governance in her master's thesis. Previously, she has worked in development cooperation, focusing on human rights and environmental due diligence obligations of European textile companies and small-scale sustainable solid waste management.

Andreas Gemählich
is an expert for international trade and development with a focus on cut-flower production in East Africa and South America. For his PhD, which he pursued in the context of the research group *Resilience, Collapse and Reorganization in Social Ecological Systems of African Savannahs*, he conducted a study on the interrelations between the social-ecological system of Lake Naivasha and the global cut-flower market.

Giulia Silvia Giberti
graduated in natural sciences, with a PhD in applied botany and forest ecology and is currently working at the National Research Council of Italy. With her research she is investigating the effects of global changes on natural ecosystems, with a perspective range over vegetation community to individual plant species to improve knowledge on conservation biology and on ecosystems and

species resilience to global changes. She is affiliated to the Institute of Research on Terrestrial Ecosystems (IRET), National Research Council of Italy (CNR).

Annalia Gminder
studies geography with a special focus on globalisation and development at the University of Bonn, Germany. Her main interests include the fields of medical, development and economic geography with a regional focus on India and Eastern Africa. Through an internship with a development organisation, she has developed a passion for the topics of WASH (water, sanitation and hygiene) and menstrual health and hygiene and the way public institutions and communities deal with it. In her current master thesis, she combines her interests by writing about the WASH infrastructure in healthcare facilities as therapeutic landscapes in India.

Marie L. Gravesen
is a researcher at the Danish Institute for International Studies (DIIS) in the unit for Sustainable Development and Governance. Her overall work pertains to conflicts and management of natural resources and climate change adaptation with a focus on East and West Africa. At DIIS, she is currently working on the FFU funded GAP programme, focusing on the political economy of funds that are channelled into climate change adaptation in Kenya and Tanzania. She undertook her doctoral work at Cologne University under the Marie Curie ITN *Resilience in East African Landscapes* (*REAL*). Her doctorate work looked at the politics of land conflicts in Kenya's Laikipia County. This work has culminated into several academic publications, including her monograph *The Contested Lands of Laikipia - Histories of Claims and Conflict in a Kenyan Landscape* published by Brill in 2020.

David Harper
is a retired freshwater scientist, formerly professor in the University of Leicester Department of Biology. His major research was the ecology of the Rift Valley Lakes for 33 years, funded by the Earthwatch Institute, British Council, student field courses and grants for local PhDs. He worked in partnership with the University of Nairobi and the National Museums of Kenya. His team initially analysed lake ecology in relation to alien species (Lake Naivasha) and degree of alkalinity (different soda lakes) but both lines of research became linked to the impacts of human activities after the turn of this century. His collaboration with Social Scientists began with NERC funding on the *EAGLO* project led by the University of Nairobi and concluded with the international authors in this volume. David Harper is a Fellow of the Freshwater Biological Association, the Linnean Society and the Royal Geographical Society.

Dominic Kimani
holds a master's degree in wildlife management from the University of Eldoret and is currently pursuing a PhD in environmental biology at Karatina University in Kenya. He is head of the Biodiversity Department at Kipeto Energy, overseeing ecological research and mitigation efforts. His research experience focuses on biodiversity conservation, natural resources conservation, resource planning, wildlife management, sustainable agriculture, indigenous knowledge systems, livelihoods, and human ecology. He is also a resident research associate in the Ornithology section within the Zoology Department of the National Museums of Kenya.

Eric M. Kioko
is an environmental anthropologist (PhD-2016, University of Cologne). He currently lectures at Kenyatta University in the Department of Environmental Studies and Community Development and at the University of Bonn where he doubles as a Principal Investigator in the Collaborative Research Centre *Future Rural Africa: Future-Making and Social-Ecological Transformation*. His area of research and teaching focuses on the dynamics of human-environment relations and his interests include environmental conflicts and peacebuilding, environmental crimes (focusing on forest crimes in Africa), green futures, large-scale infrastructures and social-ecological change, and dynamics of community-based conservancies. He has published several book chapters as well as articles in journals such as *Africa, Africa Spectrum and The European Journal of Development Research*.

Kariuki Kirigia
is an assistant professor of climate change and sustainability in Africa with joint appointment the School of the Environment and African Studies Centre at the University of Toronto. Previously, he completed a postdoctoral fellowship at the Concordia Ethnography Lab and earned his PhD in anthropology from McGill University. His research focuses on biodiversity conservation, climate change and sustainability, land governance, local and Indigenous livelihoods, environmental and epistemic justice, racial capitalism, and decolonisation in the African context and the Global South writ large. His teaching underscores decolonial methods and experiential learning that foster critical understanding of theory and practice.

Gerda Kuiper
is a cultural anthropologist based at the University of Cologne. She pursued her PhD in the research group *Resilience, Collapse and Reorganization in Social Ecological Systems of African Savannahs*. She has extensive research experience

in Kenya and Tanzania, resulting in the book *Agro-industrial Labor in Kenya. Cut Flower Farms and Migrant Workers' Settlements* (Palgrave Macmillan, 2019) and in publications in the *Journal of Eastern African Studies, Geoforum* and the *Journal of Contemporary African Studies*, among others. She is moreover dedicated to interdisciplinary teaching and international cooperation. She is scientific coordinator for the DAAD Global Climate and Environment Center *Future African Savannas*, in which she cooperates with partners in Germany, Kenya and Côte d'Ivoire.

Edward Morrison
is an applied ecologist based in Kenya. His PhD research examined the ecological restoration of papyrus wetlands at Lake Naivasha. He acts as a consultant for agribusinesses, conservation groups and development agencies working with natural resources in East Africa. His interests include agroecology, sustainable soil and water management, ecological restoration and nature-based solutions for climate change. He is a former Knox Fellow of Harvard University and an Honorary Research Fellow at the University of Exeter.

Marie Müller-Koné
is PhD-candidate in development studies at the University of Bonn, Germany, and Senior Researcher with the independent Bonn International Centre for Conflict Studies (BICC), which she joined in 2009. She obtained a Master's degree in global political economy from Sussex University, UK, in 2008. Her research foci have been resource conflicts, the politics of land access in Africa, ethnicisation and violence at frontiers, which she published on in journals such as *World Development* and *Journal of Eastern African Studies*. Her doctoral research on cross-scalar *dispositifs* of discourses of autochthony and indigeneity in Africa is part of the Collaborative Research Centre *Future Rural Africa: Future-Making and Social-Ecological Transformation.*

Detlef Müller-Mahn
is professor of development geography at the University of Bonn, and former spokesperson of the Collaborative Research Centre *Future Rural Africa: Future-Making and Social-Ecological Transformation*. His present research focuses on the political ecology of land-use change, future-making, and green development in East Africa. Recent projects addressed the "riskscapes" of climate change in Africa, pastoralism and hydro-development in Ethiopia, urban water management in Sudan, and the cut-flower industry of Lake Naivasha in Kenya.

Timothy Mwinami
holds a BSc and an MSc in animal genetics from the University of Nairobi. He is a research scientist at the National Museums of Kenya, Ornithology Section

of the Zoology Department. Currently he works as a research fellow on the *Important Bird Areas* (IBAS), a collaborative programme between the Museum and Nature Kenya.

Chigozie Nweke-Eze

holds an MA in development economics from the University of Erlangen-Nuremberg. He carries out research in the fields of energy, electricity and infrastructure development and geographies in Africa. After a research stay at the University at Bayreuth, he became a research associate at the Institute of Geography, University of Bonn, under the Collaborative Research Centre *Future Rural Africa: Future-Making and Social-Ecological Transformation*. He is currently finishing his PhD in Economic and Energy Geography, where he is exploring the financing, governance and infrastructure geographies of large-scale renewable energies development in Kenya.

Antony F. Ogolla

did his PhD as part of the Collaborative Research Centre *Future Rural Africa: Future-Making and Social-Ecological Transformation* at the University of Bonn (Department of Geography). His key interests lie in development geography and the green economy in Eastern Africa. As part of his master thesis, he did qualitative social research on the politics of fishery at Lake Naivasha in collaboration with Johannes Dittmann. In his PhD he focused on the politics of green growth in rural Kenya.

Nic Pacini

is a hydrobiologist/hydrochemist with experience in the study of Afrotropical ecosystems. He lectured at the Universities of Leicester (UK) and Calabria (Italy) and conducted research in Kenya in the Upper Tana and in the Naivasha Basin since 2009. He was international expert in institutional twinning projects concerning water quality management in Morocco and Egypt, and national Italian environmental expert during G8 and G20 negotiations. He co-authored 90 scientific articles, book chapters and reports, acts as board member of the *Integrated Environmental Assessment and Management* (SETAC) and as subject editor of the *African Journal of Ecology*. Currently he is Assistant Professor at the Department of Environmental Engineering of the University of Calabria, Italy, and Honorary Fellow at the School of Geography, Geology and the Environment, University of Leicester (UK).

Sylvain Racaud

is a senior lecturer in geography at Bordeaux-Montaigne University, researcher at the *Les Afriques dans le monde* (UMR 5115 LAM) research centre and co-director of the journal *Suds* (previously *Les Cahiers d'Outre-Mer*). He works on the geography of trade routes in Africa (low-prices goods, food) and the integration of rural-urban relations into the global economy. His research activities focus on East Africa and Cameroon.

Gaële Rouillé-Kielo
is a social and environmental geographer attached to the *Laboratoire Interdisciplinaire Sciences Innovations Sociétés* (*UMR 9003 LISIS*) research laboratory in France. Her work focuses on the circulation of models of water governance and conservation across places and actor networks. She has conducted research in Kenya, Paris and Berlin. Her doctoral research examined the translation of the concept of 'payment for watershed services' in the Naivasha region. Most of her fieldwork was carried out on the Kinangop Plateau.

Kristin Schmit
has already been able to broaden her understanding of development studies with a bachelor's degree in politics and an ongoing master's degree in development geography at the University of Bonn. More recently, she developed a keen interest for topics at the interface of development, agriculture, food systems and climate change. While preparing her master thesis on risks and drivers of maladaptation to climate change, she is striving for a rapprochement of research and practice in the field of development cooperation. She hopes that her enthusiasm for cross-cultural exchange and learning will continue to guide her academic and professional curriculum.

Megan A. Styles
is an associate professor of environmental studies at the University of Illinois Springfield. She holds a PhD in environmental anthropology from the University of Washington Seattle. She has conducted ethnographic research on the cut-flower industry in Naivasha since 2004 and is the author of *Roses from Kenya: Labor, Environment, and the Global Trade in Cut Flowers* (University of Washington, 2019). Her current research focuses on sustainable agricultural development in the United States and East Africa.

Caroline Upton
holds a PhD in geography from the University of Cambridge and is professor of environmental human geographies at the University of Leicester, UK. Her work explores environmental change and justice, resilience, sustainable livelihoods, conservation practices, and politics in diverse geographical and societal rural contexts. She has worked extensively with pastoralist communities in Kenya and Mongolia, as well as with smallholders in Kenya and South-East Asia,

including exploration of diverse understandings and ontologies of nature and environment and their relationships with identity, livelihoods and land use.

Richard Waller
received his doctorate from Cambridge University in 1979. He is now retired from university teaching but remains an active member of the Africanist scholarly community. He has been researching and writing about the Maa-speaking communities in East Africa in the nineteenth and twentieth centuries for over fifty years; some of his work is referenced in the chapter in this volume. In addition, he has written on pastoralism more generally and on the legal history of Kenya.

Silas Wanjala
is the General Manager of Lake Naivasha Riparian Association who holds a Master in Cooperation and Development from the University of Pavia and IUSS, Italy. He is a socio-ecologist who has worked around Lake Naivasha Basin for over 10 years and has vast hands-on exposure and experience on socio-ecological dynamics within the Lake Naivasha system. He has participated in and supported a number of scientific and social research programs on Lake Naivasha linked to ecological, hydrological, social, cultural and economic facades and participated in various conservation and ecological monitoring programs.

CHAPTER 1

Lake Naivasha, Kenya, in the Capitalocene

Gerda Kuiper, Eric M. Kioko and Michael Bollig

Lake Naivasha – today an agro-industrial hub, protected wetland area, tourist attraction and site for geothermal energy production – is singular for several reasons.[1] It is one of the few freshwater lakes in the East African Rift Valley, enabling a kaleidoscope of economic activities dependent on the locally available natural resources (Mavuti and Harper 2006). The economic sector that has attracted the most attention is the export-oriented flower and vegetable industry. With the establishment of this extremely successful industry in the 1970s and 1980s, the region around Lake Naivasha has transformed from an economy based on extensive livestock-keeping that developed during the colonial period into an important site of agricultural intensification (see Map 1.1). The farms' need for labour also resulted in an upsurge in human population, attracting labour migrants from different ethnic backgrounds from across Kenya and creating patches of dense unplanned urban settlements in this previously rural area. The most recent national census counted 355,383 residents in Naivasha Sub-County (KNBS 2019).

Remarkably, this labour-intensive agro-industry coexists with a large domestic and international tourism sector, a partly informalised fisheries sector, and small- and medium-scale subsistence and commercial farming in the lake's hinterland. Naivasha Town (see Map 1.1) is moreover an important transit town and stopover for trucks ferrying import and export cargo along the Mombasa-Kampala Road. In addition, the geothermal energy sites developed in the hinterland since the 1980s provide a globally relevant case study of the workings and consequences of green energy production (Harper et al. 2011; Kuiper 2019; Styles 2019).

In addition to sustaining these industries, which are of key importance to the local and national economy, Lake Naivasha and its surroundings have also gained international recognition as a site of crucial ecological importance.

1 First submission date: November 7, 2022. Acceptance date: March 13, 2023. As the process of putting this edited volume together has taken several years and as the Lake Naivasha area is transforming at a rapid pace, we have decided to include submission and acceptance dates in the chapter in order to clearly indicate at which time the draft chapters were originally written.

 | DOI:10.1163/9789004695429_002

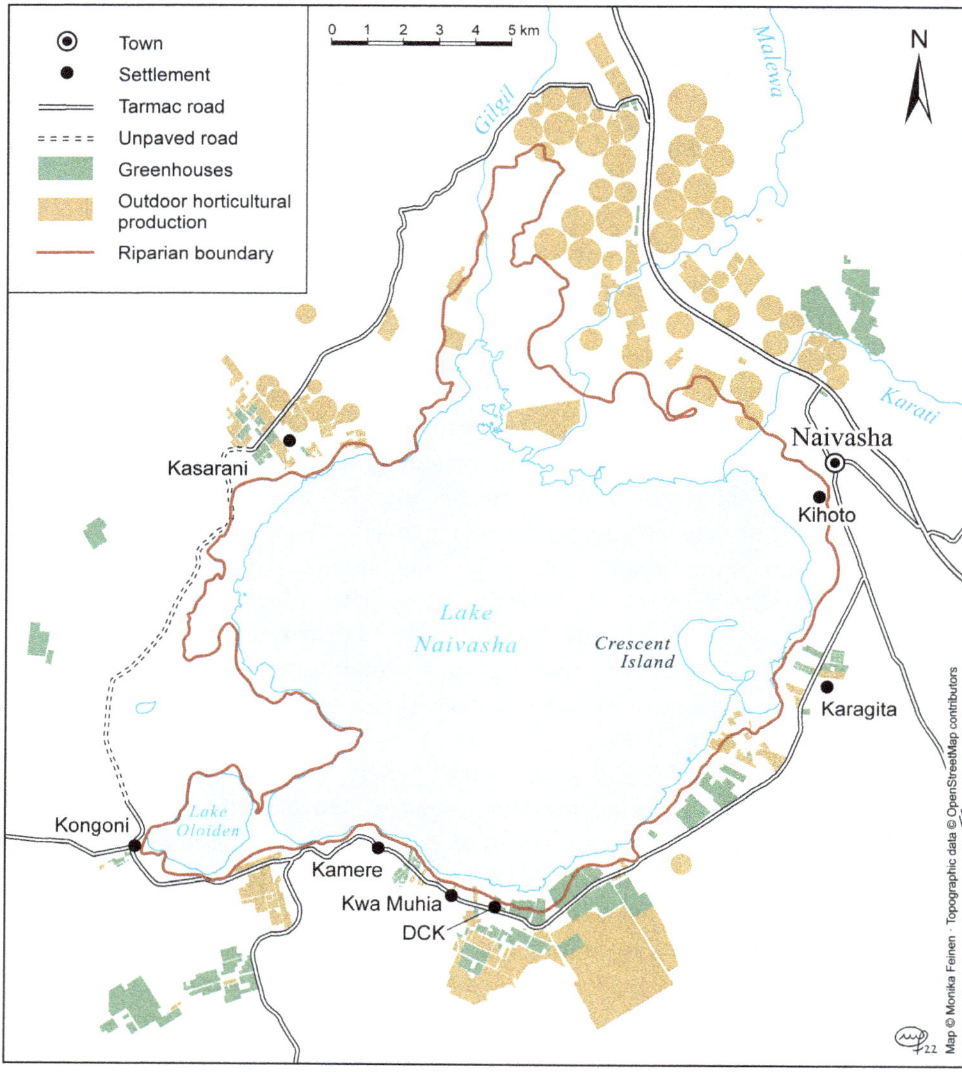

MAP 1.1 Lake Naivasha and its immediate surroundings, including riparian boundaries
CARTOGRAPHER: M. FEINEN

Several private conservancies as well as Hell's Gate National Park are located in the vicinity of the lake. The natural beauty of the lake and its (relative) abundance of wildlife and birds has figured prominently in global conservation imaginaries, underlined by its recognition as a wetland site of international importance under the Ramsar Convention in 1995 (Mavuti and Harper 2006, 30).

Thus, several major global processes and challenges – such as globalised agricultural intensification, biodiversity conservation, and natural-resource extraction – simultaneously manifest themselves in this one location. This kaleidoscope of activities has inevitably led to numerous contestations and – sometimes violent – conflicts over natural resources such as land and water. Some of these conflicts have long roots. Most notorious is the outburst of post-election violence in early 2008, which was partly rooted in, what Kanyinga (2009) called, "the ethnicised land question" in the Rift Valley. The more recent increase in diverse economic activities moreover have led to concerns about far-reaching ecological change, and have sustained large differences in wealth among the basin's residents, who all have varying aspirations for the future (their own and that of Naivasha). The area is thus characterised by an ever-present sense of social and ecological risk and crisis. At the same time, these different economic sectors and the extremely varied human population have until now managed to coexist remarkably well.

1 A "Nerve Centre" in the Capitalocene

Naivasha has been aptly described as a nerve centre. Megan A. Styles (2019, 8) elaborated on this characterisation, coined by Kenyan journalist Peter Muiruri. Styles argued that Naivasha is a nerve centre in three ways: as a politicised place that "touches nerves" in global debates on the ethicality of the horticultural industry and in national political contestations over land and other resources; as a national development hub with global connections; and as a testing ground for new forms of governance, such as the turning of private landowners into environmental stewards in the colonial period, and processes of devolution and high levels of self-regulation by the horticulture industry more recently. This characterisation of Naivasha as a nerve centre highlights its importance as a site for global capitalist expansion and neoliberalisation, but also shows that Naivasha is a complex space, providing opportunities for collaboration and for a search for alternative futures (Styles 2019, 7–8).

The concept of the "Capitalocene" can be helpful in making sense of such a manifestation of the power of global capitalism while taking into account people's agency in shaping their environment. Jason Moore (2017) has pointed out that the concept of the Anthropocene – now widely used by scholars from a wide variety of disciplines to understand the multiple, interlinked social-ecological crises of the present era – overlooks inequalities and differences between people by portraying a homogeneous "man-kind" as the culprit and by ignoring colonial histories. Moore therefore suggests to take up the concept of the Capitalocene, which he defines as "the historical era shaped by the endless accumulation of capital" (Moore 2017, 595). He moreover argues against the dualism behind the concept of the Anthropocene, which tends to portray humanity and nature as separate entities. Moore in contrast perceives humans as part of the "web of life", transforming the environment but also being shaped by the environment. Capitalism in this understanding comprises more than the economy and has revolved around producing "Cheap Nature" on a global level for many centuries.

The three aspects of Lake Naivasha as a nerve centre – its political character, its role in development and development imaginaries, and its function as a testing ground for neoliberal forms of government – all clearly relate to global capitalism's transformation of Naivasha's environment since the beginning of the twentieth century. The chapters in this volume exemplify the multiple, intertwined socio-environmental crises that consequently have played out in Naivasha in the past and the present and that continue to shape its future. We argue that the concept of the Capitalocene can be helpful in making sense of these crises, which have global origins and global reverberations.

2 Interdisciplinary Syntheses

Another remarkable feature of Lake Naivasha is that it is a true hotspot for research. The lake has attracted the attention of scientists from a wide range of disciplines. The initial focus of this body of research was on its unique ecological features, broadly categorised into fauna and flora. Studies into the lake's hydrology and ecosystem have been highly influential, both scientifically and in shaping global and local debates on Naivasha's ecological future (e.g. Becht and Harper 2002; Gherardi et al. 2011; Mavuti and Harper 2006; Stoof-Leichsenring et al. 2011; Verschuren et al. 1999). The last decade or so has seen the entry of the social sciences into an ecosystem previously dominated by the natural sciences, which has contributed to a clearer understanding of the complex and dynamic social-ecological system. Worth nothing are three large interdisciplinary research projects: *Resilience, Collapse and Reorganization in Social Ecological Systems of African Savannahs*, a Research Unit funded by the German Research Foundation (DFG) between 2011 and 2018; the EU-Funded Marie Curie Innovative Training Network *Resilience in East African Landscapes* (*REAL*), which ran from 2013 to 2017; and the ongoing DFG-funded Collaborative Research Centre *Future Rural Africa: Future-Making and Social-Ecological Transformation*. These research projects have over time brought together scientists from Europe and Africa, including geographers, anthropologists, political scientists, agro-economists, soil scientists and ecologists, all interested in the past, present and future of, among other sites, the Lake Naivasha Basin (see Map 1.2).

This volume is the first attempt to bring together interdisciplinary research from these projects, as well as from other research projects carried out in the

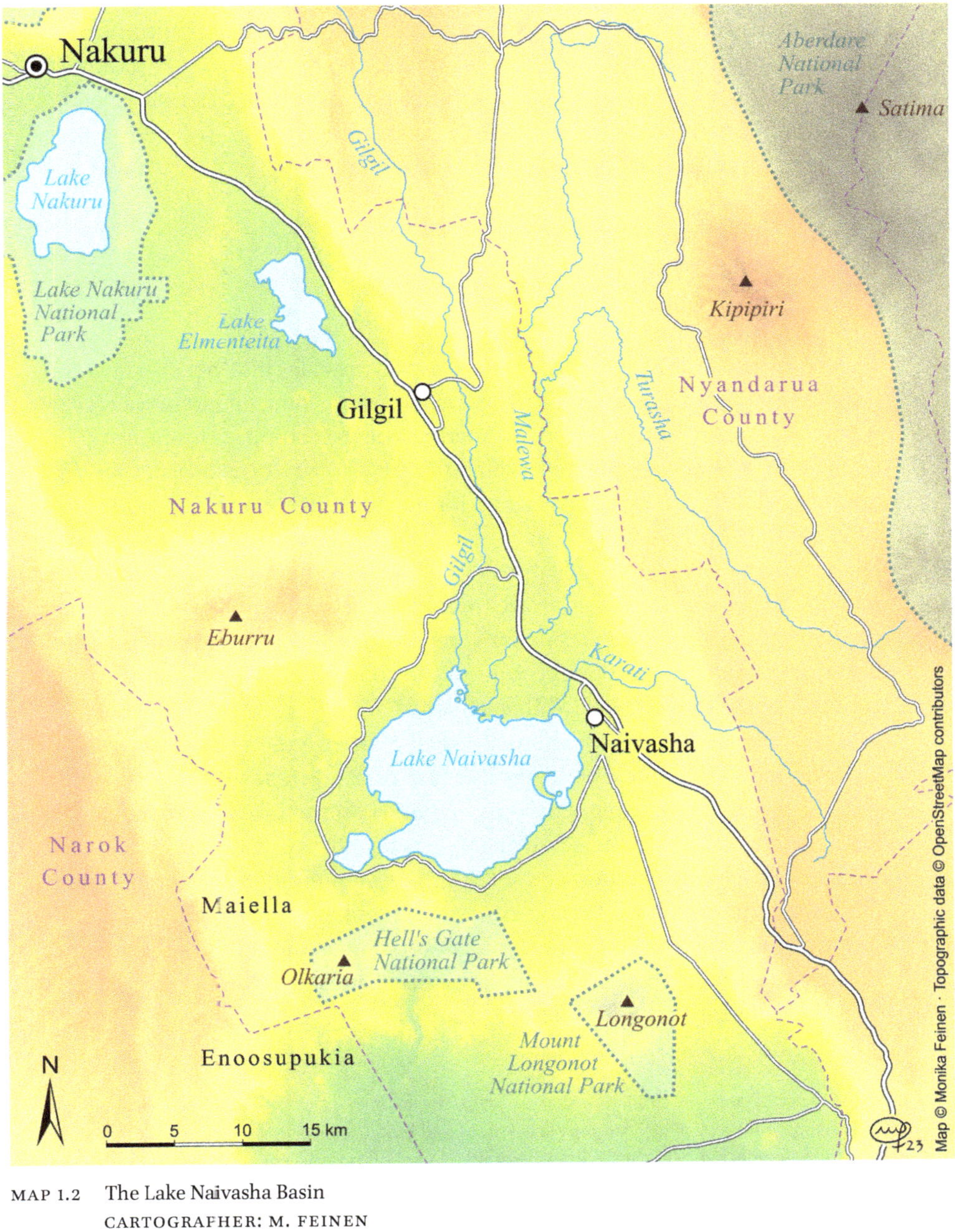

MAP 1.2 The Lake Naivasha Basin
CARTOGRAPHER: M. FEINEN

basin, into one publication. The volume connects history to the present and the future, providing an in-depth and comprehensive analysis of past and current processes of agricultural intensification and of changing use and availability of ecological resources at the unique freshwater lake Naivasha in Kenya and its surroundings. The collection thus provides insights into Lake Naivasha itself, but also into intertwined global and local processes of agricultural intensification, environmental conservation, and changing resource use. It aims to provide an example of how one can bring findings from different scientific backgrounds together in order to gain a better understanding of environmental change and its social drivers and consequences.

3 Outline of the Volume

3.1 *Naivasha's History*

The first section of the volume is historical and provides insights into how Naivasha turned from an – at first sight – largely "empty" area into a "nerve centre", where economic, social and environmental interests compete for priority.

Richard Waller's chapter on the settlement history of Naivasha between 1790 and 1912 describes the arrival and formation of the Maa community in the nineteenth century and its replacement by European settlers in the early twentieth century. The chapter illustrates that Naivasha's ecology and geographic location make it a place of strategic (political and economic) importance. The chapter moreover sets the scene for some of the subsequent chapters, which describe contemporary repercussions of the colonial removal of the Maa people and the creation of the "White Highlands".[2]

In Chapter 3, freshwater scientist David Harper and cultural anthropologist Megan A. Styles analyse consequences of the establishment of colonial settlers' farms and of later economic activities on the lake's ecosystem. They furthermore chronicle the often highly contested conservation activities in the area. These conservation activities have been dominated by foreign investors and landowners of foreign descent. The chapter thus shows that the protagonists in both the colonial and post-colonial capitalist exploitation of Naivasha's natural resources and in attempts to preserve these resources were often members of one and the same group of (non-African) people.

2 An area of some 3 million hectares, primarily located in the Rift Valley Province, was scheduled for exclusive European settlement and land use during the colonial period, and is routinely referred to as the "White Highlands" by scholars and Kenyans alike. The restriction on African land possession was lifted only in 1960 (Odingo 1971).

The final chapter on the history of Lake Naivasha, by Gerda Kuiper and Marie L. Gravesen, argues that more attention is needed to understand the role of African actors in shaping Naivasha's current landscape, which can be considered to be deeply anthropogenic. Their chapter focuses on land-buying companies that purchased plots of land after Kenya gained independence. After the establishment of the horticultural industry, these privately-owned plots of land developed into dense workers' residential areas, infamous for their cramped living conditions, poor services, and tense social relations. This chapter thus highlights the impact of both global capital (through the attraction of labour for the farms) and individual small-scale landowners on Naivasha's present-day urban environment.

3.2 *Ecological Dynamics in the Lake Naivasha Catchment*

The second section of the volume elaborates on Naivasha's ecology. The first chapter in this section is written by a multidisciplinary team of authors. David Harper, Nic Pacini, Edward Morrison, Dominic Kimani, Timothy Mwinami, and Caroline Upton highlight the lake's unique ecological features, such as its fresh water, the large fluctuations in lake water levels, and the rich diversity in flora and fauna, including local perceptions on these ecological features. The chapter furthermore discusses the influence of the introduction of alien species and of human agriculture and settlement on the lake's deteriorating ecosystem, most clearly evidenced for the period since the 1970s. The chapter urges for a better understanding and more awareness of the basic principles behind Lake Naivasha's natural resources, questioning whether its unique ecological features can survive.

The second chapter in this section – by Nic Pacini, Giulia Giberti, and David Harper – focuses on the upper catchment area , and highlights how deterioration of the ecosystem upstream impacts the provision of ecosystem services closer to the lake. The chapter shows that the impact of local climate change on evaporation and temperatures through deforestation and degradation can presently be of much more significance than the impact of global climate change. Both chapters thus illustrate the local, detrimental ecological effects of Naivasha's entrance into the Capitalocene.

3.3 *Agricultural Intensification in and around Naivasha*

The volume's third section highlights the social dynamics related to the establishment of export-oriented horticultural industries at Lake Naivasha as well as the increase in smallholder production for export in the lake's hinterlands.

The first chapter in this section, by economic geographers Detlef Müller-Mahn and Andreas Gemählich, focuses on global entanglements and local structures in the Naivasha cut-flower industry. It describes the structure of international market relations in this industry, which has changed over the years, for instance due to an increasing awareness among European consumers about the origin of the flowers. The chapter furthermore discusses their impact of these changing international market relations on the local organisation of flower production.

The subsequent chapter by Gerda Kuiper describes dynamics in labour relations within the same industry. The chapter explores the role of mutual expectations between flower-farm workers and managers – the local flower industry's "moral economy" – in maintaining the industry's relatively stable and peaceful labour relations. Kuiper observes that the industry's specific shape is largely the outcome of interactions between groups of people who all consider themselves to be only temporary residents of Naivasha. The chapter thus highlights the importance of myriad translocal and global connections for the constitution of labour relations and for the success of the horticultural industry in Naivasha.

The next chapter in this section, by Kariuki Kirigia, also focuses on horticultural farm workers. Scrutinising the assumption that global firms bring local development through providing employment, Kirigia probes the relations between the cut-flower industry and food (in)security among workers. The chapter shows the difficulties workers have with becoming food secure and describes how workers turn to their social networks to fulfil their households' food needs.

Finally, the chapter by Gaële Rouillé-Kielo, Bernard Calas, and Sylvain Racaud moves away from the lake and describes the development of export smallholder production on the Kinangop Plateau, east of Lake Naivasha, since the early 2000s. In contrast to the horticultural industry on the lakeshores, this production for export was not initiated by large foreign investors. It formed an opportunity for smallholder farmers to diversify their income-generating activities. The chapter describes how farmers try to negotiate the terms and conditions of contract farming, in a context of increasing land scarcity.

3.4 *Resource Conflicts and Co-existence in and around Naivasha*

The following section focuses on the management of natural resources and how this results in both conflict and coexistence. The first chapter in this section focuses on the highly contested resource of land. Marie Müller-Koné and Eric M. Kioko describe negotiations over land access in the settlements of Enoosupikia and Maiella, on the western side of the lake. This area is one of the main areas of food production for Naivasha's urban centre. The chapter highlights both conflicts over land and ways in which people manage to coexist in a context of increasing competition over land.

The next chapter in this section, by Johannes Dittmann and Anthony F. Ogolla, describes the highly contested fisheries industry at Lake Naivasha. The chapter interrogates the influence of the devolution of governance and the creation of counties in Kenya on the management of fisheries in Naivasha. The authors conclude that increased political interference in Naivasha's fisheries sector has led to the intensification – rather than the diminishing – of illegalised fishing practices, to an increased ethnicisation of fisheries with potential for more conflict, and to a weakened governance system.

In the final chapter in this section, Chigozie Nweke-Eze and Christine Adongo discuss the resettlement of Maasai communities south-west of the lake in the context of geothermal energy extraction. The chapter highlights how global and national visions of geothermal energy as sustainable and as a "green" alternative for the future clash with local experiences of expropriation, biodiversity loss and socio-cultural decay. The chapter highlights the inherent conflict between national-level development visions and local socio-economic and ecological consequences, which even "green" extractive industries do not seem to be able to avoid.

3.5 *The Future of the Lake Naivasha Basin*

The final section provides an outlook on Naivasha's future. The first chapter describes the recent effects of the COVID-19 pandemic on the cut-flower industry. The chapter is co-authored by Eric M. Kioko (Kenyatta University) and Silas Wanjala (scientific officer of the Lake Naivasha Riparian Association), together with four geography students – Anne Achieng, Selina Emmanuel, Annalia Gminder, and Kristin Schmit – who conducted a brief research project on the topic in September 2022. The chapter also hints at some the implications for the industry's future at Lake Naivasha.

The final chapter of the volume, also written by Eric M. Kioko and Silas Wanjala, discusses several infrastructural and industrial "mega-projects" that are currently in the planning phase or in the early stages of implementation. Examples are a large dam in one of Lake Naivasha's feeding rivers and the establishment of an industrial inland railway port. If fully implemented, these projects will have even more far-reaching implications for Naivasha's social-ecological system, on top of the consequences of the developments so far, as described in this volume.

These fourteen chapters together contribute to the ever more pressing debates on the sustainability of current economic activities in the Lake Naivasha Basin. These include debates in the Global North about the environmental and social impact of the floricultural complex, and more local debates about the impact of further development plans on the ecological state of the lake and its capacity to sustain the ever-increasing human population surrounding it. By bringing different scientific viewpoints together, the volume contributes new and vital insights to these public debates. Current development plans, outlined in the final chapter, make some of the previous crises Naivasha went through seem rather insignificant. These projected developments indicate the urgency of compiling knowledge on this place, which is singular in its diversity.

References

Becht, Robert, and David M. Harper. 2002. "Towards an Understanding of Human Impact upon the Hydrology of Lake Naivasha, Kenya." *Hydrobiologia* 488: 1–11. https://doi.org/10.1023/A:1023318007715.

Gherardi, Francesca, Robert J. Britton, Kenneth M. Mavuti, Nic Pacini, Jonathan Grey, Elena Tricarico, and David M. Harper. 2011. "A Review of Allodiversity in Lake Naivasha, Kenya: Developing Conservation Actions to Protect East African Lakes from the Negative Impacts of Alien Species." *Biological Conservation* 144 (11): 2585–96. https://doi.org/10.1016/j.biocon.2011.07.020.

Harper, David M., Edward H.J. Morrison, Michael M. Macharia, Kenneth M. Mavuti, and Caroline Upton. 2011. "Lake Naivasha, Kenya: Ecology, Society and Future." *Freshwater Reviews* 4 (2): 89–114.

Kanyinga, Karuti. 2009. "The Legacy of the White Highlands: Land Rights, Ethnicity and the Post-2007 Election Violence in Kenya." *Journal of Contemporary African Studies* 27 (3): 325–44. https://doi.org/10.1080/02589000903154834.

KNBS. 2019. *Kenya Population Census 2019*. Nairobi: KNBS.

Kuiper, Gerda. 2019. *Agro-Industrial Labour in Kenya: Cut Flower Farms and Migrant Workers' Settlements*. Cham: Palgrave Macmillan.

Mavuti, Kenneth M., and David M. Harper. 2006. "The Ecological State of Lake Naivasha, Kenya, 2005: Turning 25 Years Research into an Effective Ramsar Monitoring Programme." In *Proceedings of the 11th World Lakes Conference*, edited by Eric Odada and Daniel O Olago, 2: 30–34. Accessed September 30, 2023. https://aquadocs.org/bitstream/handle/1834/2127/WLCK-30-34.pdf?sequence=1.

Moore, Jason W. 2017. "The Capitalocene, Part I: On the Nature and Origins of Our Ecological Crisis." *The Journal of Peasant Studies* 44 (3): 594–630. https://doi.org/10.1080/03066150.2016.1235036.

Odingo, Richard S. 1971. *The Kenya Highlands: Land Use and Agricultural Development*. Nairobi: East African Publishing House.

Stoof-Leichsenring, Kathleen R., Annett Junginger, Lydia A. Olaka, Ralph Tiedemann, and Martin H. Trauth. 2011. "Environmental Variability in Lake Naivasha, Kenya, over the Last Two Centuries." *Journal of Paleolimnology* 45 (3): 353–67. https://doi.org/10.1007/s10933-011-9502-4.

Styles, Megan A. 2019. *Roses from Kenya: Labor, Environment, and the Global Trade in Cut Flowers*. Seattle: University of Washington Press.

Verschuren, Dirk, John Tibby, Peter R. Leavitt, and C. Neil Roberts. 1999. "The Environmental History of a Climate-Sensitive Lake in the Former 'White Highlands' of Central Kenya." *AMBIO: A Journal of the Human Environment* 28: 494–501.

PART 1

Naivasha's History

CHAPTER 2

Arrivals and Departures: The Maa-Speakers and Their Successors in Naivasha-Nakuru, 1790–1912

Richard Waller

This chapter looks at the settlement history of the Naivasha-Nakuru area and its environs from the early nineteenth century to the beginnings of white settlement a century later.[1] It focuses on Il Kinopop, a Maa-speaking group which included those who would later become the three northern Maasai sections Purko, Keekonyukie and Damat.[2] The chapter follows the emergence of this community and its identity during the nineteenth century and its replacement in the area by white settlers post-1904. The period of emergence is bracketed by two natural and environmental disasters: the gradual shift in the later eighteenth century from a wetter to a drier rainfall regime, culminating in a devastating regional drought around 1800, and the rinderpest pandemic in the 1890s, preceded by an outbreak of bovine pleuropneumonia in the early 1880s. The period ends, however, with a socio-political crisis caused by the arrival of white settlers and the removal of the Maasai to make way for them. Weak regulation of the allocation of land undermined government authority over white settlement. The two Maasai Moves of 1904 and 1912 accelerated the break-up of the larger Maasai community into separate territorial sections and, in effect, created "traditional Maasai society" to parallel the emergence of a landed settler society.

1 Nakuru-Naivasha and the Maasai before 1880

The Naivasha-Nakuru area has occupied a central place in the larger systems of human settlement in and around the Rift Valley for centuries. Pastoral Iron Age sites near Nakuru suggest that the area supported specialised subsistence pastoralism from early in the second millennium, centuries before the arrival of Maa-speakers (Sutton 1987). Naivasha-Nakuru may have been where "Maasai"

1 First submission date of this chapter: January 21, 2021. Acceptance date: March 13, 2023.

2 Il Kinopop (known to early administrators as the "Naivasha Masai") took its name from the Kinangop Plateau where its ceremonial centre was located.

 | DOI:10.1163/9789004695429_003

first began to emerge and from where they expanded (Galaty 1993, 68–75). Environmentally, the area was ideal for specialised pastoralism. The resource base was rich and varied, with permanent water at Naivasha, dry season grazing reserves on the escarpments above and seasonal pastures and salt licks below. In addition, its location below a bottleneck, where movement from the northern plains via the Kerio Valley and Baringo fed into a corridor between the Nyandarua and Mau Escarpments leading to the southeast plains and, distantly, Kilimanjaro, was of prime strategic importance. The combination of ecology and location explains why the area was so contested in the nineteenth century and so prized in the twentieth.

Maasai history as a sequence of oral traditions linked to successive age-sets and forming a coherent narrative, "begins" abruptly around the turn of the nineteenth century when the commonly remembered progression of named age-sets starts.[3] The first, Il Tiyioki, dates back to perhaps the 1790s, but it is not until Il Kidotu (c. 1820–1840) that particular events are associated with each set, at least in the north.[4] It can, however, be assumed, primarily from linguistic evidence, that Maa-speakers had arrived in the Rift centuries earlier and were already in occupation of the Nakuru-Naivasha corridor before 1800. Speakers of the original Ongamo-Maa branch of East Nilotic had migrated from the northwest in the seventh or eighth century, but split when the Ongamo moved south towards Kilimanjaro leaving the ancestral Maa-speakers on the plains north-west of Mount Kenya until they too moved south after 1500 (Sommer and Vossen 1993).[5] It seems likely that they did so in successive waves and that cultural differences developed between groups as they moved, dispersed and interacted with others. Some may have been agro-pastoralists.[6] In the next three centuries, Maa-speakers occupied much of what would become Maasailand, displacing or absorbing earlier South Nilotic speakers like the Tatog and Sirikwa and probably overlaying earlier Maa-speaking communities as well (Galaty 1993, 61–86).[7] By 1820, the basic ethno-linguistic map of the central Rift

3 The sequence has remained consistent since first recorded in 1883 and is supported by information from the 1850s onwards (Berntsen 1979a, chap. 2; Waller 1978, 408–14).

4 Dating is approximate at best since it depends on assumptions of regularity over a long period which cannot be checked. Mol's longer list pushes the sequence back further into the mid-eighteenth century, but without corroboration (Mol 1996, 13–14).

5 Ongamo-speakers survived into the twentieth century on the slopes of Kilimanjaro (Fosbrooke 1954, 153).

6 There are traces of other, now defunct, Maa-speaking groups, which may have been agro-pastoralists, in Maasai traditions (Fosbrooke 1948, 3–5; Waller 1978, 145–56).

7 Southern Maasai were still expanding against Tatog-speakers (*il Tatua*) in the first half of the nineteenth century and there is evidence of interaction between Southern Nilotes and

was becoming established, although on-going processes of expansion, conflict, incorporation and acculturation would simplify and to some extent obscure its original complexity.

Historians have long suspected, from oral tradition, that there had been a severe drought towards the end of the eighteenth century: recent scientific evidence, mainly derived from lake sediment cores, has put this beyond doubt. From perhaps 1790 to 1820 the entire region suffered a widespread and prolonged drought. Smaller freshwater lakes like Baringo, Naivasha, and Chala, together with the rivers that fed them, dried up completely for a time and there is evidence of abnormally low lake levels as far south as Lake Chilwa in Malawi (Anderson 2016). The impact on human and livestock communities must have been catastrophic in the short term. However, rather than seeing the drought as a single cataclysmic event, abrupt and unforeseeable – the climatic equivalent of the rinderpest pandemic a century later – it may be more useful to understand it as the outcome of a longer process of desiccation and climate change, a "long dry time", as Turkana traditions called it (Lamphear 1988, 32).[8]

Drought is essentially a process rather than an event: so too are adaptation and survival. Communities had time to adapt to harsher conditions: some, including Ateker- and Maa-speakers, did so better than others. Increasingly uncertain rainfall probably made some subsistence options, including agro-pastoralism, no longer viable and encouraged a "new" and more specialised pastoralism, based on increased mobility, smaller settlements, concentrated livestock holdings and the competitive ability to acquire and defend the resources on which long-term group survival depended. Those "new" pastoralists who had already acquired hardier, zebu, cattle were at a distinct advantage in harsher conditions (David 1982, 86–88; David 1983, 85–86, 565–66; Lamphear 1988, 32–33). However, the shift to intensive subsistence pastoralism required more than better cattle: it required rethinking community and identity and adapting social forms to meet new needs.[9]

Drought of this magnitude was socially disruptive. Communities under pressure broke up, rebuilt themselves and formed new alliances as those who had lost resources sought refuge with those who had retained or gained them.

Maa-speakers in both the Loita Hills, where there are "Tatua"- associated sites, and Trans-Mara, where *Lomoot* (Omotik-speakers), were absorbed into an emerging Siria community (Lamprey and Waller 1990, 17–20; Waller 1984, 247–48). The Sirikwa, putative ancestral Kalenjin, who had occupied the plateau area southeast of Mount Elgon, were replaced by the Maa-speaking Uas Nkishu before 1800 (Sutton 1973, 26–32).

8 Signs of change include previous droughts in northeast Uganda and the decline of irrigation agriculture in northern Tanzania (Herring 1979, 294–305; Sutton 1998).

9 For a cyclical model of pastoral adaptation and identity formation, see Bollig (2016).

Research by Lamphear and Sobania, amongst others, has revealed a complex and shifting pattern of multi-lingual and multi-ethnic alliances in the early-nineteenth-century northern Rift, involving communities with differing modes of subsistence and based on networks of kinship and complementarity (Lamphear 1976, chap. 5; Sobania 1980, chap. 2–3). Conditions further south were possibly similar, but, in the absence of evidence, we can only speculate about the changes that began to "make the Maasai" (Waller 1985, 83–133). Identities were layered and shifting, associated with specific pools of resources but structured by webs of kinship or clanship which gave a sense of belonging. "Clans" were the building blocks of community; locality provided the resources on which they depended. Claims of kinship, fictive or otherwise, provided pathways between communities and the basis for wider alliances (Bassi 2011; Schlee 1989). Berntsen (1979a: 36–43) describes Maasai sections as "clan clusters" and this emphasis holds true for other Maa-speakers, although landscape and territory always played an important part in investing group identity with memory and meaning (Waller 1978, 55–58, 156–64; Straight et al. 2016, 168–88).[10] Identity formed the basis of both inclusion and exclusion. A strong ethos, like "Maasai", attracted incomers and made their incorporation and acculturation easier. Asserting difference could deny access to resources or control the interactions with agricultural trading partners on which subsistence pastoralism relied (Spear 1993). During the nineteenth century, defeated Maa-speaking groups were displaced and then "othered" as "*Iloikop*" by the winners, so creating the supposed distinction between Maasai and Iloikop which has caused much confusion in the historiography.[11]

Lamphear has shown that Turkana made changes in their age-organisation to increase their military competitiveness and to incorporate newcomers after 1800, possibly influenced by contact with Maa-speakers living south and east of Lake Turkana. The logic he suggests would apply equally well to Maasai, although there is no hint of this in tradition (Lamphear 1988, 35–37). Since Maasai age-organisation is structured by social time rather than by external factors and driven by internal demographic pressures, it is flexible and can respond quickly to crisis or opportunity. Harsher conditions after 1800 probably increased the level of regional conflict: communities able to mobilise larger numbers of fighting men for attack or defence were thus at an advantage. The Maasai age-set system which had developed by mid-century, with its overlapping age-sets and *sirits* ("warrior companies"), provided for a constantly available and continuous supply of *murran* ("warriors"), organised by

10 Maasai sections did not become primarily territorial until after 1900.

11 The debate, which is both semantic and historical, is beyond the scope of this chapter, but see Berntsen (1980) and Jennings (2005).

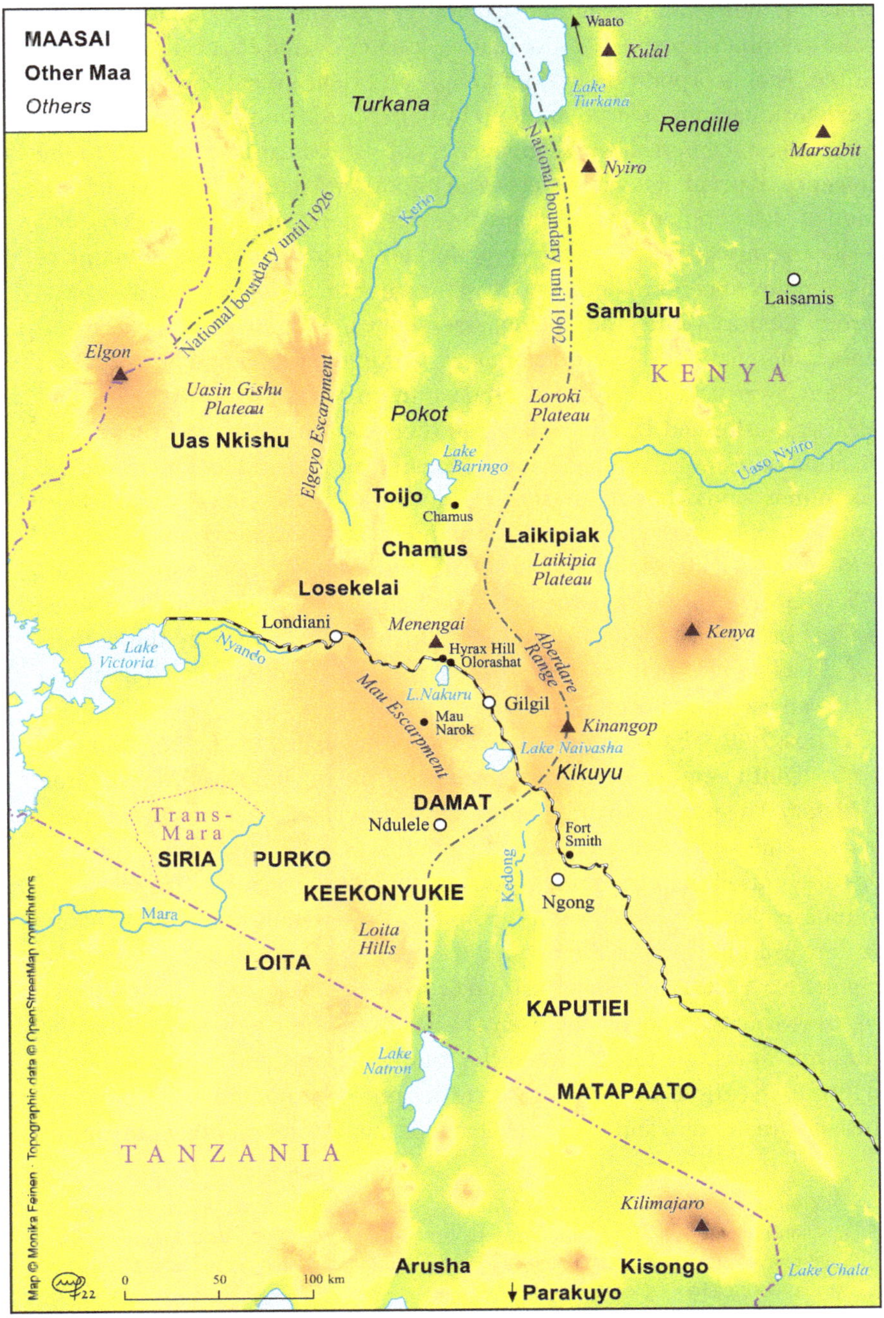

MAP 2.1 The location of different Maa-speaking groups in the central part of the Rift Valley in the late nineteenth century
CARTOGRAPHER: M. FEINEN

locality but more widely synchronised between groups. Induction into an age-set and membership of a *sirit* also made a greater degree of social integration possible. The incorporation of both people and their livestock was central to the expansionary momentum built up by both Maasai and Turkana (Berntsen 1979a, chap. 2; Lamphear 1998; Spencer, 1998). Lamphear also pointed to the emergence of prophets/war leaders as an important factor in Turkana success. A similar development can perhaps be seen among the Maa-speakers, but how far the authority of a *laibon* (prophet) extended beyond the directing of raids remains uncertain. Berntsen (1979a) argues that *laibons* acted as a focus of group identity – the group being, essentially, the following of a successful prophet able to mobilise *murran*. *Laibons*, including Supeet and his successor Mbatiany, were established among the southern Maasai by at least 1830 and both Uas Nkishu and Laikipiak also had their own prophets, but there is no indication of a similar presence among the northern Maasai before the 1880s. Jackson met "one of Mbatian's sons ... and two of his brothers" west of Lake Naivasha in 1889, possibly in search of a following.[12] Although major figures like Mbatiany wielded considerable authority and could generate strong bonds of allegiance, prophets remained entrepreneurial and potentially dangerous outsiders best kept at a distance, their influence constrained by the authority of elders and by their own professional rivalries (Lamphear 1998; Berntsen 1979b; Waller 1995; Spencer 2003, chap. 5).

The possibility that age-organisation was remodelled at the beginning of the nineteenth century leads us to reconsider the beginning of Maasai time and history. The initial explanation for the abrupt emergence of age-set traditions was that some major catastrophe around 1800 had wiped out communities and erased their memories (Sobania and Waller 1989).[13] This view still has some explanatory value, but it can be supplemented by newer approaches. One derives from work on the impact of state-imposed dating systems in the ancient Near East. Rulers demonstrated power or celebrated change by creating new calendars using themselves or a founding event as the starting point, thus drawing a line between the "outdated" past and a "new" present and future. Over time, the new reckoning became hegemonic and the past was either marginalised or shifted to a category of myth. Remembering the former

12 Jackson to De Winton, September 16, 1890, The National Archive [PRO]: FO 84/2264. Jackson also met the "lybon medicine man of the district" living near Kinangop. This might have been Lolgos, a migrated Uas Nkishu prophet who had settled near Eburru but was later killed by disgruntled Purko *murran*.

13 Other Maa-speakers had their own age-sets, but it is generally not possible to construct comparable sequences of names, except in Samburu where age-sets may go back further (Simpson and Waweru 2012: 184–85).

past could even become subversive (Kosmin 2018). Maasai did not have rulers, but they did have the experience of transformational events. "Climbing the Escarpment", the only Maasai story to refer to the drought, suggests that the "beginning of time" marked the beginning of a new identity. People found themselves trapped by drought at the foot of an escarpment – said to be the Kerio Escarpment. They noticed a bird flying high above with a leaf in its beak and, realising that there must be vegetation somewhere above, they decided to climb the escarpment to find it, knotting together strips from the hides of their dead cattle to make a rope to drag people up the steep slopes. Some reached the top, but the rope broke leaving the rest at the bottom. The lucky ones then set off to find water and grazing: they were the ancestors of the Maasai. This story of ethnogenesis describes a rupture, literal and metaphorical, and celebrates a new beginning, as "Maasai" climb from the drought stricken Kerio Valley to the highland pastures above.[14] Another approach considers adaptation rather than disaster or commemoration and, again, it has indirect support from the form of tradition. If the age-set system was indeed adjusted, the new structure provided an ideal frame for the Maasai narrative of expansion which has since informed Maasai collective memory. The progression of age-sets in the nineteenth century expressed movement forward both in time and space and thus added a powerful momentum and a moral meaning to events as one set exploited the military success of its predecessor, supported by the prestige and authority of the senior elders of the set before that (Waller 1978, 401–8).[15] Together, the three perspectives offer a plausible but still speculative answer to the problem of the "beginning of time".

By 1830, rain had returned to the Rift.[16] However, conditions remained harsher than they had been fifty years earlier and Maa-speaking communities fought to control the commanding heights of the new environment. In 1883, Thomson was told that the "war[s]" (*enjore* or *enara*)[17] between "Maasai" and

14 The story is old and well-known in many versions; there is also a comparable Samburu tale (Hollis, 1907, 103–4; Waller 1978, 398–99; Sobania 1980, 78). The Escarpment story is neither "history" (*enkatinyi* – "time") nor "myth" (*enkiterunoto* – "beginning") but is believed to reflect real events that happened "long ago". Age-sets do not appear in the story, but, in some versions, founding clan ancestors do.

15 For a useful discussion of the construction of moral narratives, see Tonkin (1992).

16 There is evidence for "drought-breaking flood events" in the Sabaki River catchment area c.1822–1826, and lakes were beginning to fill again after 1830. The development of irrigation agriculture at Leabori, the original Chamus settlement south of Lake Baringo, in the 1830s suggests that rivers had also started to flow reliably again (Buckles et al. 2016, 1255; Verschuren et al. 1999, 499; Anderson 1988, 244–51).

17 *Enjore* can also denote a major formal raid. It is likely that the "wars" consisted of discontinuous but directed raiding until attrition and stock losses tipped the balance to one side.

"Wa-Kwafi" ("Iloikop") had begun about 1830 with conflict between Kisongo and what later became Parakuyo south of Kilimanjaro (Thomson 1887, 240–42). By the late 1830s, fighting had spread north to eastern Maasailand. The "Iloogolala War" with Il Kaputiei, the eastern Maasai confederation that included the (modern sections) Kaputiei, Matapaato, and Loodokilani, ended in the mid-1850s with Iloogolala driven out of what is now Kajiado District. Whether Loogolala were part of the same larger grouping as Parakuyo further south is unclear; the war may have been in part an internal conflict which left the "winners" as Maasai and the "losers" as Iloogolala.[18]

Il Kinopop in the north do not seem to have been involved until the time of Il Nyangusi age-set (c.1851–71). Their first recounted "war" was with Ilosekelai, a Maa-speaking group who occupied grazing around Londiani. In the 1840s, Losekelai had begun to expand into Baringo, driving the Maa-speaking Il Toijo, who had grazed the scrublands around the lakes and were already under pressure from Turkana and Pokot, northwards. In the 1850s, Laikipiak occupied the Loroki Plateau and added to the pressure on Il Toijo. By 1860, they had abandoned their grazing areas and "become Chamus", joining the settlement founded by a diverse group of post-drought refugees at Leabori (Anderson 2002, 28–29; Petek and Lane 2017, 44–45).[19] Having secured control of the northern end of the Baringo bottleneck, Losekelai then turned their attention to the southern end around Nakuru, where they may previously have shared grazing with Il Kinopop. Their movements were probably perceived as a threat to both Il Kinopop and Laikipiak grazing and transhumance routes; friendly relations broke down.[20] The "Losekelai War" can be dated to the first half of the 1860s. Caught between Il Kinopop and Laikipiak, possibly acting together, Losekelai were surrounded near Menengai and forced to surrender. Most were incorporated into Il Kinopop as a subclan, but others fled to Uas Nkishu and to the Nyando Valley where they joined emerging Nandi communities (Waller 1978, 386–87; Berntsen 1979a, 144–46, 152–53).

The dispersal of Il Toijo and Losekelai and the gradual retreat northwards of Samburu removed the buffer between Il Kinopop and Laikipiak in Baringo

18 Thomson suggested that conflict had erupted after Parakuyo were weakened by disasters but does not mention drought. A summary of the "Iloikop Wars" can be found in Waller (1987, 114–20).

19 Il Toijo were probably part of the Maa-speaking "Kor" peoples encountered by the Turkana earlier in the century who then associated with the emergent Samburu (Lamphear 1988, 31–33).

20 Unreliable rainfall in the 1850s may have contributed to the worsening of relations (Weatherby 1967, 138).

and conflict was likely as relations between the two erstwhile allies soured.[21] Laikipiak now controlled a broad swathe of grazing from the Laikipia and Loroki Plateaux eastwards as far as the lower reaches of the Uaso Nyiro; and from the base of Mount Kenya to Laisamis. They faced little competition. The Samburu still held the Ndotos and Mounts Nyiro and Kulal, but had shifted their main centres away from Lake Turkana to the rich grazing areas of Wato, east of the Omo Delta and along the Ethiopian border. Their Rendille allies occupied Marsabit. ILaikipiak was probably a loose federation, somewhat like Il Kinopop, of Maa-speaking "Kor" groups, held together by shared age-sets and clans and by the cooperative exploitation of resources (Sobania 1980, 97–101; Dundas 1910, 55–56; Waller 1978, 146–51). It is not clear when and how fighting first broke out. Maasai associate the "Laikipiak War" particularly with the age-set IlAimer (c.1866–1886) but say that it began when Il Nyangusi were still *murran* (retired c.1871) and ended after Il Dalala had been initiated (c.1881). The likely timespan, then, is from the late 1860s to the beginning of the 1880s. According to tradition, confirmed by recent archaeological evidence, Lekeper, the second Chamus settlement (founded c.1860), was burned to the ground in the later 1860s by a group of Maasai, posing as friends. Since the settlement had close links with Laikipiak, it is possible that it was razed as a message or a declaration of war. Significantly, nearby Leabori was left alone (Petek 2018, 109–110, 166). In Maasai accounts, the war began when Laikipiak pushed down into the Nakuru-Naivasha area and began to drive Il Kinopop away from their main pastures and waterpoints eastwards into the less welcoming Kedong Valley. Tradition gives no reason for this aggressive move, but pressure from further north can probably be discounted. As Sobania points out, the Samburu settlements closest to Laikipia/Loroki were too weak and the main force was too far north (Waller 1978, 387–89; Sobania 1980, 136).[22] More likely is a combination of drought and disease in the north. There is some indication of both occurring in the 1860s and clear evidence of low lake levels in the central Rift in the 1870s. The need to replenish the herds and to acquire alternative grazing resources would certainly be enough to set a migration in motion; rain failure in the central Rift in 1876/7 ensured that conflict would be bitter (Waller 1978, 387–88; Sobania 1980, 143; Berntsen 1979, 263; Tiki, Oba, and Tvedt 2013, 40;

21 One account traces this back to a dispute over cattle taken in joint raids against Uas Nkishu near Ndalat, west of modern Eldoret. The spokesman of Il Nyangusi told his *murran* to seize and brand all the cattle, leaving none for the Laikipiak. Angry at being tricked, the Laikipiak vowed revenge (Berntsen 1979a, 262–3).

22 According to Stigand, Laikipiak migrated southwards *en masse* because they feared attack from the north (Stigand 1913, 207).

De Kort et al. 2013, 77–79; Gooding 2022, 210–14).[23] Once established northwest of Nakuru, Laikipiak began to raid widely south and southeastwards. This drew the eastern Maasai into the conflict and it was probably they who used their contacts to approach the *laibon* Mbatiany in Kisongo for assistance. Tradition credits the prophet with organising and blessing the Maasai coalition that ultimately defeated Laikipiak and with supernaturally destroying the power of his rival, the Laikipiak prophet Koikoti ole Tunai, and his *murran* (Waller 1978, 389–90; Berntsen 1979a, 265–72). The later stages of the war can be dated to c.1876–8. The German traveller Hildebrandt, setting off from the Coast in January 1877, found the interior trade routes already in turmoil: some caravans had been attacked and others were blocked by "Masai hordes" and waiting for conditions to improve. When he reached Ukambani, he decided not to venture further west into what was clearly an active war zone (Hildebrandt 1877, 448–53). The climactic battle took place at Olorashat near Lanet and ended, apparently, with the complete defeat of Laikipiak and the eclipse of their prophet, who took refuge with his followers among the Siria Maasai.[24] Large numbers of Laikipiak followed their herds into the camps of their conquerors as the defeated had done before them. Il Kinopop sent raiding parties north to harry the retreating Laikipiak and to pick up undefended livestock as well as women and children who were then married or adopted into their captors' families.[25] However, Laikipiak were far from "finished", as the Maasai narrative has it. As Sobania argues, Laikipiak continued to be a powerful factor in the north for the next decade and were capable of putting up stiff resistance to Maasai attempts to take over grazing areas and thus consolidate their victory. Their resilience, together with continuing erratic rainfall and the onset of cattle disease – bovine pleuropneumonia in 1883 and rinderpest in 1891 – denied Il Kinopop full control of Laikipia until a new opportunity to make it theirs arose in 1904 (Sobania 1993; Rockel 2022).

---

23 In the 1870s, cattle could be herded across Lake Naivasha to Crescent Island (Sikes 1936, 82–84).

24 One of the caravans, led by the trader Mbaruk, was destroyed north of Elmenteita. Six years later, Thomson passed the remains of two large encampments near Gilgil, where, he was told, the opposing *murran* had gathered before the final battle. By 1883, however, fighting seems to have ended in the central Rift (Thomson 1887, 199, 200, 242).

25 Amongst those picked up and adopted were boys who later became influential Maasai spokesmen like Parsaloi ole Galishu (named for Barsaloi where he was found) and Lasiti ololKeri (Waller 1978, 362–70).

2 The Decades of Disaster: 1880–1900

At the beginning of the 1880s, with the decades of post-drought confusion ended and identities and boundaries now more clearly defined, the Maasai appeared to be in unopposed control of their Rift Valley grazing areas. To the northwest, Uas Nkishu were weakened, though they had not yet disappeared. Purko and Laikipiak raids on the southern part of the Uasin Gishu Plateau and pressure from "Kakesira", Ateker-speaking raiders from north of Mount Elgon, had contributed, but internal conflict between two moieties, Nkaji Narok and Nkaji Mugie, and the withdrawal further west of the former was probably the precipitating factor in their gradual decline (Waller 1984, 250–51). Yet, despite success and consolidation, Maasai were facing something of a crisis of control: they were too weak to occupy effectively all the grazing areas they now claimed. The new threat came not from the centre but from the periphery, where newly emerging agro-pastoral groups were infiltrating into poorly defended or empty borderlands and building their herds. North of Nakuru, these groups were Kalenjin-speakers, whose separate primary identities were emerging but had not yet crystallised into the neatly defined ethnicities enshrined in colonial ethnography: like the Maa-speakers before them, they were now in the process of "becoming".[26]

Early in 1891, rinderpest struck the Maasai herds almost without warning and with devastating force. It had been preceded by pleuropneumonia brought in by cattle raided from the north, where it may already have become endemic. The suddenness and completeness of *Emutai*, the disaster (lit. "the finishing off"), were imprinted on memory: herds were destroyed in a matter of days and there was neither cure nor succour. By the end of the year, most of the cattle were dying or dead and their owners faced destitution. Under such circumstances, the usual recuperative strategies, including organised raids on outside communities, were ineffective: there were only three options for survival. One was to seek refuge with neighbouring agriculturalists, using trading partnerships or other links as an entrée or simply offering labour and marriageable girls; the second was to take to hunting and foraging in the bush, perhaps joining a hunter-gatherer community to learn how; the third involved the seizure of livestock from whomever and wherever, a desperate expedient which might violate the norms of community and risked retaliation. Towards the end of 1892, an epidemic outbreak of smallpox, striking first at vulnerable

26 Interactions on the periphery are discussed more fully in Waller (1985, 119–25). These clan-based communities were primarily agriculturalists who were developing pastoral "wings".

refugees gathered close to agricultural settlements where the disease was endemic, added to the misery of starvation and complicated the pattern of mortality (Waller 1988).

Emutai was relatively short-lived, however. The initial pandemic soon burned itself out, although disease outbreaks recurred later in the decade and both rinderpest and pleuropneumonia became endemic in the region. The herds recovered quickly, partly due to natural regeneration – pasture and water were plentiful and under little grazing pressure – and partly because raiding brought in new stock. Stocking rates probably returned to pre-1891 levels within the decade (Waller 1988, 101–3, 83–84). The Maasai had also found new patrons who needed spearmen to expand their power and were willing to pay for them in looted stock. Some were local "big-men", but these were soon joined by officials of the newly established colonial administration who found in Maasai *murran* a cheap and effective addition to their meagre and unreliable supply of troops. It was a mutually beneficial arrangement in which parties of *murran*, under their own spokesmen, accompanied the British on what amounted to joint raids on common enemies and took the livestock they captured to replenish their family herds. At first, the British selected the targets, without necessarily dictating the outcome, but, over time, the balance of power shifted. The British exerted tighter control over the fighting and kept more of the looted stock. The Maasai were no longer allies but subordinate "friendlies" (Waller 1976; Lonsdale 1977).[27]

Surviving *Emutai* exacted a high price and threatened social cohesion. *Osotua* ("peace"), the shared understandings that held the wider Maasai community together, had been broken. Survivors had seen their familiar and familial worlds destroyed; they had done things that they would prefer to forget; and they had experienced the humiliations and uncertainties of displacement. The imperatives of survival and the obligations of dependence could be harsh. The pandemic was regional, but survival was intensely local. Disaster struck unequally. Some communities and their herds suffered more severely from disease and raiding than others; some families had better networks or more social resources than others; some were lucky, others were not. This variable experience of loss shaped the context in which identities were affirmed, denied or re-evaluated in the aftermath of disaster and recovery.

27 Protectorates over Uganda and East Africa were formally proclaimed in June 1894 and July 1895.

By the 1880s, primary identities within the larger Maasai community, based on locality as well as clanship, had emerged.[28] In eastern Maasai (Kaputiei), Thomson (1887, 23) described "native districts" like Matapaato and Loodokilani, where different clans still shared resources, but generally lived separately. *Emutai* and its aftermath probably stimulated this development. After 1890, these sectional identities began to appear in official reports; and by 1904 Hollis was able to provide a table of Maasai "districts" and "sub-districts" which combined location and social identity. The signatories to the 1904 Treaty were each identified – generally accurately – by section (*olosho*) not clan.[29] *Emutai* undermined the cohesion of larger confederations like Il Kinopop and Kaputiei in several ways. Need and opportunity could widen internal cleavages and justify violence, not just against known external "enemies" (*il mangati*) but even between groups once linked by the bonds of *osotua*. "Breaking the peace" within the Maasai community was deeply unpropitious and Maasai were right to worry about its long-term implications (Waller 1988, 78; Waller 1978, 122–24). In 1892, Damat turned on Keekonyukie and seized their remaining stock; Purko retaliated, driving the Damat southwards to the Loitai, who had already "broken the peace". The *murran* of all three Il Kinopop sections had previously "shared the ox" at joint ceremonies. Purko also took the livestock of Il Dalalalekutuk who were associated with Il Kinopop. Late in 1895 at the crowded refugee camp at Ft Smith, Kaputiei and Matapaato *murran*, whose fathers had perhaps fought together against Laikipiak, had to be separated by a bayonet charge (Waller 1988, 80; Waller 1978, 537). The greatest damage was done by nearly a decade of intermittent raiding which pitted the Loitai and their allies, including the Maa-speaking Arusha, against both the eastern Maasai in Kajiado and the northern Maasai in Naivasha.[30] Conflict deepened when it fused with a succession dispute between Senteu and Olonana, the sons of Mbatiany. Loitai supported Senteu, while Olonana moved north to Ngong and built a following there among the eastern Maasai refugees. Fighting ended in 1902, when Senteu, harried by the Germans and abandoned by most of his elders,

28 Latent sectional identities were probably present earlier, but neither Krapf nor Erhardt gave any equivalent term in their vocabularies, although Erhardt did suggest that named localities existed (Krapf 1854; Erhardt 1857; Erhardt 1856, 19–32; Waller 1978, 98–136).

29 *Report by Sir A. Hardinge on the Conditions and Progress of the East Africa Protectorate from its Establishment to the 20th July 1897*, C8683, London: H.M. Stationary Office (HMSO), 1897, 22; copy of the 1904 Agreement in Sandford (1919, Appendix 1); Hollis 1905, 260–61. The modern Maasai sections did not fully form until later.

30 The fullest account of this complex "war" is in Berntsen (1979a, 301–33). He notes that it was the *murran* of Il Dalala who made the decision to "break the peace", against the wishes of their elders. For British and German involvement, see Waller (1976, 540–46).

finally crossed the Anglo-German border with two thousand of his following and "surrendered" to Olonana.[31] In autumn 1903, a great Eunoto ceremony, attended by "thousands" and blessed by Olonana, was held on the Kinangop for the Il Dwati age-set of Purko, Keekonyukie, Il Dalalalekutuk, Damat and Loitai who now became senior *murran*. This was in part a ceremony of reconciliation and unity, re-incorporating the Loitai and Damat and reaffirming Il Kinopop. The *murran* took a shared age-set name, Il Merisho, and acknowledged a senior spokesman chosen from Purko and a ritual leader chosen from Keekonyukie.[32]

Dispersal broke up communities, but it also led to the creation of new and improvised communities, especially when refugees found alien patrons like Hall at Ft. Smith (Waller 1988, 99–100).[33] Service in punitive expeditions against outsiders helped refugees rebuild their herds, but also increased the levels of violence and dislocation. The weak could once have expected to be absorbed by the strong in a spiral of accumulation that concentrated and increased wealth in people and animals, but now refugees were "lost", if only temporarily: "othered" by disaster. Those able to retain their cohesion as a group fared better. Purko had been hit by the pandemic, but had enough livestock and people, acquired through the defeat of Laikipiak, to survive without dissolving. They also avoided attacks from neighbouring communities and by Loitai raiders, unlike the weakened groups in the east. The destruction in November 1895 of a government supply caravan, which had been foolish enough to annoy the occupants of a *murran* camp in the Kedong Valley, and the subsequent peace discussions with officials, showed that the "Naivasha Masai" still had the power to dominate the route through the Central Rift and the leadership cohesion to negotiate from strength (Waller 1976, 543).[34] An official report drew a contrast between the Maasai around Naivasha, "an entire tribe", and the "remnants" around Ngong, "a mere helpless and broken fragment of a once redoubtable people". Success against Senteu and the Loitai then brought the former more

31 Ainsworth to Jackson, 10 July 1902, PRO: FO 2/573. The rest of the Loitai remained in the south, beyond Senteu's authority.

32 Bagge 1904, 167–68; *Report on the East Africa Protectorate for the Year 1903-4*, Cd 2331, London: HMSO, 1905, 6. Damat attempted to re-establish their own separate identity, but Purko refused to allow them to leave with stock captured on joint raids (Sandford 1919, 37).

33 When established by Hall in 1893, the camp at Ft. Smith sheltered about a thousand refugees; two years later it had quintupled in size and included nine hundred *murran* under their own spokesmen – Ainsworth to Hardinge, 20 Jan. 1896, PRO: FO 107/50.

34 Some of the Kikuyu porters were said to have been wearing ornaments stripped from *murran* killed earlier in an abortive raid.

raided stock and dependents. They were also able to exploit the differences between insecure administrators, divided by the Uganda/East Africa border, to maintain their own freedom of action, at least until 1902, by which time their herds and pastures were seemingly secure.[35] In 1904, Maasai in Naivasha Province reportedly had 35,000 head of cattle and 500,000 small stock (Sandford 1919, 20; Waller 1976, 547–50).

The decision to transfer Uganda's Eastern Province, including Naivasha, the Western Highlands, the Mau and part of what became the Masai Reserve in 1913, to the East Africa Protectorate in April 1902 was taken for reasons of administrative convenience – it fixed a clearer border between the Protectorates and brought the entire railway, now completed from Mombasa to Kisumu, under one administration – but had far wider consequences. It was challenged by Uganda officials like Jackson who resented transfer to what they saw as an inferior administration and were concerned about its impact on "their" Maasai, who would also be taken over.[36] Their concerns were justified: the transfer, by consolidating control over what would become the "White Highlands", both necessitated and facilitated the removal of the Maasai from the Central Rift corridor two years later.

3 Colonial Settlement and Its Consequences: 1900–1920

Constructing the railway had served imperial strategic purposes, but now that it was operational, it required freight to defray its running costs. The volume of African production seemed insufficient, so Eliot, Governor since February 1901, hoped to attract white settlers who would pay rent and create traffic for the railway. Once the right of the Crown to dispose of land in the Protectorate had been legally established, settlement could begin. Large tracts were potentially available since Crown possession assumed that African communities did not "own" land, either individually or collectively and thus had no legal rights: their use of land could, however, be "reserved" or otherwise safeguarded.[37] Colonial

35 Hardinge to Salisbury, June 11, 1896, FO 107/53; Bagge to Jackson, March 26 & July 29, 1903, PRO: FO 2/713 & /847.

36 H.H. Johnston, "Confidential Report", August 9, 1901 and Jackson to Hill, September 3, 1901, FO 2/463.

37 Land rights in East Africa were ultimately vested in the Crown (standing for the state), but their disposition was delegated to the local government concerned, with oversight from London. Land could be alienated to private ownership (under conditions set by government and specified in the lease), retained for official purposes or reserved to "native" occupation (Sorrenson 1968, chap. 3).

ideas of land use and ownership severely disadvantaged pastoralists, whose patterns of seasonal transhumance left large areas apparently unoccupied or underutilised. The area now occupied by the northern Maasai was especially desirable, its prime grazing a magnet for speculators with an eye for land.[38] Eliot looked to South Africa first for his pioneers and sought out entrepreneurs willing to use their local connections to find suitable settlers for Kenya in return for land grants for themselves. However, Eliot, who had no previous experience with land settlement, was no match for South Africans who knew exactly what they wanted and how to get it. His failure to establish adequately staffed Land and Survey Departments to handle applications invited disaster (Sorrenson 1968, 62–69).[39] The Foreign Office was equally inexperienced, although it was familiar with metropolitan investment syndicates operating overseas. To London, concessionaires seemed a better bet than settlers when it came to developing a new territory. Accordingly, the Foreign Office backed the claims of the East Africa Syndicate for 320,000 acres of grazing near Gilgil and demurred over Eliot's proposed leases of 32,000 acres each to Chamberlain and Flemmer, the South African settlement agents. Differences of opinion between London and Nairobi over both means and ends led ultimately to Eliot's resignation in May 1904.[40]

By the end of 1904, applications for 58,000 acres in Naivasha Province had been approved, together with large grants to the East Africa Syndicate and to Lord Delamere (100,000 acres); more were pending.[41] The growing demand made a settlement of the "Masai Question" a matter of urgency. Between the departure of Eliot and the arrival of his successor, Stewart, in August, senior officials, notably Bagge, Hobley and Jackson, all of whom knew the Maasai, worked to find one. The crux of the problem was summarised by Lansdowne in his dispatch approving the Agreement and endorsing the reasoning behind it. Put simply, how could land in the Rift be opened to settlement without infringing on Maasai grazing rights? The two were incompatible (Sandford, 1919: 22–24).[42] Despite their differences, Eliot and Lansdowne were agreed on this.

38 Eliot to Cranbourne, May 15, 1901, FO 2/447; Linton to Eliot, February 4, 1904, FO 2/836; Hobley to Lansdowne, July 13, 1904, FO 2/838.

39 In 1903, there were only two surveyors and two assistants available to process 117 applications.

40 *Correspondence Relating to the Resignation of Sir Charles Eliot and to the Concession to the East Africa Syndicate*, Cd. 2099, London: HMSO, 1904.

41 Figures from *Returns of Concessions …*, Cd. 2100 and *Returns of Grants of Land …*, Cd. 2902, London: HMSO, 1903 & 1904. The majority were grazing leases.

42 Jackson, "Memorandum on the Masai Question", February 22, 1904, and "Memorandum on Masai Negotiations", March 8, 1904, both in FO 2/843; Lansdowne to Eliot, February

Eliot initially accepted the view that the Maasai were important and dangerous, but rejected the idea of creating reserves, believing that coexistence and contact with Europeans would speed the process of "civilisation". The Foreign Office agreed – "segregation" had not done well in Southern Africa. Later Eliot changed his mind on both points and shifted towards separation and removal. Again, London agreed, while insisting on an "absolute guarantee" of Maasai rights once they had been moved.[43] By the middle of 1904, even administrators who had earlier defended Maasai rights in the Rift Valley had come to accept that moving the Maasai, with their agreement, was the only practicable solution. They now believed that Maasai did not require all the available grazing and were concerned that, if not given land, settlers would simply seize it.[44] When Stewart arrived, he had only to meet the Maasai spokesmen and then preside over the signing of the Agreement. Reassured that Maasai agreement was unhurried and uncoerced, Lansdowne was happy to accept what "appear[ed] to offer every possibility of a satisfactory solution to a difficult problem".[45]

Two Reserves were established under the Agreement: one in the North, including the Laikipia and Loroki Plateaux, for the northern Maasai and one in the South, south and southwest of Ngong, for the eastern Maasai. Additionally, the Maasai were to "retain control" of the Kinangop ceremonial site at Ngorinitu northeast of Naivasha and a right of way connecting the two Reserves would be created. In the event, the compensation for relinquished grazing rights urged by Hobley was omitted; the right of way did not materialise; and both movement between the Reserves and access to Ngorinitu were later prevented by quarantine regulations (Sandford 1919, 24–25; Sorrenson

23, 1904, FO 2/832; Lansdowne to Stewart, July 8, 1904, FO 2/833; Lansdowne to Stewart, October 21, 1904. FO 2/833.

43 *Report by His Majesty's Commissioner on the East Africa Protectorate*, June 1901, Cd 769, London: HMSO, 1901, 5; *Report ... on the East Africa Protectorate*, April 1903, Cd 1626, London: HMSO, 1903, 8; Eliot to Lansdowne, October 10, 1903 and reply, November 27, 1903, FO 2/716 & /708; Lansdowne to Stewart, October 21, 1904.

44 Memo by Bagge, 8 March 1904, FO 2/847; Ainsworth to Jackson, March 31m 1904, FO 2/843; papers in Hobley to Lansdowne, July 22, 1904, FO 2/838. See also papers printed in *Correspondence Relating to the Resignation of Sir Charles Eliot*, 9–15. Jackson, especially, had a very low opinion of settlers and worried that their "South African [racist] attitudes" would poison the atmosphere in the Protectorate – Jackson to Hill, May 25, 1903 and January 10, 1904, FO 2/720 & /843.

45 Stewart to Lansdowne, August 16 & September 5, 1904, FO 2/842 & /839; Lansdowne to Stewart, October 15 & 21, 1904, FO 2/841 & /833.

1968, 196–7).[46] The Maasai were left to move themselves and most had apparently done so by the beginning of 1906. Some were allowed to stay temporarily to keep pasture sweet and to provide "friendlies" for government and herders for incoming settlers; other squatted on the unoccupied East Africa Syndicate concession, adjacent to the Northern Reserve boundary.[47] Maasai reappeared briefly when they were moved across their old grazing lands on their way to their new reserve, south of the Mau, in 1912; only those who herded the livestock of others on settler estates in order to build up their own family herds elsewhere would remain. We cannot know when and how Maa-speakers first entered the Central Rift, but we do know when and how they left.

The Move marked another stage in the break-down of wider identities. Hobley hoped that, if "wandering" between the two reserves was discouraged, the Maasai would "practically grow into two distinct tribes". The forced reconfiguration of space encouraged this trend.[48] It was assumed that Il Kinopop and Loitai would go north, while the eastern Maasai remained in the south, but Loitai, Damat and about a third of the Purko stayed south of the Mau; and some Keekonyukie, later joined by families returning from Laikipia, remained in Naivasha until moved south in 1910.[49] The loss of the Kinangop site disrupted ceremonies which partly defined Il Kinopop. The 1909 Eunoto ceremony was to be held there, but was halted by Olonana, who, probably to assert his authority, insisted that it be held at Ngong. Permission to use Ngorinitu for the next ceremony, in 1915, was refused and, two years later, Purko and Keekonyukie held separate ceremonies at Mau Narok and Ndulele. 1903 had been last time that the *murran* elected joint officials and took a common age-set name. Il Kinopop had effectively ceased to exist as a corporate group. The eastern Maasai maintained their cohesion for longer, holding joint Eunoto ceremonies at Ngong in 1904, 1909 and 1918 with shared officials and age-set names, but tensions between Kaputiei and Matapaato *murran* exploded in violence in 1918 and the elders then decided to hold ceremonies separately (Sandford 1919, 28–29, 69–71).[50]

46 Hobley to Lansdowne, July 13, 1904, FO 2/838. The right to Kinangop was eventually exchanged for land elsewhere – Masai Province, *Annual Report*, 1926, Kenya National Archives (KNA) PC/SP 1/2/1.

47 Hobley to Sadler, January 4, 1906, PRO: CO 533/11; Sadler to Elgin, June 18, 1906, CO 533/15.

48 Hobley to Sadler, January 4, 1906. The 1912 Move reunited the Maasai but created a new set of strains.

49 Hemsted Memo, February 14, 1914, in *Ngong Political Record Book*, Part A, KNA: DC/KAJ 1/2/1; Hollis, "Memorandum on the Masai", July 5, 1910, in Girouard to Harcourt, April 18, 1911, CO 533/86. Neither Damat nor Loitai had been represented at the Agreement.

50 *Ngong Political Record Book*, Part B, Native General vi, KNA: DC/KAJ 1/2/2.

Six months after the Agreement, a senior official in Nairobi asked a pertinent question – if only to offer a self-serving answer. Why had the Maasai been so willing to acquiesce in the loss of their grazing lands? The reason lay in the government's "wise and considerate policy", which convinced them that its proposals must be to their ultimate benefit.[51] We might address the same question more critically, even though evidence for Maasai thinking and intentions is scanty. The Maasai were still in a state of transition. Their unity was fragile and their sense of identity uncertain. They had already acknowledged, with whatever degree of disingenuousness, that their land "belonged to government"; and they were aware, from service on punitive expeditions, of what the government could do. A "military demonstration" at the 1903 Eunoto was a reminder. Early in the negotiations, northern elders were clear about what grazing areas they needed to retain – Lake Naivasha and its surrounding pastures – but in the end were willing to settle for what they could get. The eastern Maasai had far less to lose.[52] There were other factors. One was gaining Laikipia – and moving away from Europeans. The Purko and Keekonyukie who moved north included some who had fought to win it, including signatories to the Agreement, and others who had been born Laikipiak: Laikipia was doubly theirs. In the Maasai narrative, it was Olonana who drew a "boundary", *ol polosiei l'Olonana*, dividing the land with the Europeans; and his authority was accepted. The government version likewise acknowledges Olonana's decisive influence.[53] Celebrating Olonana, however, underestimates the decision-making power wielded by the elders, with whom officials worked in 1903/4. From 1893 onwards, Olonana lived with his personal following at the foot of the Ngong Hills and established his influence over the eastern Maasai. His position in Naivasha was more ambiguous. In the mid-1890s, the three "big men of the [Naivasha] Masai" were Terere ole Naisho and Leposo ole Nkomea, chief Purko spokesmen of Il Dalala and IlAimer respectively, and "Kugarie" [unidentifiable].[54] The Naivasha elders seem to have remained their distance from the

51 Monson, "Memorandum on the Removal of the Masai …", February 3, 1905, in Stewart to Lansdowne February 9, 1905, FO 2/915. A minute on the dispatch conveys London's indifference now that the problem was solved: "no one shows the slightest curiosity in the Masai Move – no need to publish".

52 Johnston to Salisbury, October 13, 1899, FO 2/204; *Report on the East Africa Protectorate 1903–4*, 6; Bagge to Jackson, July 29, 1903, FO 2/847; Jackson, "Memorandum on Masai Negotiations", March 8, 1904. See also Hughes (2006, 35).

53 Olonana, "lybon of all the Masai", heads the list of signatories to the Agreement: he was treated as a paramount chief by the British (Sandford 1919, 3).

54 Jackson to Berkeley, January 23, 1896, FO 2/112. A decade earlier, Fischer had mentioned Terere, Leposo and Lesingo ole Kool as the most important spokesmen for Il Kinopop (Fischer 1885, 27).

laibon, despite his attempts to build a following there by summoning *murran*: they signed for their own reasons, not because they were directed by Olonana. After the two reserves had been created, his authority in the North began to diminish.[55]

With Eliot gone and the Masai Question apparently solved, the Colonial Office, which had taken over the Protectorate in April 1905, moved to bring order to the "happy go lucky" operations of the Land and Survey Offices. Competent heads were appointed, and staffing was increased (Bennett 1963, 20; Sorrenson 1968, 85–90).[56] Re-organisation took time, partly because of the sheer volume of applications to be processed and partly because the Colonial Office, drawing on its previous experience of handling white settlement in South Africa and Australasia, was determined to ensure that control over land settlement lay with government not settlers. The ensuing arguments between Nairobi and London over such matters as the procedures for allotment and the terms of leases did not finally end until a new Crown Lands Ordinance was enacted in 1915. Early mistakes could not easily be rectified, however. Settlers had initially been allowed to select land for themselves, file claims, occupy and then wait for their land to be surveyed and registered. The best land was secured early and cheaply, before the Land Office was properly established (Sorrenson 1968, chap. 5–6). Naivasha, one of the first areas opened, was particularly affected. The Kinangop site promised to the Maasai by treaty was mistakenly allocated for settlement and leased in 1908; and no provision was made for reserving riparian rights at the lake (Sorenson 1968, 197; Styles 2019, 35–37). In the long term, early confusion and the lack of a coherent policy allowed the settler community to establish a position within an important area of government from which they could not subsequently be dislodged.

By 1912, most of the available land in Naivasha Province had been allocated and there was no more officially for sale (Sorrenson 1968, 101–2).[57] Naivasha and Nakuru had been gazetted as townships in 1904 and a settler community began to form. In 1912, when the Naivasha Farmers Association held its inaugural meeting in October with twenty-five members, there were already over two hundred white residents. Settlers required labour and the removal of one group of "natives" – the Maasai – made space for another. By 1916, there were

55 Ainsworth to Craufurd, June 15, 1896; Johnston to Hardinge, October 14, 1900, FO 2/293; Hollis, "Memorandum on the Masai". Olonana had first contacted the Purko in 1889 and had later sought to move permanently to Naivasha; Hall to Pigott, March 14, 1895, FO 2/97.

56 A committee of enquiry appointed in early 1905 reported that there was "absolute confusion" over land settlement (Sorrenson 1968, 84).

57 *Report on the East Africa Protectorate for 1911/12*, February 1913, Cd 6007, London: HMSO, 1913, 45.

over three thousand Kikuyu in the district, mostly farm squatters (Styles 2019, 33–35).[58] Naivasha was being re-colonised.

Acknowledgements

The author wishes to acknowledge important assistance from Neal Sobania and John Berntsen, who graciously allowed him to draw on their unpublished Ph.D. theses, and from John Lonsdale and Thomas Spear who kindly read the manuscript and made very helpful suggestions.

References

Anderson, David M. 1988. "Cultivating Pastoralists: Ecology and Economy Among the Il Chamus of Baringo, 1840–1980." In *The Ecology of Survival: Case Studies from North-east African History*, edited by Douglas H. Johnson and David M. Anderson, 241–260. Boulder: Westview Press.

Anderson, David M. 2002. *Eroding the Commons: The Politics of Ecology in Baringo, Kenya 1890–1963*. Oxford: James Currey.

Anderson, David M. 2016. "The Beginning of Time? Evidence for Catastrophic Drought in Baringo in the Early Nineteenth century." *Journal of Eastern African Studies* 10 (1): 45–66. https://doi.org/10.1080/17531055.2015.1134532.

Bagge, S. 1904. "The Circumcision Ceremony among the Naivasha Masai". *Journal of the Royal Anthropological Institute* 32: 167–9.

Bassi, Marco. 2011. "Primary Identities in the Lower Omo Valley: Migration, Cataclysm, Conflict and Amalgamation, 1750–1910." *Journal of Eastern African Studies* 5 (1): 129–57. https://doi.org/10.1080/17531055.2011.552280.

Bennett, George. 1963. *Kenya: A Political History: The Colonial Phase*. London: Oxford University Press.

Berntsen, John Lawrence. 1979a. *Pastoralism, Raiding and Prophets: Maasailand in the Nineteenth Century*. PhD diss., University of Wisconsin-Madison.

Berntsen, John Lawrence. (1979b): "Maasai Prophets and Prophetic Leadership 1850–1910." *Africa* 49 (2): 134–46.

Berntsen, John Lawrence. 1980. "The Enemy Is Us: Eponymy in the Historiography of the Maasai" *History in Africa* 7, 1–21. https://doi.org/10.2307/3171654.

58 Naivasha District, *Annual Reports*, 1910/11–1912/13, 1915/16, KNA: DC/NKU 1/2. For a description of the Naivasha farms in 1908, see Playne and Gale (1908-9, 275–95).

Buckles, Laura K., Dirk Verschuren, Johan W. H. Weijers, Christine Cocquyt, Marteen Blaauw, and Jaap S. Sinninghe Damsté. 2016. "Interannual and (Multi-)Decadal Variability in The Sedimentary BIT Index of Lake Challa, East Africa, Over the Past 2200 Years: Assessment of The Precipitation Proxy." *Climate of the Past*, 12 (5): 1243–62. https://doi.org/10.5194/cp-12-1243-2016.

Bollig, Michael. 2016. "Adaptive Cycles in the Savanna: Pastoral Specialisation and Diversification in Northern Kenya" *Journal of Eastern African Studies* 10 (1): 21–44. https://doi.org/10.1080/17531055.2016.1141568.

David, Nicholas. 1982. "Prehistory and Historical Linguistics in Central Africa: Points of Contact." In *The Archaeological and Linguistic Reconstruction of African History*, edited by Christopher Ehret and Merrick Posnansky, 78–95. Berkeley: University of California Press.

David, Nicholas. 1983. "The Archaeological Context of Nilotic Expansion: A Survey of the Holocene Archaeology of East Africa and the Southern Sudan." In *Nilotic Studies: Proceedings of the International Symposium on Languages and History of the Nilotic Peoples, Cologne, January 4–6, 1982*, edited by Rainer Vossen and Marianne Beckhaus-Gerst, 35–107. Berlin: Dietrich-Reimer.

De Kort, Gijs, Ilse Bessems, Edward Keppens, Florias Mees, Brian Cumming, and Dirk Verschuren. 2013. "Late-Holocene and Recent Hydroclimatic Variability in the Central Kenya Rift Valley: The Sediment Record of Hypersaline Lakes Bogoria, Nakuru and Elmenteita." *Paleogeography, Paleoclimatology, Paleoecology* 388: 69–80.

Dundas, Kenneth R. 1910. "Notes on the Tribes Inhabiting the Baringo District, East Africa Protectorate." *Journal of the Royal Anthropological Institute of Great Britain and Ireland* 40: 49–72. https://doi.org/10.2307/2843144.

Erhardt, James. 1857. *Vocabulary of the Enguduk Iloikob as Spoken by the Masai-tribes in East Africa.* Ludwigsburg: Ferdinand Riehm.

Erhardt, James. 1856. "J. Erhardt's Memoire zur Erläuterung der von Ihm und J. Rebmann zusammengestellten Karte von Ost- und Central-Afrika." *Petermann's Mitteilungen* 2: 19–32.

Fosbrooke, H. A. 1954. "A Note on the Ngasa." *African Studies* 13 (3–4): 153–4. https://doi.org/10.1080/00020185408706942.

Fosbrooke, H. A. 1948. "An Administrative Survey of the Masai Social System." *Tanganyika Notes and Records* 26: 1–50.

Fischer, Gustav A. 1885. "Das Maasailand." *Mitteilungen der Geographischen Gesellschaft* 27: 189–237.

Galaty, John. 1993. "Maasai Expansion and the New East African Pastoralism." In *Being Maasai: Ethnicity and Identity in East Africa,* edited by Thomas Spear and Richard Waller, 61–86. London: James Currey.

Gooding, Philip. 2022. "Climate Change and Political Instability in Equatorial Eastern Africa, 1876–84." *International Journal of African Historical Studies* 55 (2): 207–29.

Herring, R. 1979. “The View from Mount Otuke: Migrations of the Lango Omiro.” In *Chronology, Migration and Drought in Interlacustrine Africa*, edited by J. B. Webster, 294–305. London: Longman & Dalhousie University Press.

Hildebrandt, J.M. (1877). “On his Travels in East Africa.” Trans. C.P. Rigby. *Proceedings of the Royal Geographical Society* 22, 446–53. https://doi.org/10.2307/1799846.

Hollis, A.C. 1905. *The Masai.* Oxford: Clarendon.

Hollis, A. C. 1907. “The Masai and their Traditions.” *London Quarterly Review* 108: 103–4.

Hughes, Lotte. 2006. *Moving the Maasai: A Colonial Misadventure.* Basingstoke: Palgrave Macmillan.

Jennings, Christian. 2005. “Beyond Eponymy: The Evidence for Loikop as an Ethnonym in Nineteenth-Century East Africa” *History in Africa* 32: 199–220. https://doi.org/10.1353/hia.2005.0012.

Kosmin, Paul J. 2018. *Time and Its Adversaries in the Seleucid Empire*, Cambridge: Belknap Press.

Krapf, Johann L. 1854. *Vocabulary of the Engutuk Eloikob.* Tübingen: Lud. Fried. Fues.

Lamphear, John. 1976 *The Traditional History of the Jie of Uganda.* Oxford: Clarendon Press.

Lamphear, John. 1988. “The People of the Grey Bull: The Origin and Expansion of the Turkana.” *Journal of African History* 29 (1): 27–39. https://doi.org/10.1017/S0021853700035970.

Lamphear, John. 1998. “Brothers in Arms: Military Aspects of East African Age-Class Systems in Historical Perspective.” In *Conflict, Age and Power in North East Africa: Age Systems in Transition,* edited by Eisei Kurimoto and Simon Simonse, 79–97. Oxford: James Currey.

Lamprey, Richard and Richard Waller. 1990. “The Loita-Mara Region in Historical Times: Patterns of Subsistence, Settlement and Ecological Change.” In *Early Pastoralists of South-west Kenya,* edited by Peter Robertshaw, 16–35. Nairobi: British Institute in Eastern Africa.

Lonsdale, John M. 1977. “The Politics of Conquest: The British in Western Kenya, 1894–1908.” *The Historical Journal* 20 (4): 841–70.

Mol, Frans. 1996. *Maasai Language and Culture Dictionary.* Lemek: Maasai Centre Lemek.

Petek, Nik. 2018. “Archaeological Perspectives on Risk and Community Resilience in the Baringo Lowlands, Kenya.” PhD diss., Uppsala University.

Petek, Nik, and Paul Lane. 2017. “Ethnogenesis and Surplus Food Production: Communitas and Identity Building Among Nineteenth and Early Twentieth-Century Ilchamus, Lake Baringo, Kenya.” *World Archaeology* 49 (1): 40–60.

Playne, Somerset, and Frank H. Gale, eds. 1908. *East Africa (British): Its History, People, Commerce, Industries, and Resources.* London: Foreign and Colonial Compiling and Publishing Co.

Rockel, Stephen J. 2022. "A Forgotten Drought and Famine in East Africa, 1883–1885." In *Droughts, Floods and Global Climatic Anomalies in the Indian Ocean World*, edited by Philip Gooding, 289–343. Cham: Palgrave Macmillan.

Sandford, George R. 1919. *An Administrative and Political History of The Masai Reserve*, London: Waterlow & Sons Ltd.

Schlee, Günther. 1989. *Identities on the Move: Clanship and Pastoralism in Northern Kenya*. Manchester: Manchester University Press.

Simpson, George L. and Peter Waweru. 2012. "Becoming Samburu: The Ethnogenesis of a Pastoral People in Nineteenth Century Northern Kenya." *Journal of the Middle East and Africa* 3 (2): 175–97.

Stigand, Chauncy H. 1913. *The Land of Zinj: Being an Account of British East Africa, Its Ancient History and Present Inhabitants*. London: Constable & Company Ltd.

Sikes, H. L. 1936. "Notes on the Hydrology of Lake Naivasha." *Journal East Africa and Uganda Natural History Society*, 13 (1–2): 73–84.

Sobania, Neal. 1980. "The Historical Tradition of The Peoples of the Eastern Lake Turkana Basin c.1840– 1925." PhD diss., School of Oriental and African Studies London.

Sobania, Neal. 1993. "Defeat and Dispersal: The Laikipiak and Their Neighbours at the End of the Nineteenth Century." In *Being Maasai: Ethnicity and Identity in East Arica*, edited by Thomas Spear and Richard Waller, 105–19. London: James Currey.

Sobania, Neal, and Richard Waller. 1989. "Oral History and the End of Time." Unpubl. paper, SOAS, London.

Sommer, Gabriele, and Rainer Vossen. 1993. "Dialects, Sectiolects, or Simply Lects? The Maa Language in Time Perspective." In *Being Maasai: Ethnicity and Identity in East Africa*, edtied by Thomas Spear and Richard Waller, 105–19. London: James Currey.

Sorrenson, M.P.K. 1968. *Origins of European Settlement in Kenya*. Nairobi: Oxford University Press.

Spear, Thomas, and Richard Waller, eds. 1993. *Being Maasai: Ethnicity and Identity in East Arica*. London: James Currey.

Spencer, Paul. 1998. "Age Systems and Modes of Predatory Expansion." In *Conflict, Age and Power in North East Africa: Age Systems in Transition*, edited by Eisei Kurimoto and Simon Simonse, 168–85. Oxford: James Currey.

Spencer, Paul. 2003. *Time, Space and the Unknown: Maasai Configurations of Power and Providence*. London: Routledge.

Straight, Bilinda, Paul Lane, Charles Hilton, and Musa Letua. 2016. "'Dust People': Samburu Perspectives on Disaster, Identity and Landscape." *Journal of Eastern African Studies* 10 (1): 168–88. https://doi.org/10.1080/17531055.2016.1138638.

Sutton, John E.G. 1973. *The Archaeology of The Western Highlands of Kenya*. Nairobi: British Institute in Eastern Africa.

Sutton, John E.G. 1987. "Hyrax Hill and the Sirikwa: New Excavations on Site II." *Azania: Archaeological Research in Africa* 22 (1): 1–36.

Sutton, John E.G. 1998. "Archaeological Sites in East Africa: Four Studies." *Azania* 33: 1–37.

Styles, Megan A. 2019. *Roses from Kenya: Labor, Environment, and The Global Trade in Cut Flowers*. Seattle: University of Washington Press.

Tiki, Waktole, Gufu Oba, and Terje Tvedt. 2013. "An Indigenous Time-related Framework for Reconstructing the Impact of Disasters on Ancient Water Systems in Southern Ethiopia, 1560–1950." *Journal of Historical Geography* 41: 33–43. https://doi.org/10.1016/j.jhg.2012.12.001.

Thomson, Joseph J. 1887. *Through Masai Land: A Journey of Exploration Among the Snowclad Volcanic Mountains and Strange Tribes of Eastern Equatorial Africa*. London: S. Low, Marston & Co.

Tonkin, Elizabeth. 1992. *Narrating Our Pasts: The Social Construction of Oral History*. Cambridge: Cambridge University Press.

Verschuren, Dirk, John Tibby, Peter R. Leavitt, and C. Neil Roberts. 1999. "The Environmental History of a Climate Sensitive Lake in the Former 'White Highlands' of Central Kenya." *Ambio* 28 (6): 494–501.

Waller, Richard. 1976. "The Maasai and the British 1895–1905: The Origins of an Alliance." *The Journal of African History* 17 (4): 529–53. https://doi.org/10.1017/S002185370001505X.

Waller, Richard. 1978. "'The Lords of East Africa': The Maasai in The Mid-Nineteenth Century (c1840–c1885)." PhD diss., Cambridge University.

Waller, Richard. 1984. "Interaction and Identity on the Periphery: The Trans-Mara Maasai." *International Journal of African Historical Studies* 17 (2): 243–84. https://doi.org/10.2307/218606.

Waller, Richard. 1985. "Economic and Social Relations in the Central Rift Valley: The Maa-Speakers and Their Neighbours in the Nineteenth Century." In *Kenya in the Nineteenth Century*, edited by B.A. Ogot, 83–133. Nairobi: Bookwise.

Waller, Richard. 1985. "*Emutai*: Crisis and Response in Maasailand 1883–1902." In *The Ecology of Survival: Case Studies from Northeast African History*, edited by Douglas H. Johnson and David M. Anderson, 73–112. Boulder: Westview Press.

Waller, Richard. 1995. "Kidongoi's Kin: Prophecy and Power in Maasailand." In *Revealing Prophets: Prophecy in Eastern African History*, edited by David M. Anderson and Douglas H. Johnson, 28–64. London: James Currey.

Weatherby, J. M. 1967. "Nineteenth Century Wars in Western Kenya." *Azania* 2 (1): 133–44. https://doi.org/10.1080/00672706709511444.

CHAPTER 3

Environmental Conservation and Sustainability: Rhetoric and Reality

David Harper and Megan A. Styles

1 The Beginnings of Environmental Knowledge of the Lake and Its Hinterland[1]

The conservation of the natural resources of Lake Naivasha and its hinterland is not, as some observers might imagine, a very recent activity brought about by the enormous growth of horticulture around the lake and its perceived impacts. Rather, it has a history of over a hundred years, associated initially with the negative environmental impacts of European settlement and early gazettement by the colonial government of riparian land for white settlers. It has developed momentum since Kenya's independence, initially because of relocation of formerly landless Kenyans in the catchment in the decades after independence, and latterly through the immigration of tens of thousands of citizens to work in horticulture and agriculture, living on subdivided riparian land that had been sold to developers, in settlements with no paved roads, water or sanitation.

At several points in the twentieth and twenty-first centuries, conservation advocates have argued that matters of development in the lake area should be brought under centralised control. Unfortunately, each of these attempts to create a coordinated plan for managing the lake (or the water resources available for agriculture and other land uses) has faltered. Efforts to gain control over conservation in the lake area came at times when new land uses (e.g. tourism and sportfishing, industrial floriculture, then geothermal power generation) each posed new threats to both the ecological integrity of the lake and the power dynamics within the local community. Although none have been successful at preventing the ongoing ecological degradation of the lake, each has introduced new regulatory strategies or partnerships in governance that have resulted in some positive changes.

1 First submission date of this chapter: April 19, 2021. Acceptance date: March 13, 2023.

 | DOI:10.1163/9789004695429_004

Prior to the coming of Europeans in the final two decades of the nineteenth century, the hydrologically fluctuating Lake Naivasha, with its surrounding wetlands, was an important watering point and dry-season refuge for the nomadic pastoralists of the Rift Valley floor – for several millennia. It had been a freshwater lake in the previous 300 years, probably greater in extent than any experienced in the twentieth century, but with a short period in the early- to mid-nineteenth century when its water level was very low, to the extent of probably being saline and confined to the small Crescent Island lagoon (Van der Meeren et al. 2019). It is likely that trading convoys of the Sultan of Zanzibar passed Naivasha on their way to the interior for slaves and ivory for several hundred years before the first record-keeping explorer, followed by missionaries from the nineteenth century onwards. The first map showing the approximate shape and location of the lake was that of H.M. Stanley in 1874 (Stanley 1878) but, since he did not visit the lake, it must have been visited and mapped by some of those travellers. The earliest written record appears to date from 1883, when Joseph Thomson (1887) reached Naivasha, on the Rift Valley slopes level with Mount Longonot, seeking a shorter route from the coast to Lake Victoria than was previously known through what is now Tanzania (Thomson's route was taken by the Uganda railway shortly after).

With hindsight, early explorers made an eloquent case for the conservation of both the lake and the adjoining lands, without realising the existence or value of any drawdown zone (the zone of vegetation between high and low water levels). The birds and mammals were prolific – the lake, in 1883 "was one moving mass of ducks with ibises, pelicans and other aquatic birds" (Thomson 1887, 193) with "… thousands of waterfowl screaming up into the air. The amount of birdlife on the lake and its shores was prodigious". Frederick Jackson (1930, 189), appointed Administrator in 1895, wrote "… the whole surface was thickly dotted with coots, thousands of them … an outer border of small lots of yellow-billed ducks, pochards and pink-backed teal … the whole so totally unconscious of danger, and strangers to a shot gun, that it not only appeared to be, but was downright barbarous to molest them". Yet, only a few years later Johnston (1902, 10) wrote: "The shores of lake Naivasha are in many places belted thickly with papyrus. The water-birds are not quite as abundant as formerly owning to the extent to which they have been shot at by Europeans".

The grasslands on the eastern side of the lake held "… thousands of zebras … two large herds … It was a magnificent sight to watch these beautiful animals thundering along in the grass in great squadrons" (Thomson 1887, 184) and "All along there is sweet, short grass, kept low by the grazing of innumerable gazelles and Masai [*sic*] sheep and cattle" (Johnston 1902, 7) and in 1910 "game of all kinds is plentiful; antelope, zebra, ostrich, and birds of a large size or beautiful plumage" (Bland-Sutton 1911, 21). A few years later, however, Hobley wrote "The wealth of game ... up to about 1905 ... was remarkable. It is to be feared that nowadays most of it has vanished beyond recall" (Hobley 1929).

Settlement by Europeans became inevitable after 1885 when the Berlin Conference divided up Africa for European colonial invasion and the British proclaimed an East African Protectorate. They quickly declared "waste and unoccupied land" to be "Crown Land", although Figure 3.1 illustrates that large parts of these lands were not, in fact, unused. The Uganda railway (often known as the "Lunatic Express" because of the difficulties involved in building it from Mombasa to Kisumu) commenced at Mombasa in 1896 after a decade of land surveys, and constructed a station at Naivasha in 1900, reaching Kisumu a year later. Sir Charles Elliot, the then High Commissioner, encouraged settlers from Europe and South Africa. Maasai were moved away from the area around the lake to allow European settlement as part of the (now controversial) treaties their leaders signed with the British East African Protectorate, initially in 1904 (see Waller, this volume). Evidence of how modern Maasai migrate with their

FIGURE 3.1 Image of a Maasai kraal above Naivasha, taken by explorer Ernest Gedge on September 10, 1889. Gedge noted in his diary that the kraal had been "deserted" as the inhabitants moved to Laikipia, north-west of Naivasha. These moves were seasonal

cattle in harmony with wildlife, in tune with the seasons, suggests that there would have been no negative environmental impact from their uses of the lake for water and food prior to the European invasion (Kariuki et al. 2021). In official correspondence, colonial officials themselves recognised that Maasai activity in the Lake Naivasha area "had over time greatly improved the grazing and made it sweet" (Charles Hobley, quoted in Hughes 2006, 29). Some officials recommended restricting white settlement between Naivasha and Nakuru in recognition that these were prized Maasai grazing lands. Ultimately, fears that Maasai cattle would spread disease to white livestock and concerns that settler "belligerence" would lead to ongoing conflict between the two communities led officials to proceed with a campaign of legal and physical displacement of the Maasai from the Central Rift Valley.

European settlement – with the construction of fences and more intensive agriculture –accelerated in the next few years up to the First World War. In addition to land expropriated for the establishment of white agriculture, a "colonial conservation estate" (Murombedzi 2010)[2] was created at Naivasha: all of the land surrounding the lake passed into private ownership, except for a few "lake corridors" designated for livestock watering; the exact number of these access points is uncertain, but it is understood to have been about a dozen (Morrison 2013), legally retained as government land.

The earliest settlement at the beginning of the twentieth century happened at a time when the lake was at a very high stand compared to anytime subsequently; it rose a further four vertical metres approximately in the first decade of the twentieth century and then receded from that highest recorded level. Settlers after 1910 could thus follow the lake as it receded, cultivating the additional new land that they had acquired. This decline from the highest lake stand led to the realisation among white colonial settlers that there had to be a method of demarcating newly acquired riparian land and arbitrating disputes, so they formed an association in 1929, and negotiated an agreement with the

2 Based on European ideologies that cast native populations and their subsistence practices as threats to "wild" nature, land throughout African colonies was actively expropriated for conservation. Native populations were forcibly excluded from these lands and/or their subsistence activities were outlawed. In many cases, significant, ongoing use of force was required to remove native peoples and "defend" these areas from what was perceived as human encroachment (but was actually customary use of traditional territory). This is also known as "fortress conservation" (Brockington 2002). In Naivasha, the assumption was that colonial landowners needed to be similarly empowered to defend the riparian zone adjoining their properties from encroachment by native populations (e.g. agricultural "squatters" and pastoralists) and colonial officials who might try to subdivide and sell land newly exposed when the lake level dropped.

colonial government that was gazetted into law in 1932. Among other restrictions, the 1932 Riparian Agreement banned owners from constructing any permanent buildings on riparian land, defined as the contour of the lake level in 1906. The agreement guaranteed landowners that their property boundaries would extend to the lake level, even when this fluctuated. The lake's riparian area would remain Crown land, but adjoining landowners would be recognised as caretakers of this area. Landowners were thus allowed to cultivate in the riparian area and also retained the right to build pumphouses for irrigation purposes.[3]

Debate over public access to the lakeshore emerged in the 1920s as settlers gained formal title to land adjoining the lake. In 1927, Captain Archie Ritchie, head of the Game Department, wrote to the Colonial Secretary complaining that there were only four passable outspans (access points with enough width to allow access to the lake and to water, and to provide enough room for an ox-cart to turn). He suggested that the government formally retain the right to three of these four locations for use by the Game Department. White landowners objected strongly to this plan, noting that this could set a precedent for the government essentially seizing private land and that this could allow "Asiatic" and other non-white community members to access this land and cause a "nuisance".[4] Ritchie's persistence on behalf of the Game Department helped establish the locations of the few public access points that still exist today (some of which remain contested, see Kuiper and Gravesen, this volume).

The legal corridors when originally gazetted had no consideration of wildlife value, although they were used for access to the lake water, as was any unsettled riparian land when lake levels were low. The gazettement and subsequent settlement of land around the lake took no notice of mammals and birds, which survived in reduced numbers, especially on the ranches with greater land area stretching away from the lake.

2 Management and Conservation Initiatives

The arrival of the first scientific expeditions to study Rift Valley lakes in 1928 and in 1932 provided a baseline documentation of the organisms living in the

3 Correspondence related to settler's concerns about foreshore rights on Lake Naivasha and the negotiations that led to the founding of the Lake Naivasha Riparian Owners Association and the 1932 Riparian Agreement can be found in folio AG/43/39, "Foreshore Rights: Lakes – Naivasha, Elementeita, and Nakuru: General (1928–1944)" in the KNA. A summary can also be found in Styles (2019, 35–36).

4 Summarized in Styles (2019, 35–36) based on correspondence included in folio BN/44/1 found in the KNA.

lake. Fish Warden R.E. Dent had surveyed the lake in 1926 and found that "there were no fish in the lake except *Haplochilichthys antinorhii*, a minnow-like fish of no value to anyone" (Worthington and Worthington 1933). In the same year, Dent introduced *Tilapia nigra* (from the Athi river near Thika) to create a feeder-fish population for a future sport fish. He then imported fingerlings of the (predatory) American black bass from the UK, and introduced those few that had not died on the voyage to the lake at several points in February 1929. The introduction of this alien species to Lake Naivasha captured the negative attention of the trustees of the British Museum, who felt that a more comprehensive survey of the lake was necessary. The matter was also debated by members of the Linnean Society of London in November 1929. Ultimately, the colonial government supported Dent's decision, noting that a sport fishery in Naivasha would be of great economic benefit.[5]

Europeans fishing the lake with rod and line caught both species, plus rainbow trout which had washed down from stocks in the upper Malewa River. By 1931, the Lake Naivasha Riparian Owners Association (LNROA) was considering calls to create a commercial fishery, and Dent reported to a meeting that there was "very considerable illicit netting of fish by whites and natives and sale of about 1,000 fish a day".[6] The LNROA voted at that time not to proceed with a commercial fishery; a small artisanal fishery was eventually opened in 1950 by the government. Fourteen years later came the first record of biodiversity decline, as *H. antinorhii* could no longer be found by 1964.

The introduction of fish to Lake Naivasha and increasing subdivision and settlement of land by white settlers in the early 1930s triggered the first official call for centralised conservation measures in the lake area. In 1932, Captain Ritchie of the Game Department wrote the first of many impassioned letters to colonial authorities voicing his concerns about rapid development in Naivasha.[7] He suggested that a centralised Board of Control be established and lobbied the government for additional resources and personnel. He wrote,

5 Debate surrounding the introduction of the bass to Lake Naivasha can be found in folio CO533/388/2, "Proposed Introduction of Specimens of American black bass into Lake Naivasha (1929)" in the Public Records Office, London, UK. This correspondence is also summarized in Styles (2019, 37).

6 Recorded by LNROA member Dacre Shaw in an unpublished diary, quoted in the unpublished manuscript "Lake Naivasha: An Extended History" (1976) by Mervyn Carnelley (private collection of the Carnelley family, Naivasha, Kenya).

7 Ritchie's letters pleading for control over development can be found in the folio titled Lake Naivasha KP/4/8, "Lake Naivasha (1927-1935)" in the KNA. This period is also summarized in Styles 2019, 37-38). This quote comes from a letter from Ritchie to the colonial secretary dated April 27, 1932. His letter to the colonial secretary advocating again for a Board of Control is dated February 2, 1933.

"I feel strongly that Lake Naivasha should be placed under firm control; and that, with the least possible delay. Damp clay can be moulded; but dried, it can only be chipped". At the time, LNROA members opposed this proposal on the grounds that they should not have to share control over the lake with parties from Nairobi, the Game Department, and the County Council. On the basis of their strong objections, no Board of Control was established; however, the colonial government worked closely with the LNROA on matters regarding lake-area governance. For example, the LNROA was sent drafts of new fishing and boating regulations for review and comment. Although the LNROA would later become much more involved in conservation initiatives, its early decades after its foundation were inward-looking about land-rights concerns, desires for local self-regulation, and resistance to external (or strongly centralised) regulatory control, especially regulations that would impede landowners' abilities to develop their properties commercially. Throughout most of the colonial era, the LNROA mainly arbitrated disputes between members about property boundaries, many of which arose due to changes in the lake level and absentee landownership (which left land open for use by "squatters"). Concerns about squatters became especially heated during the Mau Mau uprising between 1952 and 1960.[8] In 1961, as the terms for the end of British rule in Kenya were being negotiated, one lakeside landowner suggested that the 1932 Riparian Agreement might give owners of lakeside plots "the right to secede from the rest of Kenya and remain under the Crown as a Colony, unless the Crown would agree with the plot holders to compensate them for a change of ownership". LNROA leadership did not seriously pursue this suggestion, but they did write to the Settlement Board on behalf of members,[9] offering more than 10,000 acres of surveyed land and more than 2,000 acres of riparian land for purchase as a block.[10] Colonial authorities did not pursue this offer, and to some extent, the area surrounding Lake Naivasha remained a uniquely white space in Kenya after independence.

The LNROA dealt with several other management issues during the decades either side of independence with similar management pressures – a proposal

8 The Mau Mau was an insurrectional movement, strongly fueled by land claims, which led to the establishment of a state of emergency and undermined the colonial power in place.

9 The Kenya European Agricultural Settlement Board was established in 1946 and constituted by ordinance in 1948. It consisted of members appointed by the colonial governor to provide advice on matters of European settlement in the Kenyan Highlands.

10 This summary is based on private letters and meeting minutes in the possession of the Lake Naivasha Riparian Association (LNRA). The letter proposing that lakeside landowners might have a right to secede from Kenya was written by C.T. Todd and dated March 16, 1961.

to cull hippopotamus because their nocturnal grazing damaged crops was turned down; farmers stated that there were "over 1,500" but surveys by scientists a decade apart (1976 and 1987) found a stable population of about 230 by boat and road censuses (Smart 1988). A decade later however, an aerial census did find closer to 1,000 (Douglas-Hamilton, unpublished data). There was also a concern that duck shooting on the lake was being restricted by the competitive effect of coot upon the duck. A report commissioned in 1968 found about 40,000 coot and 15,000 duck, but no evidence of any competition. That report also recommended that riparian owners should create a management plan for the lake and create clear zones where specific activities would be allowed to take place (Parker 1968). During the 1960s and 70s, the LNROA voted to ban duck hunting and passed a number of regulations focused on boating and fishing.

The LNROA became a much more "conservation-oriented" organisation as a result of these management issues, but starting in the 1960s it also had to take a stance in response to members' concerns about invasive species and a period of rapid growth in the tourism industry. A number of new campgrounds and hotels, including the large Marina Club, which stored speedboats, opened at the lakeshore in the 1960s and 1970s. Around the time of independence in 1963, the first damaging (at least visually) alien species appeared that needed control – the coypu. This large South American rodent had escaped from a fur farm in Nanyuki and travelled along the rivers – and was controlled by shooting. In the late 1960s, *Salvinia molesta*, a South American floating fern that had spread around the globe and arrived at Naivasha (said to be keeping the bait moist in a tin brought by a bass fisherman from Nairobi) was considered for control by aerial and hand spraying using the herbicide Paraquat. *Salvinia* is controlled more, however, by wave action as a result of the afternoon storms which the lake experiences with winds from the surrounding highlands than by any human intervention; the possibilities for biological control were discussed but never implemented (Kongere 1979). Its impact on the lake became considerably reduced over time, probably as a consequence of competitive exclusion by the invasive flowering water hyacinth, another global invasive that spread from South America because of its beautiful flowers, which were sought for garden ponds. It was first recorded in the 1980s but became invasive in the lake after 1993.

The years immediately after Kenya's independence saw a second burst of scientific studies at the lake – encouraged by an LNROA Secretary, Roger Mennell, who hosted visitors and made available a small laboratory, which he had earlier set up to support his export of geranium oil for perfume from flowers he grew on riparian land. The important role of the beds of papyrus in sequestering sediments and nutrients from incoming rivers, as well as the high flowering-plant biodiversity caused by the "drawdown zone" of this uniquely fluctuating lake, were established for the first time by John Gaudet, a US academic attached to Nairobi University, and his PhD students (Gaudet 1977, 1979; Gaudet and Muthuri 1980). These accommodation and laboratory facilities were maintained for several years after Mennell's death until about 1989.

Almost as soon as the botanical diversity of the lake was fully documented, it began to disappear, with the destruction of submerged plant beds and beautiful water lily pads by yet another alien, deliberately introduced by the Kenya Fisheries Department into the fishery in 1970 – the Louisiana crayfish (Parker 1974). This initially seemed to be an economic benefit when crayfish were exported to France and Sweden for about 5 years, as well as a social benefit (to Europeans) when international film-makers Joan and Alan Root showed a delicious barbecue they made, cooked in a hot-air balloon, of dozens of crayfish that had "dripped off" water lilies caught on their anchor! This latter event was in May 1974, yet by September 1977 "there were only isolated lilies left when searching for small lily pads to transplant into the pond near the house" (Joan Root, Harper personal communication, April 2004). The crayfish had eaten the lilies and were almost certainly also responsible for major losses of invertebrates: two species of leeches, the larger of which formerly caused havoc amongst sailors and fishermen – "so bad that one simply did not go wading in the water" (Ian Parker, Harper personal communication, January 1995) – had also disappeared. The full impact of crayfish only became apparent when the elimination of all native submerged and floating vegetation from the lake was scientifically recorded from 1982 onwards (Harper 1992; Harper, Adams, and Mavuti 1995; Gouder de Beauregard et al. 1998; Smart et al. 2002), with the exceptions of Lake Oloidien and Crescent Island Lagoon (see Map 1.1) which remained crayfish-free for about 5 years and were thick with submerged plants and water lilies because they were more isolated. Crayfish populations themselves collapsed once the plants had gone, through starvation and due to their becoming easily accessible to visual predators. This created an oscillation of high plant density with low numbers of crayfish and *vice versa* that would have had about a decadal frequency. The oscillation did not continue for more than one decade, however, because the new alien invasive, Water Hyacinth, proliferated as dense littoral mats in the 1990s, far larger than and more disruptive than *Salvinia* had been two decades earlier. Crayfish lived in the hyacinth roots, sheltered from predators by day, dropping out to feed on detritus and suppressing regrowth of native water plants by eating germinating shoots at night. The hyacinth-crayfish partnership lasted until the turn of the century, when it was broken by yet another alien, deliberately introduced by the Kenya Agricultural

Research Institute to control hyacinth, and submerged plants returned for a short time (full details in Harper et al. 2011).

These new threats posed by tourism and invasive species led conservation advocates in the lake area to propose that it be brought under centralised control, echoing Captain Ritchie's 1932 plea for a Board of Control. In an article published in the *East African Standard* (MacDonald 1970), Ian Parker, Director of the consultancy Wildlife Services Limited, advocated for the "revolutionary idea" of setting up a lake authority "to supervise all future development around the lake and coordinate the activities of both official and private interests". Without such a central body, Parker predicted that "[the lake area community] will just move from crisis to crisis". He envisioned this as a body that would "take some of the control at present exercised by both local and central government departments" and involve private entities in decision-making as well. A *Daily Nation* article dated May 30, 1971, titled "Master Plan Needed for Lake Naivasha", summarised the results of a limited ecological survey conducted by I. Parker, T. Singh and R. Murray Watson and commissioned by the East African Wildlife Society following reports from members of the Naivasha Gun Club citing declining duck populations (Parker 1968). The survey cited concerns about agricultural chemicals entering the lake, the spread of Salvinia, conflicts between birdwatchers and hunters, exploitative levels of gill-net fishing, and no boating speed limits. Watson, Singh and Parker (1970) again suggested the need for a "lake authority board" that would establish "a machinery to assign priorities of exploitation" and integrate "the great number of conflicting human interests in the lake".

In 1971, LNROA members discussed a formal proposal to create a Naivasha Authority to supervise lake area development.[11] A majority of members supported the initiative, but the LNROA leadership decided to postpone any formal action. Roger Mennell, LNROA Secretary, later wrote to one of the plan's staunchest supporters to explain the executive board's decision not to act. He explained that, while the membership had voted "overwhelmingly" in support of this, the board felt that it was more important to protect members' rights to cultivate riparian land and not draw attention to the fact that this was government land in light of "a good deal of talk in certain quarters about using this land for settling Kikuyu who have no plots".[12] Once again, lakeside

---

11 This discussion is recorded in the LNROA minutes from 1971 and in documents on file in the organisation's private archive. Mennell explains this outcome in a letter to Colonel C.M. Cowie dated December 4, 1971.

12 See Kuiper and Gravesen (this volume) on shifting land tenure arrangements around Lake Naivasha in this period.

landowners chose not to create a central authority that could steer development in Naivasha because they feared that this would disrupt local control (by surrendering this to the central government and other "outsiders") and restrict their own ability to commercially develop their plots. This time, concerns about the seizure of white-owned land for native resettlement were also paramount.

In the early 1990s, the LNROA was reinvigorated by a new Chairman, Andrew Cole (later Lord Enniskillen) in response to a new perceived threat to socio-political and ecological order in Naivasha – the rapid growth of commercial floriculture. Under his leadership, efficiently supported by a new Secretary, Mrs. Sarah Higgins, the organisation was renamed the LNRA or Lake Naivasha Riparian Association (dropping "Owners" as a pre-qualification), opening membership to anyone with an interest in the lake.

The emergence of multi-party politics in Kenya at this time also gave LNRA leaders a sense that there could be new opportunities for working with the government to advocate for change; however, they worried that rising property values in the lake area might inspire a new wave of government corruption and land-grabbing. They wanted to make Lake Naivasha visible on an international stage to put pressure on officials to control development and protect their land rights (Styles 2019, 51).

The LNRA initiated three successive consultants' reports about what was known scientifically of Lake Naivasha (LNRA 1993), leading to pressure on the government's conservation agency, Kenya Wildlife Services (KWS), to declare Lake Naivasha as the county's second protected site under the international Ramsar Convention for Wetland Conservation in 1995. This paralleled wider environmental concerns across the country, which led to the enactment of the Environment Management and Coordination Act 1999 (EMCA) and the establishment of the National Environment Management Agency (NEMA) to regulate and oversee the management of the environment.

The LNRA also wrote the first Management Plan for the lake using the consultants' reports, which was promoted by IUCN to the Kenya Government and was gazetted under the EMCA in 2004. The LNRA was awarded the Ramsar prize in 1999 for this (Ramsar 2005). At the same time, the LNRA began a collaboration with European and Kenyan research scientists and the UNESCO International Hydrological Programme (IHP) to host Naivasha's first international conference – "Science & Sustainability of Freshwaters" – at the recently constructed KWS Training Centre, on a hill overlooking Lake Naivasha (Harper et al. 2002). That collaboration has lasted until the present day through UNESCO IHP's sub-programmes of "HELP" and "Ecohydrology", contributing to the academic theory of sustainable water management in parallel with the community-based practical management espoused by Ramsar (Harper and Zalewski 2001; Wanjala et al. 2018).

As the new century unfolded, negative issues of conservation and sustainability were becoming more obvious. A major concern of lake management for conservation had long been the papyrus fringe, which had been shown to be so important in filtering nutrients and sediment from the land, preventing them enriching lake water, yet so easily destroyed by farmers for planting or landowners such as hotels for access and uninterrupted views of the lake. Increasingly, conscientious horticulturalists used the ecological principles of papyrus functioning, together with other wetland plants, to retain and metabolise contaminants from their operations in wetland waste-water treatment plots. The largest companies, especially those which sell direct to supermarkets such as Homegrown (now Finlays) and Oserian, publicised the fact as part of their sustainability initiatives, using it towards gaining "Fair Trade" and other certifications. Other large sources of drainage waste, particularly from settlements, have no such treatment and have increasing and measurable impacts, some of which are also visible, such as a large sediment peninsula below the settlement Kamere created by erosion from its trackways (Boar and Harper 2002) and high phosphate measurements close to where Naivasha town sewage effluent enters the lake (Kitaka, Harper, and Mavuti 2002).

The 1999 conference had produced evidence for the over-exploitation of lake water by the rapidly growing horticultural industry (Becht and Harper 2002), but no evidence of chemical contamination of the water by flower farms. This did not prevent newspaper articles at this time from reporting the unsupported opinions of local residents that fertilisers and pesticides from flower farms were having severe impacts on lake water (e.g. IOL 2003). There had been some evidence from studies which showed that first-generation pesticides banned in most countries – such as DDT, Dieldrin, and Atrazine – were still measurable in the environment (Onyango, Kruzinger, and Nzula 2015). No evidence existed then, or has been found since, which has shown any negative impact on the lake ecosystem, however (Pacini et al. 2021).

Physical degradation of the riparian zone in places had been occurring for several decades and was shown, for example, in population declines of the most charismatic species native to Lake Naivasha in this modern age – the African Fish Eagle (Harper et al. 2002). The species needs mature trees overhanging or close to the lake edge, and damage had been visually obvious for some years, in a few riparian properties, in the cutting of mature trees close to the lake edge and the construction of "permanent" buildings and walls to hold back water on riparian land, both illegal under the Riparian Agreement of 1932. The LNRA had tried unsuccessfully to take action against individual owners.

Despite enormous investments of time, money, and international attention into conservation initiatives in the lake area, the LNRA-led efforts to establish multi-stakeholder agreement for their plan to manage Lake Naivasha had begun to falter by 2005, in part because the growth of floriculture fundamentally shifted power relations in the area. Between the mid-1990s and 2005, the LNRA had worked closely with representatives from fifteen different stakeholder groups – including from the flower industry, the tourism industry, local and national government agencies, fisherfolk, and Maasai pastoralists – to broker a comprehensive Lake Naivasha Management Plan. The plan was officially gazetted into law, but a lawsuit was then filed in the Kenyan courts to block its implementation, and an interim injunction was granted, pending further court proceedings. The lawsuit was filed by a prominent Maasai lawyer, who also worked as the human-resources manager for a Dutch-owned flower farm. The lawsuit named three plaintiffs – two community groups and a local lakeside landowner.

According to those involved in the management-plan process and the legal case surrounding it, the initiative unravelled for several reasons (Styles 2019, 51–56). Representatives from the cut-flower industry, arguably the most influential group in the region until geothermal power development shifted the balance of power again around 2014, became more heavily invested in their own floriculture-centric management processes, especially the development of localised water governance (through Water Resource Users Associations) after the passing of the new Kenyan Water Act in 2002. They preferred an approach to environmental management in the lake area focused more acutely on water resources. LNRA leaders also became embroiled in a very public conflict with the manager of one prominent flower farm, demanding that the farm remove a manager's residence and greenhouses built below the legal riparian line. More than any other factor, this dispute may have led directly to the legal challenge that stopped implementation of the plan.

The LNRA also struggled to recruit and meaningfully incorporate pastoralist representatives into the initiative; while pastoralist representatives voted in favour of the plan, it was clear that the steering group disagreed about whether and where grazing should be allowed and the legitimacy of pastoralist claims to the lake as ancestral territory. Elected and appointed officials involved in the process also harboured concerns about formally surrendering regulatory authority to community stakeholders, and ongoing government restructuring made the mandates and relative power of different agencies perpetually unclear. In this instance, attempts to establish a central plan for managing Naivasha once again fell victim to concerns about loss of control and limits on development, but following major shifts in political and economic power in the lake area, these worries came from the cut-flower industry, pastoralists, and government agencies (rather than the LNRA's "traditional" members – lakeside landowners involved in tourism and mixed farming). This time around, LNRA members were willing to share power and authority, but for all intents and purposes, they had already lost both.

As the new century began, the first evidence of "toxic algal" blooms appeared in the lake, quite frequent by 2005 (Harper 2006). The first fish kill occurred in the shallow waters of the south-western part of the lake in early 2010, caused by depletion of oxygen in lake water by blooms combined with the bacterial demand of sediment washed in and stirred up by heavy rainstorms (Njiru et al. 2015). This highlighted the struggling commercial fishery, which had been closed due to over-exploitation in 2001–2002, yet now was maintained by a highly dominant, recently arrived (and unexpected) species, the Common Carp (Pacini et al. 2021).

A rapidly declining water level through the first decade of this century resulted in increasing hectares of bare mud, fishermen unable to reach their landing stages, the construction of new urban settlements and hotels following the shrinking lake downwards through riparian land along with tourist enterprises, and distressed hippopotamus in the reduced water area by 2009, leading one journalist to write that the lake was facing a "nightmare reality" (Riungu 2009). This led to greater national activity and attracted greater international attention to the lake. Kenya Wildlife Services, ambiguous in its approach to Naivasha (which is a Ramsar site, but is neither a National Park (NP) nor a Reserve, as all other Kenyan Ramsar sites are) commissioned its own Management Plan, a decade after the first. The new Water Act 2002, gazetted in 2005, giving water users a say in water management in partnership with the government water agency, now began to build up structure. The Naivasha catchment was divided into 12 Water Resource User Associations (WRUAs), partnerships between local users and the Water Resources Management Agency (WRMA).

All these developments brought more funding into environmental conservation activities and more international involvement from NGOs. The Dutch government funded the international charity WWF to develop a training programme for WRUAs and WRMA, as well as its own technological university, ITC, Enschede (now University of Twente, Faculty of Geo-information) to conduct further research into water sources, links between ground and surface waters and availability. The German government's aid agency began to fund several initiatives in water and catchment conservation, as did several large European supermarkets directly. These international supermarkets, which were now buying directly from the larger horticultural and vegetable enterprises rather than the Amsterdam flower auctions, were suffering from customer pressure because their products were not sustainably sourced, despite some farms having "Fairtrade" status (Kuiper and Gemählich 2017). The larger horticultural enterprises were now owned by, or had grown into, international businesses, and these worked with their European customer companies to generate international initiatives, such as Alliance for Water Stewardship, Payment for Ecosystem Services (PES), and Water & Sanitation for Urban Poor (WSUP).

The Lake Naivasha Water Resource Users Association (LANAWRUA) was fast becoming the focus of conservation and environmental governance activity in the lake area. Chaired by a company manager from one of Naivasha's largest flower companies, the committee began developing new policies focused on water abstraction and permissions in the lake area. At the time, stakeholders involved in the creation of the LANAWRUA viewed this as preferable to the LNRA-led process because it was legitimised in Kenyan law (the Water Act 2002) and was structured as a public-private partnership. Government officials viewed this as less of a challenge to their regulatory authority, and local stakeholders (including active pastoralist representatives) initially viewed this management process as more inclusive (Styles 2019, 132–145).

In 2009, while water levels were at their lowest for two decades and capturing national attention, Kenyan Prime Minister Raila Odinga met with the UK's Prince of Wales at a forest-sustainability conference in Norway. The Prince's newly formed International Sustainability Unit (2010) was almost immediately sent on a fact-finding mission to the lake. It realised that there was a lack of direction, without a single coordinating body that had the power to make recommendations that produced real changes. Its report after a few months suggested the formation of a new umbrella group, to be called Imarisha (Swahili for "arise") Naivasha, which was set up under the Prime Minister's Office so it had national influence. Imarisha, joint-funded nationally and internationally, was gazetted in May 2011. For half a decade, in partnership with other organisations within the catchment, particularly the WRUAs through an umbrella organisation of all 12, it made theoretical (if not necessarily practical) headway. It commissioned yet another Management Plan, and produced a 5-year plan for achieving sustainability from 2012–17, building upon the three previous management plans. It completed, with the WRUAs, the first survey of all water users throughout the lake and catchment areas (including groundwater), showing a high proportion that were unlicensed; it persuaded abstractors to agree to a "traffic-light" system of abstraction against water-level change;[13] it brought together and sought to coordinate the researchers

13 This traffic-light system specified when permitted abstractors could draw water from the lake based on current lake levels and climatic conditions. Under "red light" conditions, for

from many nationalities and institutions into a single research group to try to achieve synergy. It also resulted in the establishment of a three-year Water Allocation Plan (WAP) 2010–2012 to help regulate abstraction. Although the WAP's lifeline ended in 2012 and has not been reviewed to date, the plan is still used to regulate abstraction.

By 2014, however, much of the initial energy surrounding Imarisha Naivasha had waned. National elections held in 2013 again shifted the political landscape, abolishing the Prime Minister's office (which had been created as part of the power-sharing agreement that ended a prolonged period of violence and tension following the December 2007 Presidential election). It became unclear which government office would continue to champion this initiative. Another major land use – geothermal energy production – by then had also experienced tremendous growth in the Naivasha area. Supported by funds from the World Bank and private-sector investment, Kenya aspires to generate 5,000 megawatts of geothermal power by 2030. Much of this development is centred on the Olkaria region – in the middle of Hell's Gate NP, just to the south of Naivasha. Geothermal power production requires large amounts of water, and although it is marketed as "green" power, drilling to release steam to generate electricity can produce negative environmental effects and cooling requires large volumes of water from the lake – which is not recorded as abstraction. At the time that this volume is being written in 2022, it is likely that the rising prominence of the geothermal industry will once again shift the balance of power in Naivasha, creating new challenges (and possibly new opportunities) for coordinated conservation efforts. If history provides any precedent, however, new conservation initiatives in Naivasha will be not be allowed to interfere with the development of industry.

All of this has occurred, however, against a background of sustained, heavy rainfall throughout Kenya, causing rising river flows and lake water levels over 5 years, such that by the middle of 2020 the height of the lake was greater than it has been since the mid-1930s (see also Harper et al., this volume). Many buildings in the riparian zone (all illegally constructed) had been flooded, thus all the initiatives of Imarisha and its partners became largely unnecessary as water became surplus to both domestic and industrial needs. The lake-water ecosystem achieved better "health", with greater transparency and lower nutrient concentrations, because of its greater volume, despite nutrients being brought into the lake by flooding pit-latrines. Papyrus regerminated and submerged plants reappeared in a limited number of places. Wildlife-conservation

instance, users could draw only small amounts of water designated for critical uses and/or abstraction was disallowed completely.

issues remained, as terrestrial animals confined to riparian grasslands – including hippopotamus feeding there at night – were concentrated in smaller land areas, but commercial fisheries remained buoyant with a greater lake area and yet another new species, Catfish, despite the very high offtake by unlicensed, shoreline fisherman using fine-meshed seine netting (offtake compared in Pacini et al. 2021).

Lake Naivasha has had four management plans written about how its resources, its "natural capital", should be sustainably managed. Ten years ago, it was clearly suffering from the plethora of organisations each managing part of the resource for its own benefit. No single one could coordinate the others; Imarisha was formed to do just this. There is evidence of increased management of water use, mostly from the WRUAs coordinated by LANAWRUA, but little evidence of any holistic management of the ecosystem by Imarisha or any other organisation. There is increased research and monitoring of the fishery by the Kenya Marine and Fisheries Research Institute (KMFRI) (e.g. Keyombe et al. 2018) but little regular attention paid to the ecosystem that sustains it, or monitoring of its status as a Ramsar site (Harper and Mavuti 2006).

The lake is an ecosystem, dependent upon the health of its inflowing rivers (Pacini, Giberti, and Harper, this volume). That includes the riparian zone, but like ecotones all over the world, the lake also connects with its adjacent ecosystems, those that are land-based. We have seen in the quotations above that early explorers found a wealth of terrestrial wildlife, coexisting with Maasai pastoralists, which soon became depleted by European firearms. The continuing survival of populations of most mammals – although generally excluding large and dangerous herds of elephants, lions, and wild dogs – was made possible by the large size of ranches back from the lake, on the higher ground to the north, south and west, where the land was unsuitable for cultivation and there was continuity of these ranches with the forests on the south (Longonot), east (Mau), north (Eburru) and west (Aberdares or Nyandarua Mountains, including Kipipiri).

Forests used to be seen as separate from their recipient lakes, even though they have always provided the "sponge" of soil and vegetation which maintained water flows and water quality in their downstream rivers. The colonial government in Kenya gazetted forest reserves, and conservators looked after gazetted forest reserves which were selectively exploited, but not clear-felled. De-gazettement of forest reserves in the upper parts of Naivasha catchment began after independence, with the settlement of landless families; for example, the forest between Kipipiri mountain and the Aberdares on the north-east side of the catchment was cleared and settled in the 1970s; the forest of the Bahati Highlands in the upper Gilgil and the land between Eburru mountain and the Mau Escarpment on the west side of the lake was cleared and settled in the 1980s. Settlement led to cultivation – often down to the stream edges – needing irrigation in the dry season and suffering erosion in the rains, leading to high sediment and nutrient loads in rivers. The resulting loss of flows and increase in erosion, which was repeated all over rural Kenya by the beginning of the twenty-first century, led elsewhere to infilling of reservoirs and disruption of urban water supplies (Odawa and Seo 2019). Efforts to deal with the problem were highlighted by the creation of the "Water Towers" concept. Initially, this showed the interrelated importance of the Kenyan economy with five upland areas which supplied 75% of all water to its cities – Mount Kenya, the Cherangani Hills, Mount Elgon, the Mau Forest complex and the Aberdares; the latter two in the east and west respectively of the Lake Naivasha catchment. Initially the Kenya Forest Service (KFS) was charged with protecting these areas, but in 2012 a separate government agency was set up, Kenya Water Towers, which has subsequently declared 18 more water towers that are of more local importance in the country.

The conservation of biodiversity in forests, as a vital, but less economically obvious ecological service, has only recently become critical because of the loss of said biodiversity through legal and illegal destruction. The northern and eastern edges of the Naivasha basin belong to the Eastern Arc Mountains, part of the Eastern Afromontane Biodiversity Hotspot, recognised to be globally important for species conservation (Mittermeier et al. 2005; Birdlife International 2012). The flagship species of the Eastern Arc Mountains is the bongo; this rare forest antelope is believed to still exist in small numbers within Eburru, and plans have been made to enhance that gene pool with reintroductions once the forest is secured by its fencing (Church 2015).

Reasonable savannah wildlife populations on the Rift Floor of Naivasha's catchment survived throughout the twentieth century, coexisting with livestock, many visited by tourists; indeed, most of the ranches provided tourist accommodation by the mid-late twentieth century. Hell's Gate, the former outflow of the very large Pleistocene lake, was a particular focus of tourism, because its cliffs supported rare birds of prey such as the Lammergeier, with large populations of species of other vultures roosting and eagles nesting and self-sustaining ungulates on the plains of the gorge. The cliffs in the gorge and the volcanic plugs of Fischers Tower were attractive to rock climbers, with some conflict breaking out from time to time between climbers and vultures. The descent of the gorge was attractive to walkers, as was the nearby Longonot crater, because the crater rim could be completely circumnavigated with spectacular views down onto intact forest and savannah inside, although few could venture down because of the vertical sides. These two areas were declared separate National Parks in 1984. Far greater problems than rock-climbers, however, were looming for Hell's Gate NP; geothermal power had already been found to be economically viable in the middle of the park (Nweke-Eze and Adongo, this volume).

Wildlife within that part of Hell's Gate NP above its gorge is provided with waterholes using piped water from Lake Naivasha, linked to a supply that had been built some years previously to supply the ranch land at the bottom of the gorge, further south. There was thus limited need for individual animals to move back to the shores of the lake. Movement did occur however, through those few riparian estates that were not intensively farmed, such as the property of Joan and Alan Root, now Kilendegwa Conservancy, along the south shoreline, which otherwise rapidly developed with flower farms after 1980. Animals from the hilly scrubland around Olkaria volcano in the west of the park would move to and from the south-west side of the lake, to Lake Oloidien (Figure 3.2). This was encouraged by the expanding flower company, Oserian, which purchased the ranch running along the western side of Hell's Gate NP and then opened a private wildlife corridor from the Park's north-west boundary down to Oloidien. This created a large private game ranch for high-end tourism and enabled the general public to watch wildlife crossing South Lake Road as they travelled along it. This corridor was (and is still) seen as critical during the dry seasons; it is part of a much larger unit that includes Mount Longonot NP (KWS 2013, 2017), and to be sustainable, the corridor requires agreements with Kedong Ranch and the design of an innovative solution to address wildlife crossing South Lake Road.

In the area north of the lake, wildlife from the Eburru Mountains and the very large ranches such as Loldia and Marula were able to access the lake more easily, as development was slower further away from Naivasha Town and the main Nairobi-Nakuru highway. The past decade, however, has seen greater and greater subdivision of more ranches into small plots, with a concomitant reduction in ease of movement by wildlife. Such subdivision and settlement is happening all over Kenya, with wildlife dispersal being threatened by the shrinking of migration routes under the impact of intensifying human activities. This is exacerbated by the fact that a large majority of the country's wildlife species reside outside national parks and are undergoing severe decline (Ottichilo et al. 2000). Artificial wildlife corridors are now being designed to protect wildlife-migration routes and to overcome the isolation created by protected areas, several of which have been or are being enclosed by electric fencing. Since 2009, a 400 km-long fence has isolated the Aberdare NP, and now Eburru is being enclosed to conserve bongo (Church 2015). The creation of corridors requires lengthy negotiations with private landowners and

FIGURE 3.2 Image of Lake Oloidien and Lake Naivasha, taken between 1935 and 1956 by T.D. Simon
REPRODUCED BY KIND PERMISSION OF THE COPYRIGHT HOLDERS AND BY THE SYNDICS OF CAMBRIDGE UNIVERSITY LIBRARY, ITEM REFERENCE CODE RCS/Y304R

local communities that must agree to restrict some of their customary uses of the land within the corridor area. Their effectiveness is questioned by some observers (Green et al. 2018) and in many cases it is too soon to obtain a full assessment; nevertheless, the creation of wildlife corridors is a major initiative in the Naivasha catchment. Its legal basis is the 2010 Kenya Constitution, which cites wildlife corridors and dispersal areas within tracts of land that can be acquired by the state in the name of national benefit.

North of Lake Naivasha, a seven km-long corridor is being developed through a collaboration between Rhino Ark, KWS, KFS and the LNRA, and private landowners to link the Eburru National Forest Reserve to the lakeshore via Loldia Ranch. This idea was conceived in 2013 with the commissioning of a feasibility study (Acacia Consultants Ltd. 2013). As the fencing of Eburru forest neared completion under the support of Rhino Ark and the Mpesa Foundation, it became clear that it was hampering wildlife movement out of the forest to the lake and back. "Corridor 1" was conceived, through Loldia Ranch, to include an overpass on North Lake Road allowing animals to cross safely. Operations started in 2015, with on-site observations of wildlife movements using field surveys and camera traps. It is expected that the corridor will be fully operational by the end of 2022. A second corridor is planned, "Corridor II", to link Eburru Forest to the lake through Olsuswa, Korongo and Wileli farms and the Crater Lake Wildlife Sanctuary. Feasibility studies are now underway to monitor wildlife movement and map ideal parcels of private lands for this.

A larger, north-south wildlife corridor (the "LANNECA Corridor") is planned to secure free movement of wildlife between Nakuru Park and Lake Naivasha by connecting key community, private and public lands. It was conceived in 2009 to extend through Lake Naivasha's north swamp and Marula ranch to the Soysambu Conservancy surrounding Lake Elmenteita (Combes 2009). Recent census results (KWS 2017) show this area features the highest densities of remaining ungulates in the central part of the Rift Valley.

The success of these bio-corridors is enormously dependent upon landowner agreement, their acceptance by the county and national authorities, and their adoption into their Spatial and Integrated Development Plans. These plans are critical in planning, zoning and regulating future land use. The Naivasha Integrated Sustainable Development Plan (ISUDP) 2014–2034 has become a crucial document, which zoned these areas for ecotourism and conservation use.

At the time of completion of this chapter, the sustainability of Lake Naivasha and its linked ecosystems (lake – rivers – riparian – catchment) seems precarious because, despite the strong theoretical basis for sustainability-management plans – supported by a considerable volume of scientific publications – there is once again no single body promoting it. There is adequate water, and many people seem not to remember as far back as 2009, when the lake levels were the lowest they had been for sixty years. Imarisha has waned in influence, through the elimination of the Prime Minister's office, which provided oversight and funding. At the same time, the LNRA has lost its two strong leaders; the chairman retired and moved away, the Secretary sadly died in 2017. Sustainability is no longer a major issue for a government now committed to developing an inland railhead for all the containers brought from Mombasa and destined for central Africa; this will be constructed just to the south of Naivasha and will need a large and reliable water supply. That supply is proposed to come from a new reservoir in the Malewa catchment, so large that just filling it will interrupt the supply to the lake for two years, and its operation will remove the unpredictability that maintained "drawdown zones" and their high biodiversity (see also Wanjala and Kioko, this volume). Glossy urban-development proposals show promenades along a beautiful lakeshore without a reed in sight. The reality could be quite different – a very dirty lake without the self-cleansing benefits of drawdown that, despite recent proposals to rehabilitate the town's sewage-treatment plant, still receives effluent from tens of thousands of un-sewered residents. Centralised control and management strategies that keep in mind the ecological limits to growth and development are necessary if this lake and the many human and non-human communities it sustains are to continue to thrive.

Acknowledgements

David Harper wishes to acknowledge his funding from the University of Leicester and the Earthwatch Institute between 1982 and 2007; and the EU Tempus programme, the British Council and the UK Natural Environment Research Council between 2008 and 2016. He thanks the hundreds of Earthwatch volunteers and students from both Leicester and Nairobi Universities who studied with him. His successive Kenya Government research on permits were granted through collaborations with Nairobi University and the National Museums of Kenya. He is particularly grateful to the friendships of several staff in those two institutions and the many people around Lake Naivasha he worked with. This chapter is dedicated to Joan Root, Jill Simpson, John Caba and Velia Carn, who helped him but died before his work was finished.

Megan Styles wishes to thank the Wenner Gren Foundation for Anthropological Research, the University of Washington, and the University of Illinois Springfield for funding this research. She is also eternally grateful to the many Kenyans who shared their stories with her and granted her access to historical documents during the course of this research.

References

Acacia Consultants Ltd. 2013. *A Feasibility Study on the Conservation of the Wider Eburru Forest Ecosystem through the Establishment and Management of the Wildlife Corridors and Dispersal Areas*. Report to the Ministry of Environment and Mineral Resources. Nairobi.

Becht, Robert, and David M. Harper. 2002. "Towards an Understanding of Human Impact upon the Hydrology of Lake Naivasha, Kenya." *Hydrobiologia* 488: 1–11. https://doi.org/10.1023/A:1023318007715.

BirdLife International. 2012. *Eastern Afromontane Biodiversity Hotspot: Ecosystem profile*. Critical Ecosystem Partnership Fund. January 24, 2012. https://docslib.org/doc/3281616/eastern-afromontane-biodiversity-hotspot.

Bland-Sutton, John. 1911. *Man and Beast in Eastern Ethiopia: From Observations Made in British East Africa, Uganda, and the Sudan*. London: Macmillan and Co.

Boar, Rosalind, and David M. Harper. 2002. "Magnetic susceptibilities of Lake Sediment and Soils on the Shoreline of Lake Naivasha, Kenya." *Hydrobiologia* 488: 81–88. https://doi.org/10.1023/A:1023357910441.

Brockington, Dan. 2002. *Fortress Conservation: The Preservation of the Mkomozi Game Reserve, Tanzania*. Bloomington: Indiana University.

Combes, Kathryn. 2009. "The Lake Nakuru National Park to Lake Naivasha Project". *Soysambu Conservancy* (blog). November 13, 2009. Accessed September 7, 2023. https://soysambuconservancy.org/blog/the-lake-nakuru-national-park-to-lake-naivasha-project/.

Church, Colin. 2015. "Mau Eburru: The 'Smoking Mountain Regenerates'." *Swara* (April–June): 44–48.

Daily Nation. 1971. "Master Plan Needed for Lake Naivasha." May 30, 1971.

Gaudet, John J. 1977. "Natural Drawdown on Lake Naivasha, Kenya, and the Formation of Papyrus Swamps." *Aquatic Botany* 3: 1–47. https://doi.org/10.1016/0304-3770(77)90002-X.

Gaudet, John J. 1979. "Seasonal Changes in Nutrients in a Tropical Swamp: North Swamp, Lake Naivasha, Kenya." *Journal of Ecology* 67 (3): 953–81. https://doi.org/10.2307/2259223.

Gaudet, John J., and Francis M. Muthuri. 1980. "Nutrient Regeneration in Shallow Tropical Lake Water." *Internationale Vereinigung für theoretische und angewandte Limnologie: Verhandlungen* 21 (2): 725–9. https://doi.org/10.1080/03680770.1980.11897077.

Gouder de Beauregard, Anne-Christine A., David M. Harper, Francoise Malaisse, and Jean-Jacques Symoens. 1998. "Dynamique recente et cartographie de la vegetation aquatique (1960–1996) du Lac Naivasha (Rift Valley, Kenya)." *Bulletin des Seances, Academie Royale des Sciences d'Outre-Mer* 44 (3): 373–89.

Green, Siân E., Zeke Davidson, Timothy Kaaria, C. Patrick Doncaster. 2018. "Do Wildlife Corridors Link or Extend Habitat? Insights from Elephant Use of a Kenyan Wildlife Corridor." *African Journal of Ecology* 56: 860–71. https://doi.org/10.1111/aje.12541.

Harper, David M. 1992. "The Ecological Relationships of Aquatic Plants at Lake Naivasha, Kenya." *Hydrobiologia* 232 (1): 65–71. https://doi.org/10.1007/BF00014613.

Harper, David M. 2006. "The Sacrifice of Lake Naivasha." *Swara Mag.* (*J. East Afr. Wildlife Soc*) 29 (April-June): 27–37.

Harper, David M., C. Adams, and Kenneth M. Mavuti. 1995. "The Aquatic Plant Communities of the Lake Naivasha Wetland: Pattern, Dynamics and Conservation." *Wetland Ecology & Management* 3: 111–123. https://doi.org/10.1007/BF00177693.

Harper, David M., Edward H.J. Morrison, Michael M. Macharia, Kenneth M. Mavuti, Caroline Upton. 2011. "Lake Naivasha, Kenya: Ecology, Society and Future." *Freshwater Reviews* 4(2): 89–114. https://doi.org/10.1608/FRJ-4.2.149.

Harper, David M., and Kenneth M. Mavuti. 2006. "The Ecological State of Lake Naivasha, Kenya, 2005: Turning 25 Years Research into an Effective Ramsar Monitoring Programme." In *Proceedings of the 11th World Lakes Conference Vol. 2*, edited by Eric Odada and Daniel O. Olago, 30–34. Nairobi: AquaDocs. Accessed September 9, 2023. https://aquadocs.org/handle/1834/2127.

Harper, David M., and Maciej Zalewski, eds. 2001. "Ecohydrology: A Summary of the Projects Presented to the Conference on Science and the Sustainable Management of Tropical Waters, Naivasha, Kenya, 11–16 April 1999." *Technical Documents in Hydrology* 46. Paris: UNESCO.

Harper, David M., Maureen M. Harper, Munir A. Virani, Andrew Smart, R. Brooks Childress, Rachel Adatia, Ian Henderson, and Bernard Chege. 2002. "Population Fluctuations and Their Causes in the African Fish Eagle, (*Haliaeetus vocifer* (Daudin)) at Lake Naivasha, Kenya." *Hydrobiology* 488: 171–80. https://doi.org/10.1023/A:1023390800872.

Hobley, Wiliam C. 1929. *Kenya: From Chartered Company to Crown Colony; Thirty Years of Exploration and Administration in British East Africa.* London: H.H. & G. Witherby.

Hughes, Lotte. 2006. *Moving the Maasai: A Colonial Misadventure.* New York: Palgrave Macmillan.

IOL. 2003. "Flower Farms May Kill Lake Naivasha." November 4. Accessed September 7, 2023. https://www.iol.co.za/business-report/international/flower-farms-may-kill-lake-naivasha-774827.

Jackson, Frederic. 1930. *Early Days in East Africa.* London: Edward Arnold and Company.

Johnston, Harry H. 1902. *The Uganda Protectorate.* New York: Dodd & Mead Company. Accessed September 7, 2023. https://www.wdl.org/en/item/2561/.

Kariuki, Rebecca W., David Western, Simon Willcock, and Robert Marchant. 2021. "Assessing Interactions between Agriculture, Livestock Grazing and Wildlife Conservation Land Uses: A Historical Example from East Africa." *Land* 10 (1): 46. https://doi.org/10.3390/land10010046.

Keyombe, James L. A., Ruth M. Lewo, Edna Waithaka, Alice Mutie, and Priscilla Boera. 2018. "Has the Latest Fish Introduction in Lake Naivasha Improved Income of Fishermen? The Economics of Nile Tilapia (Oreochromis niloticus) in Lake Naivasha." *Fisheries & Aquaculture Journal* 9 (3): 250. https://doi.org/10.4172/2150-3508.1000250.

Kitaka, Nzula, David M. Harper, and Kenneth M. Mavuti. 2002. "Phosphorus Inputs to Lake Naivasha, Kenya, from Its Catchment and the Trophic State of the Lake." *Hydrobiologia* 488: 73–80. https://doi.org/10.1023/A:1023362027279.

Kongere, P.C. 1979. *The Salvinia Infestation on Lake Naivasha in Kenya.* Swedish Fund-in-Trust, Project TF-RAF 112(SWE), Suppl. 1, 143–53. Rome: FAO.

Kuiper, Gerda, and Andreas Gemählich. 2017. "Sustainability and Depoliticisation: Certifications in the Cut-flower Industry at Lake Naivasha, Kenya." *Africa Spectrum* 52 (3): 31–53. https://doi.org/10.1177/000203971705200302.

KWS (Kenya Wildlife Services). 2013. *Proposal for Securing the Naivasha Wildlife Sanctuary – Lake Naivasha – Hells Gate NP – Mount Longonot.*

KWS (Kenya Wildlife Services). 2017. *Hell's Gate-Mount Longonot National Parks Ecosystem Management Plan, 2017–2027.*

LNRA. 1993. "A Three Phase Environmental Impact Study of Recent Developments around Lake Naivasha." John Goldson Associates, Nairobi. Accessed September 1, 2023. http://ftp.itc.nl/pub/naivasha/PolicyNGO/LNROA1993.pdf.

MacDonald, Peter. 1970. "Lake Naivasha – Centre of Crisis or National Asset?" *East African Standard*, December 2.

Morrison, Edward. 2013. "Ecological Restoration of Papyrus Wetlands at Lake Naivasha, Kenya: Social and Ecological Considerations." PhD diss., University of Leicester. Accessed September 7, 2023. http://hdl.handle.net/2381/28183.

Murombedzi, James C. 2010. "Agrarian Social Change and Post-Colonial Natural Resource Management Interventions in Southern Africa's 'Communal Tenure' Regimes." In *Community Rights, Conservation and Contested Land: The Politics of Natural Resource Governance in Africa*, edited by Fred Nelson, 32–51. London: Earthscan.

Mittermeier, Russell A., Patricio R. Gil, Michael Hoffmann, John Pilgrim, Thomas Brooks, Cristina G. Mittermeier, John Lamoreux, and Gustavo A. B. da Fonseca. 2005. *Hotspots Revisited: Earth's Biologically Richest and Most Endangered Terrestrial Ecoregions.* Chicago: University of Chicago Press.

Njiru, James, George Morara, Edna Waithaka, and James Mugo. 2015. "Fish Kills in Lake Naivasha, Kenya: What Was the Probable Cause?" *International Journal of Fisheries and Aquatic Studies* 3 (1): 179–84.

Odawa, Simon, and Yongwon Seo. 2019. "Water Tower Ecosystems under the Influence of Land Cover Change and Population Growth: Focus on Mau Water Tower in Kenya." *Sustainability* 11 (13): 3524. https://doi.org/10.3390/su11133524.

Onyango, Joel, Norbert Kruzinger, and Kitaka Nzula. 2015. *Pesticides Residues Contamination in Lake Naivasha Catchment, Kenya.* Saarbrücken: AV Akademikerverlag.

Ottichilo, Wilber K., Jesse Grunblatt, Mohammed Y. Said, and Patrick W. Wargute. 2000. "Wildlife and Livestock Population Trends in the Kenya Rangeland." In *Wildlife Conservation by Sustainable Use*, edited by Herbert H. T. Prins, Jan G. Grootenhuis, and Thomas T. Dolan. Conservation Biology Series, vol. 12. Dordrecht: Springer. https://doi.org/10.1007/978-94-011-4012-6_10.

Pacini, Nic, M. Baxa, M. Kosík, J. Grey, O. Lepšová-Skácelová, D. K. Mbogo, Timothy Mwinami et al. 2021. "Ecohydrological Costs and Benefits of Common Carp, the Dominant Species in a 'Novel' Tropical Lake Ecosystem." *Ecohydrology & Hydrobiology* 21 (3): 467–89. https://doi.org/10.1016/j.ecohyd.2021.08.011.

Parker, Ian S.C. 1968. *Some Observations on the Relationship Between Red-Knobbed Coot (Fulica Cristata Gmelin) and Wild Duck (Anatidae) on Lake Naivasha.* Nairobi: I.S.C. Parker Wildlife Services.

Parker, Ian S.C. 1974. *The Status of the Louisiana Red Swamp Crayfish in Lake Naivasha; Report Commissioned by the East African Wildlife Society.* Nairobi: I.S.C. Parker Wildlife Services.

Ramsar. 2005. "Lake Naivasha." Ramsar Sites Information Service. Last modified January 1, 2005. Accessed September 9, 2023. https://rsis.ramsar.org/ris/724.

Riungu, Catherine. 2009. "Lake Naivasha is Dying: Government, Users Wake Up to Nightmare Reality." *East African Standard*, July 20.

Stanley, Henry M. 1878. *Through the Dark Continent: Or, the Sources of the Nile, Around the Great Lakes of Equatorial Africa, and Down the Livingstone River to the Atlantic Ocean.* New York: Harper & Brothers.

Smart, Andrew C., 1988. "Population Census of Lake Naivasha Hippopotamus *Hippopotamus amphibius*." In *Lake Naivasha Vertebrate Survey 1987*, 65–82.

Smart, Andrew C., David M. Harper, Christine A. Gouder de Beauregard, Sophie Schmitz, Stephanie Coley, and Francoise Malaisse. 2002. "Feeding of the Exotic Louisianan Red Swamp Crayfish, Procambarus Clarkii (Crustacea, Decapoda), in an African Tropical Lake: Lake Naivasha, Kenya." *Hydrobiologia* 488: 129–42. https://doi.org/10.1023/A:1023326530914.

Styles, Megan A. 2019. *Roses from Kenya: Labor, Environment, and the Global Trade in Cut Flowers.* Seattle: University of Washington Press.

Thomson, Joseph J. 1887. *Through Masai Land: A Journey of Exploration Among the Snowclad Volcanic Mountains and Strange Tribes of Eastern Equatorial Africa.* London: S. Low, Marston & Co.

Van der Meeren, Thijs, Emi Ito, Kathleen R. Laird, Brian F. Cumming, and Dirk Verschuren. 2019. "Ecohydrological Evolution of Lake Naivasha (Central Rift Valley, Kenya) During the Past 1650 Years, as Recorded by Ostracod Assemblages and Stable-isotope Geochemistry." *Quaternary Science Reviews* 223 (1): 105906. https://doi.org/10.1016/j.quascirev.2019.105906.

Wanjala, Silas, Timothy Mwinami, David M. Harper, Ed H.J. Morrison, and Nic Pacini. 2018. "Ecohydrological Tools for the Preservation and Enhancement of Ecosystem Services in the Naivasha Basin, Kenya." *Ecohydrology & Hydrobiology* 18 (2): 155–73. https://doi.org/10.1016/j.quascirev.2019.105906.

Watson, R.M., T. Singh, and I.S.C. Parker. 1970. "The Diet of Duck and Coot on Lake Naivasha". *East African Wildlife Journal* 8: 131–144.

Worthington, Stella, and E. Barton Worthington. 1933. *Inland Waters of Africa.* London: Macmillan and Co.

CHAPTER 4

Shifting Land Tenure Arrangements: Private Land-Buying Companies and the Shape of Naivasha's Anthropogenic Landscape

Gerda Kuiper and Marie L. Gravesen

1 Introduction: Naivasha's Anthropogenic Landscape[1]

This chapter describes the history of shifting land-tenure arrangements around Lake Naivasha with a primary focus on the post-independence period. We argue that an analysis of these land-related dynamics can deepen our understanding of the multiple threats to the local ecological system and the socio-economic hardships that are so characteristic of contemporary Naivasha (see e.g. Harper et al., this volume, and Kirigia, this volume). Lake Naivasha and its immediate surroundings have been heavily influenced by human activities since the late nineteenth century (see Waller, this volume). The local landscape can thus be considered to be deeply anthropogenic, "a product of human thought and action" (Gravesen 2020a, 181). Colonial settlers' land-use practices and their perceptions on what constitutes proper land tenure have had a decisive and continuous influence on Lake Naivasha's social-ecological system (Kanyinga 2009; Harper et al. 2011, 93–96; Styles 2019, 33). A conspicuous example is the establishment of the flower and vegetable industry. Only because of the massive dispossession of land in the early twentieth century (see Waller, this volume), did horticultural investors find large parcels of land ready for lease or sale some eighty to ninety years later. These agro-industrial ventures have in turn left a large imprint on Naivasha's ecological and social landscape, if only by their attraction of tens of thousands of migrant workers (Kuiper 2019). However, something that has received little attention so far is the ways in which some parcels of land returned to the hands of Africans in the intermediate period. Although Naivasha was not included in formal resettlement programmes that were set up elsewhere in the former "White Highlands" in the 1960s and 1970s, it was part of the informal "Africanisation" of land tenure, where unregulated

1 First submission date of this chapter: November 24, 2021. Acceptance date: March 13, 2023.

 | DOI:10.1163/9789004695429_005

land transactions exchanged the former white landowners with Kenyan landholders and tenants. Specifically, several plots of land were sold to private land-buying companies owned by Africans in the years after Kenya gained independence. These shifts in ownership, the agency of the different stakeholders involved, and the consequences for Naivasha's landscape and socio-economic make-up are the focus of this chapter.

We pay particular attention to the origins of Naivasha's large urban sprawls, the largest of which have developed on land that was sold off to land-buying companies in the late twentieth century. These transfers initially seemed insignificant for the development of the area as a whole. But in hindsight, they have shaped present-day Naivasha extensively, due to how the function of these plots has changed. They have turned into residential areas, housing the majority of migrant workers on whom the Naivasha horticultural economy depends. Smallholders in the area have erected rooms for rent on their plots, often next to their own houses, thereby turning the smallholder settlements into dense workers' quarters (Kuiper 2020). The plots' specific historical trajectories have determined their physical layout and socio-economic makeup. Notably, these "shanty towns" have put a stain on the international reputation of the horticultural industry and were subject to intense outbreaks of the 2007–2008 ethnicised post-election violence (*Daily Nation* 2005; Okanga 2008). In this chapter, we describe the transfer, subdivision, and current state of two of these plots, Karagita and Kihoto, by now Naivasha's most infamous low-income settlements. What is significant here is that the subdivision processes of these plots have been characterised by recurring conflicts and court cases. These conflicts influenced decisions to build rental housing (or not) and contributed to the already insecure position of tenants in Naivasha. But despite their considerable impact on the (re)shaping of Naivasha's landscape and social dynamics, the land-buying companies that founded these settlements have hardly been noticed in scholarly literature on the area. More generally, even though the majority of rental housing in African cities is provided by private landlords, the informal supply systems and the dynamics of these markets are massively understudied (Andreasen et al. 2021).

We will unpack these dynamics by briefly describing land-tenure arrangements in Naivasha in the colonial period and by introducing land-tenure transformations and the phenomenon of land-buying companies and cooperatives in Kenya in the period around independence. By bringing in the dynamics around the origin of the land-tenure relations, we seek to point out the underlying logic followed by different actors that steered how the land relations developed. Finally, we draw a brief comparison to the situation of land-buying companies in another part of what was referred to as the White Highlands,

Laikipia County, in order to be able to clarify which parts of the conflicts and tension are specific to Naivasha and which are comparable to more general tendencies surrounding land transfers and subdivisions in Kenya in the period of study. Moreover, the case of Laikipia offers a point of comparison on specific types of tension, including points of contention related to political involvement and ethnic composition in the informal Africanisation carried forward by the private land-buying companies (cf. Klopp 2001; Kanyinga 2009; Gravesen 2020a).

Overall, we aim to make three contributions. First, we seek to fill a gap in the literature on how Naivasha's socio-economic landscape developed between independence and now, a time period that otherwise tends to be overlooked in scholarship, in favour of analyses that link colonial matters with current affairs. Secondly, we argue that the agency of African land actors in this period holds immense importance for Naivasha's contemporary anthropogenic landscape. Here, the dynamics around the land-buying companies become instrumental in pointing to shareholders, politicians and tenants alike. And thirdly, we highlight how the history of the land dynamics can explain how some of the sources of tension, ethnic and class-related, predate the arrival of the horticultural industry, whereas other points of tension have taken form and intensified along with the diversification of the landscape. The chapter thus draws attention to the influence of the informal Africanisation process that the private land-buying companies introduced, and to the agency of a variety of actors in shaping contemporary Naivasha.

The description of the case studies is based on an analysis of varying archival sources, accessed in the Kenya National Archives (KNA) in Nairobi, in its provincial deposit in Nakuru, and in the archives of the Nation Media Group in Nairobi. Other sources, such as recent newspaper articles, court cases, and parliamentary discussions, have been accessed online. Findings from the archival records have been triangulated with data from interviews conducted in the period 2014 to 2016, in the context of a research project on labour relations in the Naivasha cut-flower industry.[2]

2 Land-Tenure Arrangements in the Pre-Independence Period

Let us first outline the early developments around land-tenure arrangements in Naivasha. As described by Waller (this volume), Maasai were moved out of

2 This methodology is discussed elaborately in Kuiper (2019, 6–16).

Naivasha to make space for colonial settlers at the start of the twentieth century. These settlers obtained long-term leases on large tracts of land. The average holding size was over 5,000 acres in 1960 (Odingo 1971, 45).[3] The semi-arid land around Lake Naivasha was considered unsuitable for intensive farming and was primarily used for extensive livestock ranching (cattle, sheep, and pigs) on uncultivated grasslands while some fodder crops were cultivated on irrigated fields along the lakeshore. Settlers hired resident labourers, known as "squatters", who were allowed to cultivate part of the land and, at least initially, to keep some livestock themselves. The majority of the squatters were Kikuyu, who were pushed out of their region of origin in central Kenya due to land shortage, pressure to pay taxes, and oppression by chiefs. In addition, there was a minority of labourers originating from western Kenya who would stay in Naivasha temporarily or seasonally.[4] In order to provide food rations to the contract workers, settlers purchased maize, potatoes, and beans from their squatters and from Kikuyu farmers from central Kenya (Odingo 1971; Kanogo 1987, 9–30; Styles 2019, 34–35).

In the 1940s, squatters' rights with regard to land cultivation and livestock keeping were curtailed in order to avoid competition with the settlers. This repression led to violent resistance in squatter areas across the White Highlands, culminating in the Mau Mau uprising in the 1950s. European settlers were also attacked in Naivasha and the township's police station was raided. As a repercussion, Kikuyu were evicted from settler areas and replaced by other labourers (Kanogo 1987, 127–52; Styles 2019, 38–40). The Naivasha Annual Report of 1953 described: "The Emergency brought with it many problems affecting labour (…). Kikuyu labour [*sic*] who were removed or repatriated were replaced partly by labour from other tribes, mainly Kipsigis, Maragoli and Wakamba, who were found not to be as satisfactory or hardworking as the Kikuyu".[5] Some settlers called for the return of Kikuyu – this time as contract labourers, not as squatters. The Annual Report of 1957 mentioned that Kikuyu indeed returned to the area, but "[t]he number of Africans other than Kikuyu employed in the District did not decrease drastically".[6]

3 Notably, the distribution of land in the Rift Valley was highly uneven. The Naivasha settler community comprised impoverished settlers residing on 20-acre plots, but also included influential settlers such as Lord Delamere, who was granted over 100,000 acres of land partly located in Naivasha (Sorrenson 1968).

4 In 1920, it was reported that there were 9,116 Africans in the district, of whom 6,600 belonged to Kikuyu squatter families (Kanogo 1987, 14).

5 KNA, DC/Nais/1/1/1/15, Annual Report (AR) Naivasha Division 1953, 7.

6 KNA, DC/Nais/1/1/1/45, AR Naivasha Division 1957, 11. Ominde (1968, 7) observed that Naivasha was ethnically more mixed than most other Kenyan regions. Nevertheless, whereas

In the years leading up to independence, there was much political pressure on the colonial administration to buy land from settlers in the White Highlands and to redistribute it to Africans in the form of government resettlement schemes. The most expansive and well-known was the Million-Acre Settlement Scheme, the implementation of which started in 1962 and which held a particular weight on land transitions in Naivasha's neighbouring Nyandarua District, also known as the Kinangop (Rouillé-Kielo, Calas, and Racaud, this volume). Land was purchased from settlers and reallocated to African-owned cooperatives (Odingo 1971). The process of reallocating land was however influenced by multiple interests. It served the purposes of sustaining and intensifying agricultural production by the African population after Kenya gained independence and of appeasing public pressure from especially the Kikuyu population so that an orderly transition of political power could take place and a new Mau Mau uprising was avoided (Leo 1984; Boone, Lukalo, and Joireman 2021). For this reason, the Kikuyu population were favoured in the land allocations, which disenfranchised other ethnic groups and thereby escalated the ethnicisation and politicisation of the land question (Branch 2009, 2011; Kanyinga 2009). In addition, the settler economy was sought to be protected so that the British Administration could show that they were leaving their legacy in Kenya after independence to feed a public image that Kenya had benefited from colonialism. Therefore, a cap was put on the land to be bought and reallocated through the resettlement schemes (Wasserman 1976; Leo 1981). Consequently, strict regulation was put in place to meet the balance of allocating just the right amount of land to appease the African population while maintaining good prospects for the settler economy. Among these regulations was an assessment of the land to be included in the resettlement schemes: only land that had been used by a settler for mixed farming could be included, whereas a settler who had used the land for ranching could not receive compensation. This regulation was meant to ensure the prospects for small-scale farming after subdivision in the resettlement schemes since small-scale farming was deemed unviable on the more arid lands on the ranches. Thus, despite protective intentions for the settler community, these regulations disenfranchised especially the settlers on ranches, among whom a fair number were interested in selling but were unable to because their land was not approved for inclusion in the schemes. This became even more complicated when laws on Exchange Control

among non-Kikuyu it were mainly young men and some young women who resided in Naivasha, age and gender distribution among Kikuyu in Naivasha was comparable to age and gender distribution in the former Kikuyu Reserve (Ominde 1968, 154). In other words, Kikuyu settled in Naivasha as families, while non-Kikuyu mainly resided there as temporary workers.

were introduced, as it disallowed land sales outside the resettlement schemes. The eligibility criteria regarding land use therefore had the side effect of locking many of the settler families to their land even though they might have wished to sell and leave the country (Gravesen 2020a). Also many settler farmers around Naivasha, especially those with small farms, were willing to sell their land to the resettlement board, at least if they would be paid a good price – but the board was not interested, likely because of Naivasha's reputation of being unsuitable for intensive farming (Styles 2019, 41). As illustrated by Boone, Lukalo, and Joireman (2021) and Gravesen (2020a), large areas of the former White Highlands, including all areas around Lake Naivasha and all of Laikipia, apart from its southern fertile lands, were thus barred from formal Africanisation.

Instead, a more unstructured process of Africanisation and re-settlement took place. In 1962, it was reported that plots of land had started to be given out to landless Kikuyu under the Million Acre Scheme, including in the Kinangop bordering Naivasha to the east. Nevertheless, many Naivasha Kikuyu residents did not qualify for the scheme, such as those who had only recently taken up employment in the area, and "unattached" women. Instead, some of them squatted in the township.[7] Ominde (1968, 127) furthermore noted significant out-migration to Nairobi. Africans rarely obtained tenure rights in Naivasha itself in this period. Rather, most Naivasha settlers remained, and even some other settlers arrived, when they sold their large estates elsewhere in Kenya and moved to a smaller plot in Naivasha (Styles 2019, 42). Some plots were transferred to expatriates, such as conservationists Joy Adamson and Joan Root, and the horticultural adventurers who started DCK and Oserian flower farms (Kuiper 2019, 50–52). And although gradually some (long-term leases on) farms were sold under a "willing buyer, willing seller" principle, the large plots were unattractive to African cooperatives because the subdivision of large farms was discouraged by the local government. Naivasha's District Officer for instance stated in 1977 that no approval would be given for subdivision of large farms because "subdivision lowers the economy of the country".[8] Nevertheless, a couple of cooperative farms were established in the region, but primarily in the lake's hinterland, such as Maiella Estate (see Müller-Koné and Kioko, this volume). Overall, the informal "Africanisation" of land has been a slow and gradual process, and a number of large estates remain in colonial

7 KNA AR Naivasha Division 1962.

8 KNA Nakuru ANN 1/1/21, Minutes of the second meeting of Naivasha division farmers leaders conference, 26 March 1977.

settler families' hands until today.[9] Only a handful of smaller plots in the vicinity of the lake were sold to former squatters or to land-buying companies in the years after Kenya gained independence. We argue that a more formalised resettlement process would have laid the foundation for an entirely different present-day anthropogenic landscape in Naivasha. The local economy might have been based more on smallholder farming than on agro-industry, which in turn would have produced an entirely different socio-economic composition of residents in the area and would have had differently impacted the social-ecological system around Lake Naivasha.

3 The Emergence of Land-Buying Companies in the Post-Independence Period

The informal and gradual redistribution of land allowed for the emergence of private land-buying companies with African shareholders. The agency of the different stakeholders in these seemingly communal entities stands as an essential factor for how the informal land transfers took form – in Kenya's former White Highlands in general as well as in Naivasha specifically.

The general pressure among the Kenyan population to acquire land following independence was continuous because the limited land in resettlement schemes was not nearly sufficient to saturate the demand. The high demand for land, combined with willing settler-sellers being barred from having their land included in resettlement schemes which would have allowed them to leave the country with their assets, created a window of opportunity for private-sector land markets (Wasserman 1976; Leo 1981; Kanogo 1987; Kanyinga 2009). Where the formal land transactions – the resettlement schemes – were highly controlled and functioned as vehicles for different political agendas, the informal land transactions were instrumentalised for individual gains of various types, including to further the interests of a particular ethnic group, as lucrative investments for the elite, or as opportunistic schemes to build a political base for prospective politicians (Gravesen 2020b).

Land-buying companies started to spring up, in which members were recruited as shareholders who pooled their funds in order to collectively purchase a former ranch from a settler who wished to leave. Land-buying company

9 Examples are the descendants of Lord Delamere, who was influential in cementing Naivasha's reputation as a cattle and dairy centre in the 1920s (Styles 2019, 34), and the Hopcraft family, who renewed the land grant on their estate at North Lake for another 942 years in 1963 (Kenya Gazette, Notice No. 2,483, January 1, 1964).

offices opened up in towns across the country, where clerks and company directors would enlist members and provide them with share certificates as proof of their deposits and status as shareholders. Ideally, once land had been secured, the land could be surveyed and subdivided so that each shareholder's share certificate could be exchanged into individual title deeds. However, each part of this process often turned out to be complicated. By and large, company shareholders shared ethnic ties and especially Kikuyu were dominating land-buying companies (Kanogo 1987, 175). For that reason, there may have been an expectation that cooperation would be smooth. Complications arose if the constitution of shareholders was less homogeneous, for instance when larger portions of different ethnic groups were represented in the land-buying companies or cooperatives, but also when shareholders included Kikuyu from different clans or areas. Fractions along such lines shaped the dynamics between shareholders. In that sense, the trend to acquire land through land-buying companies generated ethnic struggles as well as clan- and class-based competition. Another problem was that many of the plots were too small or unsuitable for subsistence agriculture and had been sold only out of political (not economic) consideration. If a settler agreed to sell privately, good political connections were necessary to bypass Exchange Control. For this reason, many aspiring politicians saw the founding of land-buying companies as an instrument to establish a political base of support by enabling access to land for votes. Given the political rhetoric around land acquisitions in the independence years, a politician was almost guaranteed political support if he or she was able to provide land for a potential voter base – even if the land was in fact of poor quality. In many places across the former White Highlands, the effects of the informal land transformations driven by the land-buying companies were immense: they became instruments for political opportunism, vehicles for poverty traps, and inciters of conflict by forcing competition between already established loyalties, which by default ran along ethnic, clan and class lines. These dynamics were particularly clear in Laikipia County, where legal struggles often transpired into decades of court cases and where conflicts also impacted the physical landscape. Claims and counterclaims encouraged increasing numbers of physical enclosures such as fences and trenches. In Laikipia, the political involvement in land-buying companies tended to lead to an inability to subdivide and title the land for the shareholders, and people therefore were unable to acquire bank loans to invest in their acquired land plots. This resulted in a divided landscape, with visible poverty and absent development in some parts of the county that, to this day, stand in great contrast to the wealthier tenure-secure properties and the remaining ranches. In addition, the increased tension around property and tenure insecurity encouraged a form of territorialisation where individuals of ethnic or class-related commonalities settled in clusters and renamed their settlements in such a way as to support their claims to the place. These changes in the way the landscape is shaped, named and divided stems to a large extent from the land-buying companies' involvement, their presence, and the associated increased conflicts and tension around land that was brought forth in their midst. Due to their lack of accountability and the state's lacking willingness to regulate, the emergence of the land-buying companies has immensely impacted people's own initiatives to claim and defend what they perceive as rightfully theirs in various ways (Gravesen 2020a, 2020b).

A few land-buying companies also became active around Naivasha, even though they were not as numerous as in other parts of the Rift Valley Province.[10] These companies were not as politically salient as in Laikipia, but they were similarly conflict-ridden and likewise (initially) failed to bring development. The mismanagement of these companies gave rise to regular complaints, as did the frequent issue of the slow or even lacking issuing of title deeds once a farm had been subdivided.[11] In 1998, local leaders in Naivasha Division were asked to educate citizens about the harmful effects of the prevalent wrangles in land-buying companies and the lack of title deeds on development in the area as a whole.[12] In the following year, the Naivasha MP Kihara complained in parliament that the subdivision processes of land-buying companies in his constituency were "riddled with a lot of corruption. The Government appointed liquidators in collusion with the former chairmen of these societies who are colluding to defraud the members of those areas which were reserved as common areas for the benefit of the members (…)".[13] The plots we introduce below have also been affected by such conflicts over ownership and tenure. In these instances, the numbers of people affected have even multiplied, since landowners and land claimants have started to develop makeshift housing on

10 By 1973, there were 57 land-buying entities active in Nakuru District, to which Naivasha belonged at the time. KNA Nakuru, GU/9/1/104, The Provincial Planning Officer Rift Valley Province to all members of the Nakuru District Development Committee, "District Development Plan 1974–1978", August 16, 1973.

11 KNA Nakuru, GU/1/5/41, S.B. Anunda, District Agricultural Officer Nakuru, to the DC, Nakuru, "District Development Plan", June 6, 1978.

12 KNA Nakuru, GU/1/9/100, Minutes of the Naivasha Divisional Leaders' meeting, October 19, 1998.

13 Kenya National Assembly Hansard, April 27, 1999. Accessed August 23, 2023. https://books.google.de/books?id=FKrJe3rNRv4C&printsec=frontcover&source=gbs_ge_summary_r&cad=0#v=onepage&q&f=false.

their plots to rent out to the migrant workers attracted by the horticultural industry since the 1990s.

4 Karagita: Land Wrangles and (Tenure) Insecurity

Karagita is currently the most populous residential area at Lake Naivasha, apart from Naivasha Town itself.[14] It is located in what was called "Poverty Bay" during the colonial period: an area on the eastern side of the lake where impoverished Europeans would buy small plots of about 20 acres. The plot that later turned into Karagita settlement was part of a settler's farm, which was also divided into smaller pieces to be sold off.[15] One of the company directors of the land-buying company Karagita East Africa Limited, Peter Mburu, explained that the company, founded in Nairobi in 1965, acquired this plot of 20.6 acres in 1966.[16] The company was named after the neighbourhood in Nairobi where the company had already purchased another plot. Archival sources suggest the shareholders were lower-class Kikuyu residing in Nairobi (rather than in Central Province) who were otherwise landless (*Daily Nation* 1984).[17] According to Peter Mburu, the first company members constructed houses on the plot and moved there in 1972. The main income-earning activities were selling illicit brew or firewood, and fishing. The land itself was unfit for smallholder farming (see Pacini, Giberti, and Harper, this volume). The Karagita residents had no immediate access to the lake and had to pay the settlers, who still owned the plots on the lakeshore, for jerry cans with drinking water or to pass through their land to go fishing. There was little work on the large farms in the area, which changed when the first horticultural farms were established in the late 1970s. Instead of starting to work for the farms themselves, the Karagita plot owners erected wattle-and-daub housing on their plots, where they rented out rooms to migrant workers. When an increasing number of migrants

14 The development of Karagita has also been described in Kuiper (2020), where the impact of migrant workers' practices on the shape of this settlement was interrogated. In contrast, this chapter focuses on the agency of the landowners in shaping the local landscape.

15 Interview with Sarah Higgins, the British widow of a settler descendant and former LNRA-secretary, on February 10, 2015.

16 Interview on October 25, 2014. Notably, unlike almost all other interviews conducted, this interview took place in Kikuyu (with the help of a translator), as Mburu was more comfortable with speaking Kikuyu than Swahili or English.

17 KNA Nakuru, EA/2/10/97, Minutes of the Nakuru Land Control Board meeting held on November 29, 1977; KNA Nakuru, GU/3/36/174, Chief of Hells Gate Location to Sonrahi Surveys, Naivasha, April 22, 1995, "Sub-division of LR no. 396/31".

arrived in the course of the 1980s, some plot owners constructed brick rental rooms, which they could rent out for higher prices. Nevertheless, until today, a large proportion of the houses on this specific plot are wattle-and-daub.

With the flower industry growing and ever more migrant workers coming to the area, other land-buying companies or cooperatives started to develop the plots immediately adjacent to the original plot of Karagita (EA) Limited. Since the early 1990s, the former Mirera Suswa cooperative farm (now subdivided) bordering the Karagita plot on the east is also considered part of Karagita settlement.[18] The differences in type of construction and age of the corrugated iron roofs between the original Karagita plot and the neighbouring plots to the north, south and east are clearly visible on Map 4.1. In 1995, the subdivision of Karagita into 60 commercial plots and 91 residential plots was approved, under the condition that two plots were set aside for a nursery school and a health centre

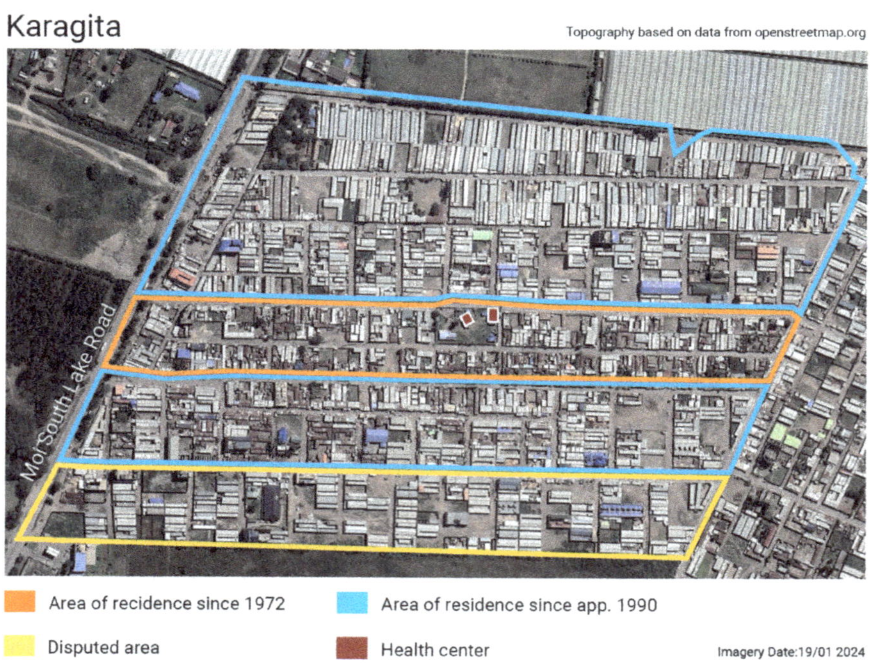

MAP 4.1 Karagita
CARTOGRAPHER: LONE RAVNKILDE

18 KNA Nakuru, GU/3/9/46 Peter Mburu, KANU Chairman and Area Chairman of the Elder, Karagita Village Elders, Naivasha, to the DO's office, Naivasha, April 29, 1991, "Dispute of plots against Kunyamarika Group and other group members".

(in an existing building that had been erected by the settler).[19] In 2004, Karagita became part of sublocation Mirera and of the location Hell's Gate, and there-with received its own chief (the lowest level of government officials in Kenya).[20] By now, many of the original subdivided plots have been sold by members to "outsiders", who do not move there themselves but who consider such plots with rental housing a form of investment. However, the company Karagita EA Ltd. continues to exist to take care of public utilities such as the health centre and is still involved for instance in the transfer of property rights in succession cases.[21]

Karagita has been plagued by numerous conflicts over land ownership, even if the plots were initially considered to be of little economic value due to their lack of farming potential. The first conflict started already at the time Nairobi-based Karagita (EA) Ltd purchased the plot. In narrating its history, company director Peter Mburu mentioned that Karagita (EA) Ltd cooperated with the Naivasha-based Ndogo Cooperative to piece together the necessary funds. However, they soon came into conflict over ownership rights. Eventually Karagita (EA) Ltd received the title deed for the whole plot. Subsequently, numerous internal conflicts between the company's directors and the shareholders arose, as company directors often acted in their own interests rather than those of the shareholders. For instance, in 1977, the company applied with the Nakuru Land Control Board for approval to mortgage the Naivasha plot. But this would be a risky venture for the many shareholders who by then resided on the plot and who had no other land to move to. The application was therefore deferred and later withdrawn, when the company chairman acknowledged that the mortgage was not in the interest of the shareholders.[22] In 1984, it appeared that Karagita (EA) limited had lost millions of Kenyan Shillings (KSh), according to a local government official due to the untrustworthiness of the company's directors. "He noted with regret that most of the [former] board members were either illiterate or semi-illiterate and were unable to keep records of the company's financial records. He called on the current board to intensify efforts to trace [*sic*] fate of the company's financial standing" (*Daily*

19 KNA Nakuru, GU/3/36/174, "Sub-division".

20 KNA/ Nakuru, GU/9/1/202, The District Officer I of Naivasha Sub-District, J. Karung'o Kamau, to the District Commissioner Nakuru, August 10, 2004, "Merging of administration boundaries".

21 Interview with Peter Mburu; *Alex Njoroge Robert & 3 others v Jane Wanjiru Kamau and another* (High Court at Nairobi) (2020), accessed April 11, 2024. http://kenyalaw.org/caselaw/cases/view/193458/.

22 KNA Nakuru, EA/2/10/93, Minutes of the Nakuru Land Control Board meeting held on October 11, 1977; KNA Nakuru, EA/2/10/97, Minutes of the Nakuru Land Control Board meeting held on November 29, 1977.

Nation 1984). In 1995, the subdivision process of the Naivasha plot also caused an internal dispute. This dispute over (again) "malpractices in the company" was settled in 1998, and in the following year, the formal processing of the leases and registration in the Nakuru Lands Registry could take place.[23] However, conflicts continued to occur, for instance about the use of the plot on which the dispensary is located. Several company officials were brought to court in 2004 over private construction there, and the management of the shared plot was transferred to other company officials.[24] These continuous internal conflicts – sometimes resulting in drawn-out court cases – illustrate the enormous interest in acquiring land and the high stakes in such land-buying ventures. The conflicts also point to the precarious position of many of the shareholders before they were finally able to secure their title deeds.

Apart from such internal conflicts, there were also conflicts between different land-buying entities. In 1993, the two plots immediately south of the original Karagita were registered as property of the Karagita Self Help Mixed Group (a different entity than Karagita (EA) Ltd), and the group received permission for subdivision into 140 very small residential parcels and a few plots set aside for specific purposes, such as sewage and a borehole. However, it turned out that the most southern 20 acres of this plot had been sold twice by the then-owner, a company called Thika River Estate Ltd, to both the Karagita Self Help Mixed Group and the Weru Wamuthurwa Investment Company. This last company had in turn sold the plot on.[25] The double sale resulted in a lengthy legal process, with numerous court cases. To make matters worse, one of the quarrelling parties had started construction on the plot and allowed hundreds of families to settle. When in 2002 a court ruled in favour of other claimants, these families and their tenants were evicted, and all buildings were demolished. Only a mosque was left standing, after the government intervened:

23 KNA Nakuru, GU/3/36/172, Abdi M. Hassan for the District Commissioner, Nakuru District, to the Commissioner of Lands, Nairobi, June 22, 1998, "Karagita E. Africa Limited LR No 396/31"; KNA Nakuru, EA/2/16/122, G. Ireri for Commissioner of Lands to the Chairman of Karagita (East Africa) Co. Ltd, February 10, 1999, "The Land Planning Act (Cap 303)".

24 KNA Nakuru, GU/3/36/166, Ten members of the "Karagita Standing Sub-division Committee of 396/31", Naivasha, to the Chief of Hells Gate Location, Naivasha, July 15, 2004, no subject; KNA Nakuru, GU/10/5/173, Meeting for members on the day of 15 September 2004 at Chief's Office Mirera.

25 *Edward K. Kimuhu & 17 others v Weru Wamuthurwa Investment Company & 3 others* (High Court at Nairobi) (2005), accessed August 24, 2023. http://kenyalaw.org/caselaw/cases/view/11050/; KNA Nakuru, GU/3/36/127 J.M. Nzukuh, Directors of surveys, to the Secretary, Central Authority, Nairobi, 12 November 1993, "The Land Planning Act (Cap. 303)"; *Karagita Self Help Mixed Group v Thika River Estate Ltd* (Court of Appeal at Nairobi) (2017), accessed August 24, 2023. http://kenyalaw.org/caselaw/cases/view/141472/.

"Police in Naivasha shot in the air to disperse a gang that had been sent to demolish a mosque built on a plot whose ownership is in dispute. Nobody was hurt during the operation on Wednesday evening. Muslims had earlier surrounded the mosque, vowing to tackle the advancing mob of about 200 before the police arrived" (*Daily Nation* 2002). The mosque remained intact, but this specific ruling was not the end of the matter and the plot's ownership remains under dispute until the present day.[26] The insecurity with regard to land ownership continues to be visible on the plot, which until today has much more vacant space than the rest of the settlement, despite its attractive location next to the Moi South Lake Road.

Yet another land issue affecting (in)security around Karagita is the lack of public corridors to the lake, which has repeatedly led to tension in the area.[27] On May 2, 2015, a crowd gathered at the gate of a private plot just north of Karagita. A parliamentary committee investigating land matters arrived and declared that the plot contained a public lake corridor. The next day, a mob of residents returned: "The demonstrators burnt tyres and used huge tree trunks to block the road used mostly by tourists and flower farm workers" (*Daily Nation* 2015). Moreover, the mob demolished the gate of the property to enable access to the lake. The two landowners, who had purchased the violated plot in 2012, subsequently sued the Naivasha MP Kihagi and a few others, claiming the damages caused by the invasion of their land. Several surveyors were called up in court, who concluded that there was no historical evidence that the plot had ever contained a public corridor. The judge stated: "Whoever declared it a public utility land, misled the public, and misinterpreted the survey maps drawn in respect of the area under dispute".[28] This court case did not mean the end of the search for a public corridor close to Karagita. After the Nakuru County Assembly passed a motion on the opening of public corridors in 2020, members of the White Eagle Self Help Group started to clear a path on a stretch of land between the leasehold plots of a flower farm and a tourist lodge. They claimed it as a public corridor, with reference to the 2003 Ndungu Report, written by a parliamentary commission that had investigated the illegal or irregular

26 There was still a ruling about this case in 2021: *Grace Wangui Mburu & another v Peter Mburu Nguri & 4 others* (High Court at Nairobi) (2021), accessed August 24, 2023. http://kenyalaw.org/caselaw/cases/view/209689/.

27 See Harper et al. (this volume) on the long history of concern about sufficient public access points to Lake Naivasha.

28 *Jackson Kariuki Kahungura & another v John Karanja Kihagi & 5 others* (Environment and Land Court at Nakuru) (2018), accessed on August 24, 2023. http://kenyalaw.org/caselaw/cases/view/151697/. See Dittmann and Ogolla (this volume) for a more elaborate description of this conflict in relation to lake access for fishermen.

allocation of public land. But also in this case, the judge concluded there was in fact no public access road.[29] These conflicts illustrate the tension between the protection of private property rights and the lack of public services, due to unregulated planning of land use transpiring from the unstructured processes of transfer in the years around independence.

5 Kihoto: A Settlement on Riparian Land

Kihoto, the workers' settlement located between Naivasha Town and the lakeshore, was sold to Kihoto Farmers Company Ltd in the 1960s. The company had 30 Kikuyu shareholders, who used to work for the settler's farm Alfafa Ltd, which sold this plot to their former employees in the 1960s. The shareholders divided it in such a way that one could access the lake from all plots, with each of them owning a small strip of land, a division that continues to be visible in the layout of the plot. The shareholders continued to farm despite the risk of flooding. Another income source was the making of charcoal from trees on their land. Finally, like in Karagita, the shareholders constructed rental houses on their plots. The parts of the plots closest to Naivasha Town have turned into dense residential areas, while landowners and land renters continue to farm on the riparian land close to the lake (even if this is officially not allowed), with more land available for farming at times when lake levels are low (Ghawana 2008; World Bank 2014).[30]

Kihoto Farmers Company Ltd has had several disputes over land rights with the government. For instance, the shareholders obtained the consent of the Land Control Board to subdivide in 1997, but they had not yet received their title deeds by 2001. An official of the Ministry of Lands asked the Nakuru District Commissioner to intervene, as there appeared to be a "misunderstanding" between the shareholders and the Naivasha Division Commissioner.[31] Since at least the year 2000, the shareholders have moreover been in conflict with the

29 *Flamingo Horticulture Kenya Limited & 3 others v White Eagle Self Help Group & 5 others* (Environment and Land Court at Nakuru) (2021), accessed August 24, 2023. http://kenyalaw.org/caselaw/cases/view/215226/.

30 Interview with a long-term Kihoto resident, June 8, 2016; *Karanja Mutitu & 19 others v Kihoto Company Limited and Naivasha Municipal Council* (Environment and Land Court at Nakuru) (2018), accessed August 24, 2023. http://kenyalaw.org/caselaw/cases/view/171036/.

31 KNA Nakuru, EA/2/19/204, E.M.O Opar for the Permanent Secretary of the Ministry of Lands and Settlements to the DC, Nakuru District, June 18, 2003, "LR No. 12079 Kihoto farmers company Ltd".

MAP 4.2 Kihoto
ANNOTATED SATELLITE IMAGE FROM GOOGLE MAPS, ACCESSED JANUARY 31, 2024. ANNOTATIONS BY MARIE GRAVESEN AND YOUSSEF EL MOUZAKKI

Naivasha Municipal Council over a 64-acre plot located between Kihoto and the main road to Naivasha Town that has been earmarked as a public park, "Kenyatta Gardens" (see Map 4.2). The Kihoto shareholders have disputed its status as public land and brought the matter to court. According to Naivasha residents, who wrote to the Nakuru DC to ask him to intervene in the conflict, the plot used to be a detention camp during the colonial period.[32] After independence, presidents Kenyatta and Moi planted "commemorative trees" on the plot in an effort to conserve the indigenous acacia, and a public primary school was constructed there. Court cases on the matter of ownership rights seem to have remained inconclusive until now (Mwangi 2005).[33] The park currently hardly looks like a park: it is scattered with trees, but usually also with litter, and is partly used as a football field. Moreover, it was for a long time notorious for robberies at night. For flower-farm workers living in Kihoto, the walk

32 Interestingly, the ten petitioners – all men with Kikuyu names – identified themselves as "the indigenous residents of Naivasha", implying both a claim on Naivasha as "Kikuyu territory" and a claim that the Kihoto shareholders were not "indigenous" residents (KNA Nakuru, EA/7/33/34, Naivasha residents to the DC Nakuru, September 10, 2000, "The position of 'The Naivasha Kenyatta Garden Public Park'").

33 KNA Nakuru, EA/7/33/34, "The position"; KNA Nakuru, EA/7/33/38 J Karung'o Kamau, District Officer Naivasha, to the DC Nakuru, September 24, 2003, "Naivasha Kenyatta Garden Public Park"; *Karanja Mutitu & 19 others v Kihoto Company Limited and Naivasha Municipal Council* (2018).

from home to the staff bus stop at the road before sunrise was a precarious undertaking, until a World Bank project installed high-tower floodlights in the settlement a few years ago (World Bank 2014).

Adjacent to Kihoto, a flower farm was established in the early 2000s, which had to close down again due to a conflict over the land rights, and after a court ordered its closure in 2011 the farm was ransacked by Kihoto residents (Kuiper 2019).[34] Notably, this plot was described as riparian land that had been illegally allocated by the Kenyan Wildlife Service in the Ndungu report, and so even the rights to the land of the party that won the court case has remained questionable (Republic of Kenya 2004, 163). Moreover, the same counts for Kihoto Settlement itself. That such occupation of riparian land is not only illegal but also dangerous became clear in 2020, when lake levels rose steeply and almost all of Kihoto flooded. The number of landowners in Kihoto had multiplied in the meantime. At the time of the flooding, the number of house owners – united in the Kihoto Homeowners Self-Help Group – had reached 500. The displaced house owners petitioned parliament for help. They explained that not only were their houses flooded, but also hippopotami and buffaloes were straying into the settlement since their habitat had shrunk. Interestingly, they asked for their land to be classified as part of the lake and for them to be resettled – explicitly asking for the land to be given back its riparian status.[35] Tenants likewise had nowhere to go: although they could try to move to a rental house in town, it had already been extremely difficult before to find an affordable place and their attempts were often in vain.[36] Both house owners and tenants thus continue to reside in Kihoto under highly precarious ecological and legal conditions.

6 Discussion: A Comparison between Land-Buying Companies in Naivasha and Laikipia

The conflict-ridden settlements of Kihoto and Karagita, initially only inhabited by members of the land-buying companies, have turned into dense sprawls, housing tens of thousands of migrant workers. As they are located

34 *Geoffrey Muhoro v Lake Flowers Limited* (High Court of Kenya at Nakuru) (2011), accessed August 24, 2023. http://kenyalaw.org/caselaw/cases/view/75475.

35 Kenya National Assembly Hansard, March 20, 2021. Accessed August 24, 2023. http://parliament.go.ke/sites/default/files/2021-03/Hansard%20Report%20-%20Thursday%2C%204th%20March%202021%20%28P%29.pdf.

36 Informal conversation on WhatsApp, 1 July 2020.

along the main roads, they are just as noticeable as features in Naivasha's landscape as the greenhouses used for flower production. They are notorious for their poor living conditions and are described in government documents as crime-ridden and prone to violence. They thus further contribute to the flower industry's global reputation as unsustainable and exploitative (Mwangi 2007; World Bank 2014).[37] However, both the extremely dense layout of and lack of infrastructure in these settlements and the tense relations between residents cannot be ascribed to the attraction of migrant workers by the horticultural industry alone. Instead, and supported by the comparison to the Laikipia land dynamics in the years around independence, these features have largely been determined by the settlements' origins as smallholder plots, owned by land-buying companies. Crucially, these smallholder settlements that hosted the workers were not "empty" or neutral terrain. They had a specific ethnic and class composition, and most migrant workers who have moved in more recently stem from a different background than their landlords.

Both Naivasha and Laikipia were so-called "extra-provincial" settler districts under the colonial dual land policy (before being included in Rift Valley Province in the 1930s), and average landholdings were large in both regions – over 5,000 acres in 1960 (Ominde 1968: 11; Odingo 1971: 45). But pressure on land in the years after independence has caused fragmentation of land use. Although this process was much slower in Naivasha than in Laikipia, the plots of the former land-buying companies are defining components of the current varied but starkly segregated landscape in both areas.

Moreover, dynamics around and within the land-buying companies in Naivasha were similar to those described by Gravesen (2020a) for Laikipia, and were equally tense: there were conflicts between founding partners; company directors and shareholders had diverging interests; there were conflicts over double ownership with other land-buying entities that caused land to remain vacant; and there were disputes with the government over public land that had been allocated to private parties. Such conflicts all enhanced insecurity and hampered development. Both in Laikipia and Naivasha, land claimants tried to force their access to land by simply physically invading the land, such as in the case of lake corridors. In addition, the Kikuyu landowners or claimants in Naivasha have often resorted to court cases. Although in a few cases they managed to get an eviction order, this strategy usually did not end insecurity as the court cases tend to drag on for decades, as in Laikipia.

37 KNA Nakuru, GU/10/5/201, Minutes of a security meeting in Olkaria Sublocation, August 5, 2005.

The Laikipia and Naivasha land-buying companies differ when it comes to levels of political involvement. In Laikipia, aspiring politicians were often tempted to sell more shares than a piece of land could support to increase their voter base (Gravesen 2020b).[38] This was different in Naivasha, where there is no evidence that the land-buying companies were instrumentalised by politicians for the amassing of voters at a large scale. However, the problem of the extensive processes of issuing titles for the land was also noticeable in Naivasha, albeit for partly different reasons. Shareholders tended to be from lower classes with little or no political connections to pull – especially as the country entered further into Moi's political rule, where Kikuyu were increasingly disenfranchised from positions in power (Mueller 2008). In addition, the land available for land-buying companies in Naivasha was limited, which may have reduced the appetite for political involvement. The lacking political involvement may have contributed to the land-buying company members' inability to advocate their cause and effectively push for title deeds.

Although land-tenure insecurity was high in both areas, the implications of such insecurity differed between Naivasha and Laikipia. The Laikipia land-buying companies were characterised by complex shareholder demographics. The lower-class shareholders who had not seen the land before investing and who had no alternative settlement options became locked to land that was ultimately unsustainable as smallholder or subsistence properties – situations that tenure insecurity only exacerbated. For these shareholders, the land-buying companies were effectively poverty traps. Lower-middle-income shareholders, who might also have intended to settle, relocated after having determined farming to be unviable. Lastly, the wealthier shareholders with no intention of settling merely assumed that an investment in land could stay undeveloped and still be sold with a profit later. What is common for the latter two is that they would never invest in developing or settling on the land, and even so, few had any intention of selling before a considerable profit could be claimed. They in turn became absentee landlords and left large parts of the landscape seemingly vacant – a trademark of the Laikipia landscape that attracted pastoralists to settle and complicate the dimensions around land rights and access further (Gravesen 2020a).

In contrast to this diversity in Laikipia, shareholders who settled in Naivasha in the 1960s and 1970s tended to belong to lower classes. To them, the plots were not an economic investment, but their new (or, in the case of the Kihoto shareholders, their existing) homes. However, their position changed drastically with extensive increases in population and in land prices. With

38 For instance, a land-buying company in Laikipia had a total acreage of 37,676 and the company directors sold one-acre shares to 53,138 shareholders (Gravesen 2020a, 157).

the influx of – often landless – labour migrants, the impoverished small-scale landowners turned into wealthy landlords, even when their plots were small and unfit for farming. They could rent out makeshift housing to migrant workers while the market value of their plots exploded. Insecurity of tenure has therefore become even more significant in recent years. This insecurity has not only impacted the smallholder landowners but also the thousands of migrant workers who reside on their plots as tenants, for instance in the case of the plot in Karagita where all housing was demolished due to a conflict over ownership rights.

Moreover, due to these transformations, the plots of Naivasha land-buying companies have also become instruments of class and ethnic competition and points of incitement, similar to the situation in Laikipia. There is no clear indication that land-buying companies in Naivasha were vehicles for bringing Kikuyu voters into this constituency, but nevertheless they solidified the Kikuyu presence. In addition, socio-economic stratification increased after the arrival of the horticultural industries (Kuiper and Greiner 2021). Whereas the Kikuyu shareholders could now accumulate wealth, their primarily non-Kikuyu migrant tenants in majority had low and insecure incomes. The relation between the Kikuyu landlords and business owners and the tenants is characterised by distrust (described in more detail in Kuiper, 2020), and is somewhat akin to the "relationships of necessity" between multi-ethnic neighbours in Laikipia as described by Gravesen (2020a).

In addition to the class- and ethnic-based distinctions, there was also a political divide, which together intensified tension and eventually led Karagita and Kihoto to become ignition points for the post-election violence in early 2008, during which migrant labourers of for instance Luo and Luhya background were targeted. This episode of violence in Naivasha was, in fact, linked to previous insecurity and struggles over land and territory in Laikipia. The violence in Naivasha was partly incited by members of the Mungiki criminal network, who at that time sought retaliation for land-related violence against Kikuyu elsewhere in Kenya, including in western Laikipia where its origins were (cf. Rasmussen 2018; Anderson and Lochery 2008). Claims on land and especially on territory were also crucial in the post-election violence, and it is not surprising that the settlements Kihoto and Karagita, which started as Kikuyu smallholder plots, were among the hotspots of violence. The violence has left a mark on neighbour relations and feelings of insecurity since then (Lang and Sakdapolrak 2015; Kuiper 2020).[39]

39 In addition, the colonial heritage still looms large and contributes to the sense of risk and crisis in the wider Naivasha Basin. The sale of former settler farms in the hinterland of Lake Naivasha to farming cooperatives created tensions and sometimes violent conflict

Finally, such broader insecurities and more particular wrangles over land also have repercussions beyond the settlements. For instance, the blockage of roads clearly hinders the transport of flowers, and such insecurities were a main reason for one flower investor to move his farm from the south part of the lake to the Flower Business Park north of town. He asserted there was much more security there.[40]

The similarities between the dynamics around the land-buying companies in Naivasha and Laikipia suggest that such informal and unregulated structures of land division and transfer are the cause of subsequent exacerbated tension, rather than such conditions rising as a product of the formation of a diverse social landscape. On these grounds, we argue that the unstructured process of land transfers in the years around independence laid the ground for the current shape and demographic composition of the Naivasha settlements on former smallholder plots. More particularly, we argue that the insecurities around tenure and land access and stakeholders' agency in manoeuvring these insecurities have greatly impacted Naivasha's anthropogenic landscape, contributing directly to the general sense of crisis and risk – as well as opportunity – in the area.

7 Conclusion

Land-tenure arrangements and land uses changed drastically in some areas of Kenya in the 1960s and 1970s. Although changes were less profound in Naivasha, some notable transfers did take place, which have had a crucial impact on the shape of the area as a whole. In this chapter, we have drawn attention to the often-overlooked agency of African land-buyers and land-claimants in shaping Naivasha's contemporary, deeply anthropogenic landscape. The country-wide frenzy around land acquisition that had driven the expansion of land-buying companies in other places in the country was not as prominent in Naivasha in the years after independence. However, the few Naivasha plots that were purchased by land buying companies were also characterised by fierce competition and collusion around securing land – processes that were riddled with various types of strategies, including physical occupations, violently enforced

between Kikuyu landowners and the Maasai who were previously dislocated from this land (Kioko and Bollig 2015; see also Müller-Koné and Kioko, this volume). But the conflict-ridden urban sprawls close to Lake Naivasha have not much been affected by these particular conflicts between Maasai and Kikuyu.

40 Interview on February 27, 2014.

land claims, legal opposition, and interference by elites, including government officials on different levels. The chapter has therewith highlighted how the permanent sense of crisis in the area not only stems from transformations during the colonial period or changes in the current globalised era, but is also deeply rooted in the manner in which changes in land tenure took form in the post-independence decades, which have continued to affect the lives and livelihoods of smallholder plot owners and their tenants up until today.

Acknowledgements

Gerda Kuiper conducted her research in Naivasha in the context of the DFG-funded Research Unit *Resilience, Collapse and Reorganization in Social Ecological Systems of African Savannahs.* Marie Ladekjær Gravesen received funding for her research in Laikipia under the EU-Funded Marie Curie Innovative Training Network *Resilience in East African Landscapes* (REAL).

References

Anderson, David, and Emma Lochery. 2008. "Violence and Exodus in Kenya's Rift Valley, 2008: Predictable and Preventable?" *Journal of Eastern African Studies* 2 (2): 328–43. https://doi.org/10.1080/17531050802095536.

Andreasen, Manja H, Gordon McGranahan, Griet Steel, and Sadaf Khan. 2021. "Self-Builder Landlordism: Exploring the Supply and Production of Private Rental Housing in Dar Es Salaam and Mwanza." *Journal of Housing and the Built Environment* 36 (3): 1011–31. https://doi.org/10.1007/s10901-020-09792-y.

Boone, Catherine, Fibian Lukalo, and Sandra F. Joireman. 2021. "Promised Land: Settlement Schemes in Kenya, 1962 to 2016." *Political Geography* 89 (August): 102393. https://doi.org/10.1016/j.polgeo.2021.102393.

Branch, Daniel. 2009. *Defeating Mau Mau, Creating Kenya: Counterinsurgency, Civil War, and Decolonization.* African Studies. Cambridge; New York: Cambridge University Press.

Branch, Daniel. 2011. *Kenya: Between Hope and Despair, 1963–2011.* New Haven: Yale University Press.

Daily Nation. 1984. "Mystery Surrounds Fate of Farm Company's Cash." July 2, 1984.

Daily Nation. 2002. "Villagers Ordered to Leave Disputed Plot." January 28, 2002.

Daily Nation. 2005. "Complaints by Workers Mar Bloom in Flower Farms." August 25, 2005.

Daily Nation. 2015. "Transport Paralysed as Locals Block Road." January 10, 2015.

Ghawana, Tarun. 2008. "Flowering Economy of Naivasha: Impacts of Major Farming Systems on the Local Economy". Thesis, University of Twente.

Gravesen, Marie Ladekjær. 2020a. *The Contested Lands of Laikipia: Histories of Claims and Conflict in a Kenyan Landscape*. African Social Studies Series 42. Leiden: Brill.

Gravesen, Marie Ladekjær. 2020b. "Twilight Institutions: Land-buying Companies & Their Long-term Implications in Laikipia, Kenya." In *Land, Investment & Politics: Reconfiguring Eastern Africa's Pastoral Drylands*, edited by Jeremy Lind, Doris Okenwa and Ian Scoones, 144–154. Woodbridge: James Currey.

Harper, David M., Edward H.J. Morrison, Michael M. Macharia, Kenneth M. Mavuti, and Caroline Upton. 2011. "Lake Naivasha, Kenya: Ecology, Society and Future." *Freshwater Reviews* 4 (2): 89–114. https://doi.org/10.1608/FRJ-4.2.149.

Kanogo, Tabitha. 1987. *Squatters and the Roots of Mau Mau*. Eastern African Studies. London: James Currey.

Kanyinga, Karuti. 2009. "The Legacy of the White Highlands: Land Rights, Ethnicity and the Post-2007 Election Violence in Kenya." *Journal of Contemporary African Studies* 27 (3): 325–44. https://doi.org/10.1080/02589000903154834.

Kioko, Eric M., and Michael Bollig. 2015. "Cross-Cutting Ties and Coexistence: Intermarriage, Land Rentals and Changing Land Use Patterns among Maasai and Kikuyu of Maiella and Enoosupukia, Lake Naivasha Basin, Kenya." *Rural Landscapes: Society, Environment, History* 2 (1). https://doi.org/10.16993/rl.ad.

Klopp, Jacqueline M. 2001. "'Ethnic Clashes' and Winning Elections: The Case of Kenya's Electoral Despotism." *Canadian Journal of African Studies / Revue Canadienne Des Études Africaines* 35 (3): 473–517. https://doi.org/10.1080/00083968.2001.10751230.

Kuiper, Gerda. 2019. *Agro-Industrial Labour in Kenya: Cut Flower Farms and Migrant Workers' Settlements*. Cham: Palgrave Macmillan.

Kuiper, Gerda. 2020. "'They Just Move in with Relatives": Translocal Labour Migrants and Transient Spaces in Naivasha, Kenya." *Journal of Eastern African Studies* 14 (2): 227–49. https://doi.org/10.1080/17531055.2020.1730562.

Kuiper, Gerda, and Clemens Greiner. 2021. "Export Horticulture and Labour Migration in Kenya: Translocality and Transiency in a Secondary Town." *Geoforum* 122 (June): 1–10. https://doi.org/10.1016/j.geoforum.2021.03.013.

Lang, Britta, and Patrick Sakdapolrak. 2015. "Violent Place-Making: How Kenya's Post-Election Violence Transforms a Workers' Settlement at Lake Naivasha." *Political Geography* 45 (March): 67–78. https://doi.org/10.1016/j.polgeo.2014.09.005.

Leo, Christopher. 1981. "Who Benefited From the Million-Acre Scheme? Toward a Class Analysis of Kenya's Transition to Independence." *Canadian Journal of African Studies / Revue Canadienne Des* Études *Africaines* 15 (2): 201–22. https://doi.org/10.1080/00083968.1981.10803973.

Leo, Christopher. 1984. *Land and Class in Kenya*. Toronto: University of Toronto Press.

Mueller, Susanne D. 2008. "The Political Economy of Kenya's Crisis." *Journal of Eastern African Studies* 2 (2): 185–210. https://doi.org/10.1080/17531050802058302.

Mwangi, Macharia. 2005. "Mzozo wa bustani sasa wachukua mwelekeo mpya [Quarrel about a garden now takes a new direction]." *Taifa Leo*, December 14, 2005.

Mwangi, Macharia. 2007. "Naivasha Town: Where Poverty and Affluence Live Side-by-Side." *Daily Nation*, August 18, 2007.

Odingo, Richard S. 1971. *The Kenya Highlands: Land Use and Agricultural Development*. Nairobi: East African Publishing House.

Okanga, Daniel. 2008. "Fujo za Naivasha zilikuwa ni kulipiza kisasi mauaji [Chaos in Naivasha was a retaliation for killings]." *Taifa Leo*, October 19, 2008.

Ominde, Simeon H. 1968. *Land and Population Movements in Kenya*. London: Heinemann.

Rasmussen, Jacob. 2018. "Parasitic Politics: Violence, Deception, and Change in Kenya's Electoral Politics." In *Violence in African Elections. Between Democracy and Big Man Politics*, edited by Mimmi Söderberg Kovacs and Jesper Bjarnesen, 176–96. Uppsala: Zed Books.

Republic of Kenya. 2004. "Report of the Commission of Inquiry into the Illegal/ Irregular Allocation of Public Land (The Ndungu Report)." Nairobi: Government Printer.

Sorrenson, M.P.K. 1968. *Origins of European Settlement in Kenya*. Nairobi: Oxford University Press.

Styles, Megan A. 2019. *Roses from Kenya: Labor, Environment, and the Global Trade in Cut Flowers*. Seattle: University of Washington Press.

Wasserman, Gary. 1976. *Politics of Decolonization: Kenya Europeans and the Land Issue, 1960–1965*. African Studies Series 17. Cambridge: Cambridge University Press.

World Bank. 2014. "Environmental and Social Impact Assessment for Naivasha." Kenya – Informal Settlements Improvement Program Project: Environmental Assessment Vol. 11. http://documents.worldbank.org/curated/en/592331468050686738/Environmental-and-social-impact-assessment-for-Naivasha.

PART 2

Ecological Dynamics in the Lake Naivasha Catchment

CHAPTER 5

The Unique Natural Resources of Lake Naivasha; Can They Survive?

David Harper, Nic Pacini, Edward Morrison, Dominic Kimani, Timothy Mwinami and Caroline Upton

1 Why Is Naivasha Unique?[1]

1.1 *Freshwater Where Saline Water Is Common*

The Great Rift Valley of Africa is a giant depression running for 7,000 km from the Red Sea in the North to Mozambique in the South, caused by the gradual separation of two of the earth's plates. The faulting, which began some 25 million years ago, uplifted mountains either side of a deep trench dotted with volcanoes; some of these are very recently active and in northern Tanzania, some, such as Ol Donyo Lengai in the north, were active up to very recently. Rain falling on the mountain masses in Kenya – Mount Kenya, the Aberdares, Mau Forest complex and Cherangani – drains down one side into the floor of the Rift Valley, where water has no escape, accumulating in depressions known as endorheic lakes – lakes with no outflow. Such shallow lakes in the tropics with no surface outflow lose water by evaporation, leaving behind a concentrated solution of the salts that dissolved from the volcanic rocks of their catchments. These are the alkaline-saline lakes, commonly called soda lakes, famous throughout Eastern Africa, because their extreme chemistry implies that very few species can live in them. Their waters are intensely green, because a single species of colonial photosynthetic bacteria reaches enormous densities, fed upon by the world's only filter-feeding bird, the lesser flamingo. Lesser flamingos can congregate in hundreds of thousands, sometimes over a million, creating what has been described as the greatest ornithological spectacle in the world.

These lakes are common along the Rift Valley in Kenya, but what is much rarer is an endorheic freshwater lake. There are only two such freshwater lakes in Kenya's Rift Valley; Naivasha is the larger of them. An endorheic lake can only be a freshwater lake if it has an outflow, with enough inflow to match and

1 First submission date: March 10, 2021. Acceptance date: March 13, 2023.

 | DOI:10.1163/9789004695429_006

to overcome the salt concentration caused by evaporation losses (see Map 5.1). Naivasha sits at the highest point of the Rift Valley floor – almost 2,000 metres, with permanent inflows from the Aberdare Mountains to its east, 4,000 metres at the highest point. Colonial scientists were aware of an underground outflow to the south-west, together with inflowing groundwater seepage from the river valleys in the north-east (Sikes 1935; Thompson and Dodson 1963), but it was not quantified until the 1970s by water-budget calculations from river inflow and evaporation, leaving a residual (estimated seepage out) of up to 20%. The river waters are naturally very dilute, and it is believed that some chemical sedimentation may also occur within the lake (Gaudet and Melack 1981); the combined effect maintains the lake water as fresh, albeit with variability,

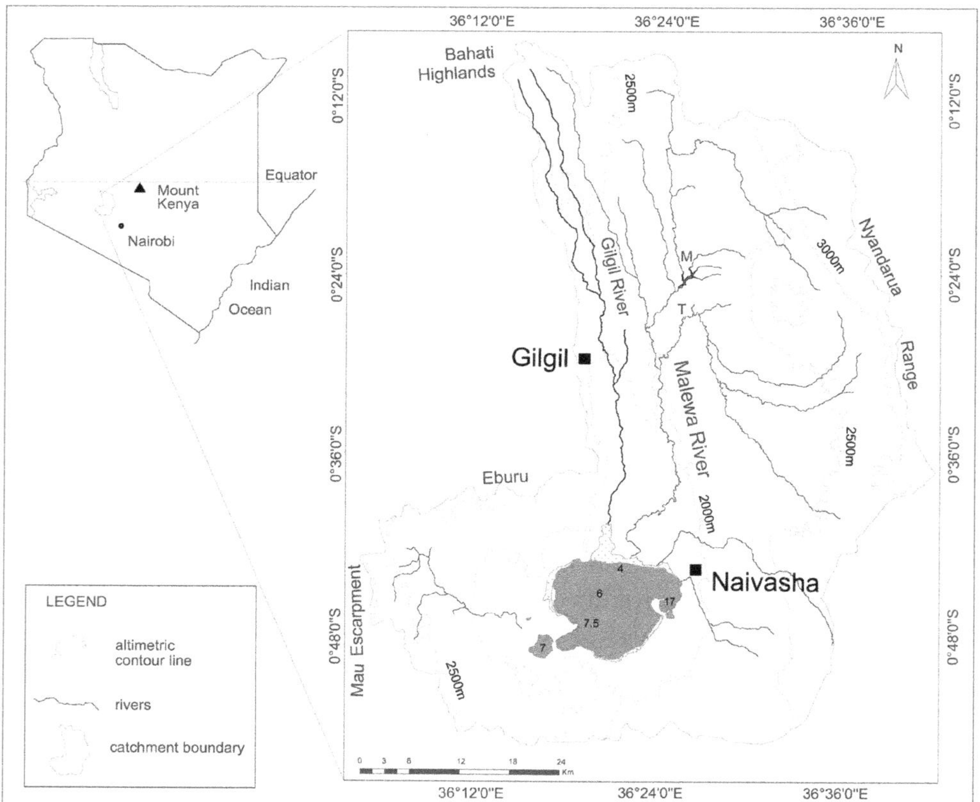

MAP 5.1 The Lake Naivasha Basin. The hydrographic network, the altimetric contour lines and the catchment boundaries were redrawn by consulting 1:25000 maps and Google Earth. The location of the Turasha Dam (built in 1992) and of the proposed Malewa Dam are represented with letters T and M
CARTOGRAPHER: NIC PACINI

depending on seasonal hydrological changes. Subsequent calculations using the mass balance of conservative ions (not biologically utilised) such as chloride, again showed the outflow as 20–30% of inflow (Ojiambo and Lyons 1996). More recently, measurements of groundwater volume, composition, and movements using piezometer measurements (Misongo 2020) and sophisticated modelling (Hogeboom et al. 2015) confirm these earlier findings.

1.2 *Irregular Amounts of Freshwater*

Outward seepage provides the Rift Valley with its most important natural resource, freshwater, which makes it important to Kenya, but does not of itself make the lake globally unique. Its global uniqueness comes from its high species diversity, paradoxically arising from its hydrological irregularity, which is greater than almost any other lake (see Figure 5.1). The supply of rainwater is very erratic, due to global events, such that the lake has fluctuated by 10 metres vertically (more than three times its current average depth), and many square kilometres horizontally (over three-quarters of its surface), since records began in 1906. This, however, is small in comparison to the prehistoric changes evidenced from geological strata and sediment-core analysis.

Examination of the evidence of former lakeshores on adjacent Rift Valley walls early in the twentieth century showed that once, at least half a million years ago, there was a mega-lake including all of present day Nakuru, Elmenteita, and Naivasha lakes. Excavation of the *Homo erectus* tool site of Kariandusi, north of Gilgil once on this lake's shoreline, has shown that obsidian – hard, black shiny vitreous volcanic rock, likely brought from Eburru (north) or Olkaria south of the present lake to be carved into tools – almost certainly arrived by boat (Leakey [1931] 1971; Nilsson 1932; Butzer et al. 1972). More recent examination of lake sediments, by mineral-isotope analysis and particularly by inspection of the biotic remains in them such as the silica shells of microscopic diatoms (lake algae) and pollen from land plants that use wind dispersal, have shown that Naivasha has been extremely unstable over the past several hundred thousand years. It ranged from a small dry soda lake to a large freshwater lake then discharging through Njorowa Gorge (now part of Hell's Gate NP) until about 7,500 years ago, during a period when rainfall is estimated to have been 65% greater than the present (Richardson and Richardson 1972). Fluctuations have not been as great in the last one thousand years or so but have still been considerable; from near-dryness 1,000 years and 200 years ago, to over 40 metres depth inside Crescent Island Lagoon about 300 years ago (Verschuren, Laird, and Cumming 2000); the depth inside the lagoon has fluctuated by around 14 metres for the past few decades (see Harper and Styles, this volume).

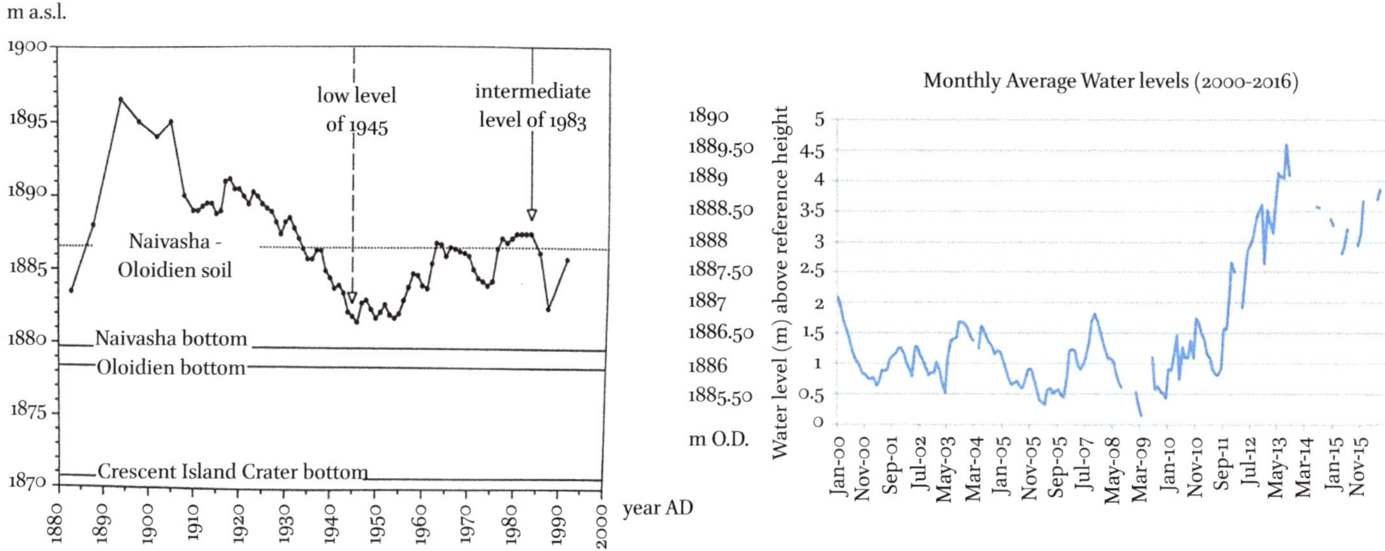

FIGURE 5.1 Fluctuations of the water level in Lake Naivasha, from 1880 to the present day. The left hand graph initially shows the probable water levels calculated from the writings of explorers and early settlers. In 1909, a datum point was established at the Naivasha town side of the lake and subsequent datum points on the eastern and southern shores levelled against that as the lake receded; the data points are thus regular and accurate (from Verschuren, 1994). The right hand graph shows the fluctuations during the first 15 years of the 21st Century, calculated from surface area of satellite imagery by Odoyo et al., (2018). Although the precise pattern of this graph does not agree with others (e.g. Fig. 3 in Harper et al. 2011, Fig. 2 in Morara et al, 2022), it does show that the decade 2005–2005 saw a low and a high that matched those of the past 70 years. Recent reports indicate even higher water level fluctuations since 2010; Hermegger quotes 7.5 metres. The 4 satellite images over the past 50 years in Fig 5.2 can be compared with the appropriate point on these graphs
GRAPHS BY DAVID HARPER

The causes of this great variation are understood in outline. On the longest time scale (tens of thousands of years), East Africa's climatic variations are linked to Milankovitch cycles caused by irregularities in the earth's orbital movements (Buis 2020) which change insolation by up to 25%. In the past thousand years, such changes in insolation that have caused major Northern Hemisphere events which have also affected the East African climate. The Medieval Warm Period (1000–1270) is associated with higher solar radiation in Europe and North America and with drought in East Africa; the Little Ice Age (1300–1850) with lower insolation and wetter periods in East Africa. This latter was quite variable, but the three very cold few decades – from 1650, 1770, and 1850 – were associated with high lake levels (Stoof-Leichsenring et al. 2011), also seen in changes identified in Crescent Island sediments by Verschuren, Laird, and Cumming (2000). Variability in Northern Hemisphere insolation impacts upon temperatures in the Indian and Atlantic oceans, which affects the magnitude and direction of monsoon winds – south-west from the Indian, north-east from the Atlantic. The winds meet over the East African highlands, dropping the rain they carry from the two adjacent oceans; the region where they meet forms an ideal boundary oscillating according to their relative strengths – the Intertropical Convergence Zone (ITCZ) – creating two wet and two dry seasons each year. The ITCZ can be pushed northwards or southwards by relative temperature changes in the two oceans, under the influence of global-climate drivers known by climatologists as the Indian Ocean Dipole and the El Nino/ Southern Oscillation. Superimposed onto this pattern, northern-hemisphere cooling pushes the ITCZ southwards, causing drought conditions in East Africa.

1.3 *High Plant Diversity Comes Out of Climate Unpredictability!*

Climatic variability and lake-level fluctuations are now not only believed to have been responsible for the diversity in early Hominid evolution in East Africa over longer timescales (Maslin et al. 2014), but have also been shown to maximise biodiversity over shorter timescales. Modern ecological theory calls this the Intermediate Disturbance Concept (Wilkinson 1999), whereby an external factor, which may be physical, such as rainfall or strong winds, or biological, such as grazing or predation, creates gaps in a biological community which can be filled by pioneer species. Disturbance thus creates opportunity for the coexistence of a larger number of species. At Naivasha, the disturbance is the water level constantly changing and either drowning terrestrial species and promoting aquatic species as it rises or creating bare mud for recolonisation and succession as it falls. A US botanist at Nairobi University as part of an exchange programme after Kenya's independence, John Gaudet, studied the vegetation at Lake Naivasha in the early 1970s and demonstrated that the lake-level fluctuations gave rise to successive plant zones, each with unique species, together leading to high species richness. Variability of water level meant that each zone was always represented on the shoreline; this, superimposed upon variation in soils, between volcanic sand and papyrus peat, ensured the formation of a complex mosaic. The zones passed from the yellow-barked acacia trees growing as far up the shore as the 1906 high-water level downwards to the submerged plants up to 5m deep in the shallow water (called the littoral zone) of the lake. One hundred and eight species from 45 families were recorded, between 1973 and 1974, in five zones (Gaudet 1977a). This is an exceptionally high family diversity, also with an average of just over two species per family. Five zones were identified – (1) shallow water (littoral) zone with submerged and floating plants; (2) wet mud recently exposed with seedlings; (3) dried oxidised mud called the *Sphaeranthus* (globe thistle) zone after the dominant genus; (4) recently dried mud with sedges dominated by *Cyperus* species (papyrus the most well-known), (5) dry soils with composites (dandelion family) dominated by *Conyza* (fleabane) species (aster family), including yellow-barked Acacia (now called *Vachellia*) woodland with other woody species. The number of species declined from land to water with 52 on dry land, 49 in the combined wet zones, 36 on mud and 25 in the water.

Twenty-five fully aquatic plant species in the littoral zone, although large in a freshwater lake, compares with 172 species of microscopic algae (including 25 blue-green bacteria) found by Hubble (2000) in the open water of the lake (see also Pacini et al. 2021). Such a high number is not exceptional; the open water of all freshwater lakes is highly species-rich, a phenomenon explained by Hutchinson (1961) that the niche of any one species was constantly changing due to water movement altering conditions of light, temperature, nitrogen, phosphorus, silica, and so on, enabling multiple niches to coexist. Later authors have found similar numbers of species in Naivasha (e.g. Ballot et al. 2009; Pacini et al. 2021) but in different proportions. Algae provide a very good indicator of change in lakes, something that will be returned to later.

1.4 *High Animal Diversity Follows Plant Diversity*

The diverse lakeshore vegetation belt is critical for maintaining a high biodiversity of animal life, for the same reason – different species are specialists in each zone. We don't have such a complete understanding of animal presence, as many species are seasonal migrants, but three examples illustrate the principle of high plant diversity supporting high animal diversity, two from yellow-barked Acacia (*Vachellia*) woodland at the top of the drawdown zone and one from the lake. A 20-day study of the bird species in the woodland (Kimani 2012) found 169 bird species of 55 families from 10 sites around the lake. This represents 15% of the total number of bird species recorded in Kenya. At the time of finalizing this chapter, one of these species, grey-crested helmet shrike *Prionops poliolophus*, is categorised as Near-Threatened on the IUCN's Red List (IUCN 2023). Two are Endangered – martial eagle *Polemaetus bellicosus* and the grey-crowned crane *Balearica regulorum*. One of these species, Rüppell's vulture *Gyps rueppellii*, is even critically endangered. We do not suggest that all of the 169 species were confined to the woodland zone, but rather that some are dependent upon it and would not be able to live outside it – such as the grey-crested helmet shrike – whilst most are mobile but need its food and shelter resources.

Insects however, tend to be more confined to one habitat or even one plant species; a study of macro-moths of the yellow-barked and other Acacia (*Vachellia*) species growing together behind the lake shoreline found 76 species in six families (Agassiz and Harper 2009), 12 new to science, with six presently described, after collecting larvae from the trees and rearing them to emergence as adults (Agassiz 2009). It is quite remarkable that species unknown to science can be found 'under our noses' at Lake Naivasha and not just in uninhabited forests or deep ocean!

The only long-term monitoring programme on the natural resources of lake Naivasha, which is also undertaken by Citizen Scientists, is the annual water bird count organised by National Museums of Kenya (NMK) in partnership with Nature Kenya (NK) and the Kenya Wildlife Service (KWS), which remarkably has been undertaken continuously since 1990 (Owino et al. 2001, 2009). This takes place on a single day, usually in January. The survey confirms the high biodiversity of the lake's resources – in this case water birds – which have shown a total of 110 species over 25 years, varying in individual years from 61 (in 1992) to 82 (in 1994, 1995, and 2002). Half these birds are carnivorous on invertebrates and small vertebrates; a quarter piscivores and a quarter omnivores. Very few (six) are herbivores. Just like other animal groups, the birds are habitat-specific – Henderson and Harper (1992) identified seven different vegetation habitats plus an eighth, open sand/mud shoreline, which supported distinct bird assemblages. A one-day census covers the whole lake, but more detailed studies which focus on individual habitats will provide greater detail and more species. Papyrus swamp also supports rare warblers, such as lesser swamp warbler, little rush warbler, and African reed warbler (Tyler 1991), within an assemblage of about a dozen species that overlap with other habitats (Nasirwa and Nalianya 2002).

1.5 *Key Plant Species Regulate the Ecosystem*

Plant biodiversity is not only uniquely valuable in its own right and in the animal species it sustains, but it also contributes to maintaining optimal quality of lake water (in the sense of water use by people) through a number of mechanisms, the most important of which John Gaudet demonstrated. This is by controlling the cycling of nutrients between land and water, smoothing out peaks in nutrient concentration that would otherwise be damaging, retaining large quantities of nutrients in plant biomass and in decomposing vegetation, and releasing them slowly. Gaudet showed in his 1977 study that papyrus (*Cyperus papyrus*) could exist, once it had germinated from seed, in several forms, the most mature being fringing swamps and floating islands. He investigated the North Swamp, which in the early 1970s covered 11 km^2 in the Gilgil and Malewa estuaries at the north of the lake, where almost all the inflowing water to the lake comes through this swamp (Gaudet 1977b; 1979). He collected water and sediment over a year on a transect from river to swamp to lake (when papyrus swamp completely covered the Malewa estuary). Analysis showed that most of the nutrients (nitrogen, phosphorus, silicon) in the rivers entering the swamp were in particulate form, whilst most of the major ions (sulphur, manganese, iron) were dissolved. The swamp itself was a large source of nitrogen, through bacteria fixation of atmospheric nitrogen in the mat, such that nitrogen concentrations within the swamp water were 600% of those entering via stream water; in comparison the mat contributed only 60% of the phosphorus (which has no gaseous form). The largest output to the lake was organic detritus from dead plant material, which in low river flow settled on the bottom of the swamp, while in high river flow was carried into the lake. Detritus contained almost all the important elements; inside the swamp it was mostly in anaerobic (oxygen-poor) conditions, but once into the lake, within the submerged plant (littoral) zone which extended to 5m depth, it was broken down by aerobic (oxygen-requiring) processes slowly and then utilised within that zone.

The shallow littoral zone beside the swamp, forming the edge of the aquatic ecosystem, is an important part of the recycling of nutrients from land back to water, because the nutrients from flooded soils are extracted by the roots of submerged plants and brought back in circulation. Gaudet and Muthuri (1981) showed that 73% of the phosphorus and 26% of the nitrogen in the lake littoral comes from recycling of the flooded soils. Other pathways of land-water recycling also exist, such as grazing by hippopotamus, which puts most of the grass they consumed on land by night back into the littoral zone as faeces by day (Grey and Harper 2002). The precise functioning of the littoral zone has not been investigated in Naivasha, but it has been well studied in other African lakes and shown to be critical to their efficient functioning in terms of fisheries and good water quality (Howard-Williams and Lenton 1975). A study of fish distribution in Naivasha in the late 1950s did show the importance of the littoral zone, with highest numbers of tilapia (*Oreochromis leucostictus*) found under papyrus and highest numbers of *Tilapia nigra* along the margins of the swamp in the littoral macrophyte beds – the "rich carpet of waterweeds". Double the number of fish were caught in nets parallel to the swamp edge compared to at right angles, showing that the movement of fish between littoral and open water was much more important than the movement along one habitat (Elder 1959). Later, Muchiri, Hart, and Harper (1995) showed that *Oreochromis leucostictus* spent 94% of its time in the littoral compared to open water, *Tilapia zillii* 67% and bass (*Micropterus salmoides*) 63%. The littoral zone with its "forest" of submerged plants up to 5m tall (deep) with the floating leaves of water lilies at the top, used to retain a very different physical and chemical environment than under the swamp (deoxygenated, detritus-dominated) or the open water (wind-mixed, dominated by microscopic-sized plant and animal plankton). It used to be very well oxygenated due to photosynthesis of the plants, despite high production of oxygenated detritus from plant leaves and epiphytic algae (living on the leaves), which sink quickly to the bottom because of the lack of currents leaving the crystal-clear water (Harper, Adams, and Mavuti 1995).

The importance of natural organic detritus, generated from plants within the ecosystem (autochthonous), cannot be overestimated, although it is often overlooked in research and monitoring on the lake. Muchiri (1990) showed that 49% of the diet of *O. leucostictus* was made up of it and 41% of *T. zillii*. The two species were able to coexist because they did not overlap in the other dietary components – *O. leucostictus* fed more on the bottom invertebrates and algae (consistent with its habitat preferences under papyrus swamp); *T. zillii* fed more on submerged plants and the bugs/beetles which lived within them (whose preferred habitat is the littoral zone). These latter, particularly one ubiquitous and abundant species of water boatman, called *Micronecta scutellaris*, were very important in the diet of the young predatory largemouth bass, which was very size-selective, moving to larger food such as frogs and small fish as it grew (Hickley et al. 1994).

2 Do We Recognise and Sustainably Use the Resources?

2.1 *Recognising the Resources*

The very high richness of plant species is dependent upon flooding and recession, and the interactions of the important ones (papyrus and submerged plants) with water quantity (hydrology) and chemistry. The cause has been expressed as "The Flood Pulse Concept" from studies of similar ecological processes that have been shown in large rivers which annually flood, such as the Amazon and the Danube, and in river-dominated large wetlands such as the Tonle Sap in Cambodia (Junk, Bayley, and Sparks 1989; Junk and Wantzen 2004). The way that plant biodiversity is maintained by hydrology, whilst plant communities then modify hydrochemistry, is also an illustration of the principles of Ecohydrology (Zalewski 2002), which is a practical concept that uses our understanding of relationships between hydrological and biological processes at different scales to recommend management measures that improve water quality, water quantity, and biodiversity (Khan, Tchadie, and Llovera 2011). The Kenyan and European research scientists working at Naivasha have been involved with the UNESCO International Hydrological Progamme since 1997 (Harper and Zalewski 2001) and the results of their work have promoted the lake as an international Demonstration Site for Ecohydrology since then (Negussie, Urbaniak, and Zalewski 2011).

It is the case in almost all scientific and social studies of communities and natural resources that the information gathered by research is communicated as rapidly as possible to funders and decision-makers (whether they use it or not is another matter!) but often does not so quickly reach the users of the resources on the ground. Natural resources globally are in decline, and measures to define the economic value of their uses more accurately have been adopted this century. The commonest of these is Ecosystem Services (Millennium Ecosystem Assessment 2005). A good example is shown by people's recognition of the benefits of papyrus. Quite apart from the moderating of water quality described above (known as a Supporting Ecosystem Service), papyrus has a very high productivity – indeed, one of the highest of any plant (Muthiri, Jones, and Imbamba 1989), and it is widely used for building material and for fuel, making it also a Provisioning Ecosystem Service (Morrison et al. 2013a). Resident communities who were surveyed around the lake (see Upton 2013) recognised a wide array of ecosystem services associated with papyrus wetlands across the four categories of ecosystem services. Clean water supply (36.1%), fishing/hunting opportunities (22.9%), and enhanced agricultural potential (21.1%) together accounted for over three-quarters (80.1%) of all perceived benefits associated with Provisioning Services. Recognition of the wetlands' aesthetic qualities, opportunities for maintaining human relationships, and values related to being satisfied with one's sense of place in the world were the majority (87.3%) of benefits assigned to Cultural Services. Influences on local temperature and precipitation (42.4%), removal of pollutants from water (27.1%), and the ability of the swamps to assist in erosion control (13.6%) were most of the (83.1%) perceived benefits of Regulating Services. The value of wetland habitat for the survival of wild plants and animals (biodiversity support) dominated Supporting Services (88.1%). Interestingly, nobody recognised nutrient regulation, photosynthesis/biomass or heritage as services provided by papyrus. In this, the lakeside communities surveyed differed from a comparable, far longer-settled, group at Lake Victoria, which had higher awareness overall, with particularly more awareness of the use of fibres (provisioning), climate regulation and erosion/hazard control (regulating), educational/spiritual/heritage values (cultural) and biodiversity (supporting). In contrast to Lake Victoria, Lake Naivasha communities in riparian settlements are generally ethnically heterogeneous, recently settled, with a high rate of immigration.

Neither resident communities at Naivasha nor visitors are fully aware of the value of the lake's natural resources, which they should carefully manage.

A specific survey of hotel managers around the lake (Vereecke 2012) representing half of the range of 35 hotels from most expensive to least expensive, showed that 57% of interviewees have only poor or average knowledge of the intrinsic benefits of papyrus, and that some interviewees were unable to identify the plant, whilst 43% had good knowledge and were able to list multiple benefits. Managers of lakeside hotels had better knowledge (60% good) than those further away. Hotel guests had even poorer knowledge, not one able to list multiple benefits and many not even knowing of papyrus' existence: "I came for the animals, not the plants", was a typical comment.

Wider studies of residents' knowledge about the natural resources of Naivasha show that understanding differs according to who they are and where they live.[2] In the informal lakeside settlements where almost everybody is on the poverty line, use of natural resources is widespread but understanding limited, regardless of their sustainability. Naivasha's most populous settlement Karagita (see Kuiper and Gravesen, this volume, on the development of this settlement) is located landward of the South Lake Road, with an "unofficial" access to the lakeside between two riparian properties (see also Dittmann and Ogolla, this volume). A borehole has been provided by a charity, Water & Sanitation for the Urban Poor (WSUP), such that few people must visit the lake for domestic water. It is rich in fluoride however, and the effects are obvious in schoolchildren, with brown teeth widespread; treated water is sold from kiosks, but untreated water is less expensive. Borehole water, lake water (a few collect directly) and rainwater (uncommon) are drunk directly, sometimes without boiling, and frequent waterborne diseases including typhoid are reported.

The settlement is surrounded by ranches, and most people rent their houses, though almost no food is grown within them; a few people keep pigs, which wander the trackways between plots. An average Karagita resident consumes fish once/twice a week, but if she/he is a Luo, fresh Lake Naivasha fish or dried omena (*Limnothrissa miodon*) from Lake Victoria will be eaten every day. It is virtually only fish poachers that supply Karagita as there is no official landing beach for the catch. Residents enjoy eating fish, and that demand is rising. In 2013, the price of fish varied between KSh 100 and 350 depending on its size. Very small fishes are sold at the price of 4 x 20 KSh, while omena costs KSh 25 for a small tin.[3]

2 The remainder of this section is primarily based on work by Caroline Upton. See for a description of the methods and more details Upton (2013).

3 At January 1, 2013, KSh 1 was equivalent to $0,01142 (source for the exchange rate: https://www.oanda.com/currency-converter/en/). Hence, in dollars, fish prices varied from roughly $1,15 to $4, while smaller fish were sold for about $0,90 per 4 and a small tin of omena for about $0,30.

These amounts are considered sufficient to provide one meal for an adult. Fresh lake fish supply is seen to be decreasing and of concern.

Most cooking is by charcoal, imported from Maella/Ndabibi highlands just to the south-west (see Müller-Koné and Kioko, this volume). The price rises from 700–800 KSh/bag during the dry season (2013 prices), up to 1,000 during the rains due to poor road transport.[4] The charcoal market is regulated by the government, which is supposed to issue permits to each dealer; the system is however riddled with corruption and it is impractical to obtain the permits without paying the controllers. Firewood is not abundant around Karagita and many residents are obliged to purchase charcoal at high prices. While demands for charcoal are increasing, the consumption per person is decreasing due to its cost. The entire system is under crisis due to unavailability of a cheaper alternative. Most residents substitute paraffin (kerosene) for cooking when there is a shortage of, and an increase in, the price of charcoal. Women would like the availability of charcoal to be increased as well as the price of paraffin reduced (in 2013 at KSh 80 per litre, currently (2023) more than double this price).

Recreation is not considered very important to the residents of Karagita; despite this, most visit the public beach at least twice a week, and every time someone comes to visit them. They go down for picnics with friends and enjoy doing so also at Crayfish Camp and at Fish Eagle Inn where they can pay something small to be allowed access and picnics are not restricted. Concerts, picnic sites, and improvised discos by the lake were more common before the COVID-19 pandemic. Besides allowing residents to have a good time, business also boomed as they could sell small items like candy, boiled eggs, etc.

The lake edge is also important for educational purposes as schoolchildren are taken there by bus and given talks on nature. In the future, locals would like the activities to increase and be varied such that entry fee is paid. There could be sale of small items of environmental interest, showing of short films, and so on. All the above activities are only possible if there is adequate access to the public beach. The residents would like the road to the beach expanded so that cars can pass through as this would increase the number of people going there, which in turn leads to booming business for small traders. Many people are baptised in the lake, and a small section should be reserved for this.

Other activities include mainly walking and watching birds; nobody seems to go out for excursions because "you need to have something in your pocket to get the morale to go for a long walk". Considering the low wage that flower farms pay their workers – according to Upton (2013) common wages were 6,000

4 This price was equivalent to \$8-\$9 during the dry season and about \$11,40 during the rainy season, see the exchange rate in the previous footnote.

KSh/month, or 150 KSh/day[5] – few residents feel concerned with recreational activities. Access to the lake represents the only recreational activity that is considered relevant and that the residents feel is worth fighting for (riots in Karagita dispersed by the paramilitary GSU with tear gas in 2011). The residents expect more access points to be opened by riparian landowners to allow people to access the waterfront far more freely. Confronted with the question of the impact that this might have on the lake, they responded that they would request the City Council authority to police the beach grounds.

Female residents interviewed underlined the poor state of the infrastructure; the market in Karagita is small and occupies the road reserve. Since this poses a problem to residents and vendors, they would like a larger and more organised market space. Despite numerous projects and citizen-collective action, Karagita has very poor solid-waste disposal facilities. This is of great concern, as the quantities of the garbage around the residential areas are increasing. They would like laws on garbage disposal and would like plot owners to place large drums for disposal of solid waste (and the municipal council to collect it).

Herbal medicine is not popular in Karagita and many would prefer to go to clinics. Local health services are insufficiently staffed or equipped, and more facilities should be set up nearby. The transport system is poor and most roads in the settlement are impassable during the rainy season.

In Kihoto, the informal settlement nearest to the town, and in Naivasha Town itself, many people can rent small plots of land to grow vegetables, as Kihoto has wide riparian access that is intensively farmed in thin strips which follow water-level changes. Their awareness of the lake's provisioning services is high, particularly water for irrigation of plots and fish for food, as well as of native plants that can be used for food or medicinally. Recognition of biodiversity was strong, but only of animals and related more to the earnings of tourism rather than anything supporting the lake's ecosystem. Such uses of ecosystem services were rare however, because few respondents acknowledged having spare time. Almost everybody felt that the lake water was not clean, believing it to be contaminated by horticulture and by municipal waste (Upton 2013).

2.2 *Sustainably Managing the Resources*

These uses of resources are not sustainable and refer to a public lake edge which is already almost completely devoid of papyrus because of the human pressure, yet most residents are unaware of any change from the natural state, except water-level changes on a few months' timescale. Papyrus loss over the entire lake edge and river estuaries has been very dramatic over the past 50 years,

5 These wage levels were equivalent to $68,50 per month or $1,70 per day, see the exchange rate in footnote 2.

with the last major period of germination and regrowth in 1988 (Harper and Mavuti 2004; Morrison and Harper 2009). Individual water-level-rise events enable small quantities to germinate, but the ecohydrological events leading to widespread creation of islands and swamps described by Gaudet (1977a) have not been repeated, and papyrus lost by grazing and clearance has not been naturally replaced.

There are many ways in which papyrus can be restored and sustainably used, but there is only a single locality around the lake where it is conserved in order to be used in this way – Fisherman's Camp (now split into two properties, the second called Carnelly's Camp). Papyrus has been demonstrably utilised for sustainable harvesting for matting and other uses of the fibre (Terer, Triest, and Muasya 2012) and for fuel from briquettes (Morrison et al. 2013a), which could theoretically be achieved on the small scale of a single location (e.g. Fisherman's Camp, Kamere). Its restoration for its larger-scale roles of water-quality maintenance of biodiversity conservation of the full species, as discussed above, is a much larger task, although evidence has been presented for the need for this (Morrison et al. 2013b; Pacini et al. 2018; Wanjala et al. 2018). Specific ideas, such as the use of papyrus in a series of artificially created wetlands in the inflowing estuaries of the rivers Gilgil and Malewa, to recreate the erstwhile North Swamp and "kick-start" its water-cleansing services were proposed as a concept twenty years ago (Harper and Zalewski 2001). The concept was investigated practically as an Environmental Impact Assessment (EIA) (Githaiga 2008), was proposed in greater detail (Morrison and Harper 2009), and then were calculations made with hydrological modelling (Cornelissen 2014). The private owner of the land, Marula Estates, had created the first wetland on the Gilgil river in 2009 (Marula 2009), but was prevented in developing the second by the rapid rise in water level which occurred in early 2010. The concept still exists, with a landowner willing to give up the necessary land. Unfortunately, no extant official document has any reference to the restoration of papyrus at this critical location or indeed, anywhere else on the lake. The Integrated Lake Management Plan (KWS 2012) which was relevant until 2022 has almost no mention of papyrus and none of ecological restoration of any kind, neither do WWF's policy proposals (WWF 2012a). Their Integrated Water Resource Action Plan (WWF 2012b) uses the word "restoration" extensively, but only in objectives, never in explaining how it will be done. The Imarisha Sustainable Development Action Plan (Imarisha 2012) which ran out in 2017 but is still in force, has no mention of how restoration will take place. Most of these documents talk about Lake Naivasha as a "papyrus-fringed lake" when it has not been like that for 50 years or so, except for the few places where the landowners have ensured there will be minimal change, such as Fishermans' Camp! These documents are obviously written from a desk and not in the lake environment!

It is possible to claim that the most heavily used natural resource, water, is moving towards sustainable management, with the new "traffic-light" system of voluntary abstraction against one's licence held (see Harper and Styles, this volume). It is difficult to claim sustainable management for any other natural resource.

2.3 *Payment for Restoration of Ecosystem Services*

A PES-programme has existed for about a decade whereby several of the large horticultural growers make a payment which helps farmers in the steep-sided upper Malewa catchment undertake sediment-control measures (see also Pacini, Giberti, and Harper, this volume). It is too small to be make a difference beyond the individual farms, or to be measured as lowered sediment/nutrient runoff into rivers, but several alternative mechanisms have been suggested to raise funds that might provide an adequate quantity and reliability of funds that would enable catchment-scale, measurable improvement in ecosystem health and services to people at some future date. Mekonnen, Hoekstra, and Becht (2012) suggested that a water-sustainability agreement should be made between major companies growing flowers that included a premium on the final bouquet of flowers sold in the supermarket that bought them in the Northern Hemisphere. They felt that this would support a Water Fund, which could be used, in their example, to implement the technology that would optimise the sustainable use of water in the flower-growing companies to minimise their "blue water" footprint (see Hoekstra and Chapagain (2007) for an explanation of water footprints) and thus improve watershed management. Mulatu, Oel, and van der Veen (2015) suggested that instead of funding direct activities in the upper catchment, horticultural enterprises at the lake would be willing to pay directly into a Water Fund for ecosystem restoration. Cook et al. (2013) went further and examined each major water user – horticultural and agricultural companies, tourist enterprises, water-service companies and geothermal generation. They evaluated whether a small levy – on top of that already paid to the government authority WRMA could be charged to support a "Lake Naivasha Basin Sustainable Development Fund" (a water fund, as in the other two studies). They found that a modest levy on commercial users of KSh 0.5–1 per m^3 and of 1% of room charge for the tourist industry could raise between KSh 70–150 million per annum.[6] This was the first calculation of the possible revenue that could be generated for ecosystem support, which would be greater than any previous ecological funds spent at the Naivasha basin, either donated or from government. It would, in theory, pay for substantial

6 Equivalent to roughly $800,000 to $1,700,000 per annum, according to the exchange rate in footnote 2.

natural-resource restoration, if the raising of such an income and its use could occur without corruption.

2.4 *Losses and Changes in Natural Resources*

In reality then, there has been a steady loss of most natural resources at Lake Naivasha for the past 75 years. Losses had started even before the seminal studies of Gaudet and his co-workers in the 1970s, but it is impossible to know how much, in terms of plant species for example, had already been lost from the drawdown zone in 1973/1974 when Gaudet studied it. There are two causes, explained in more detail below, both with multiple effects which are not easy to unravel, as well as with some overlaps. The first is alien species, both deliberate introductions and accidental arrivals. The second is the "industrialisation" of agriculture based on irrigation, together with the settlement of large numbers of immigrants as labour for farms and in support enterprises.

Some natural-resource losses cannot be quantified, others can be put together by looking at the results of successive short academic studies. Here, we describe or reference five examples; the first three of losses and the last two of changes to the resources – 1) drawdown zone plants, resampled on one of Gaudet's transects 40 years after his seminal work; 2) papyrus losses since 1960, from maps and satellite imagery; 3) submerged macrophyte losses since 1974, from repeated visits; 4) water quality changes using chemical and algae since 1974, from successive visits and publications, and 5) changes in the water-bird populations.

2.4.1 Plant-Species Loss in the Drawdown Zone

The flat riparian zone on the west corner of Lake Naivasha, outside Crescent Island lagoon, which was used by Gaudet in 1973/4 was revisited in 2009 and 2010 with similar methods used to identify plant species on a transect from within the lake to Acacia woodland. Fifty-six species in 25 families were recorded, three only at genus level (*Inigofera* sp., *Panicum* sp., and *Plectranthus* sp). Only 35 species were common between the two studies; the remaining 73 previously documented by Gaudet (1977a) were missing; a decline in species richness of 48% over 35 years. Some of these will be rarer species simply not found in the second study; for example, a study of plant species colonising floating islands of water hyacinth bound with grasses ten years earlier recorded 16 terrestrial species not found there (Harper, Adams, and Mavuti 1995; Adams et al. 2002). Another 25 species, one-third of those missing, were fully aquatic species; the most likely cause of their disappearance is described below. The remaining 48 were wet-grassland species, whose disappearance may be linked to the appearance of new species.

One obvious difference was the lack of *Cyperus papyrus* in general and, despite its emergence as the dominant species after re-flooding in 2010, its failure to persist as permanent fringing swamp. There were prominent plant species not recorded by Gaudet (1977a); for example, *Cyperus rotundus* (purple nutsedge) attained the highest relative abundance of all species in the present study. It has a widespread distribution (>100 countries) and known deleterious impacts on over 50 agricultural crops (Rao 2000); described by Holm, Plucknett, and Pancho (1977) as the "world's worst weed". A highly competitive perennial, *C. rotundus* can survive periods of environmental stress due to a prolific underground system of asexual reproductive structures (including basal bulbs, tubers and rhizomes), which favour rapid vegetative growth following disturbance. A single tuber has been shown to give rise to >100 additional tubers in <3 months, the allelopathic properties of which are known to supress germination and growth in adjacent species (Holm, Plucknett, and Pancho 1977). *Cirsium vulgare* (Savi) Ten, (spear or bull) thistle, which was also recorded for the first time during this study with a ubiquitous distribution along transects in all but the Sedge Zone, is a wind-dispersed biennial common in open, grazed or otherwise disturbed sites (Stace 1997). *C. vulgare* is unpalatable because of its sharp spines, with no significant nutritive value for livestock; hence it is avoided by grazing animals, although goats may eat it when young (Mitich 1998). Several species of *Cirsium* are known to inhibit the growth of surrounding plants; *C. vulgare* is thought to do this by nutrient immobilisation during litter breakdown (Klinkhamer and de Jong 1993).

2.4.2 Papyrus Area Losses

It was known ever since the first European explorers wrote descriptions that Lake Naivasha was fringed with thick papyrus islands (see Harper and Styles, this volume). Early scientists recorded this and collected plant specimens (Beadle 1932; Jenkin 1936) but no quantitative work was carried out until Gaudet (1977a). The earliest quantitative records of the Naivasha papyrus swamp are the Ordnance Survey maps. The Ordnance Survey map from 1960 shows an extensive "North Swamp" which together with fringing papyrus also shown, suggests a cover of about 45 km^2 and, since the map was derived from aerial photographs this probably relates to 1945–50. Anecdotal information suggests that the North Swamp was thick at that time, because it was used as a hideout by Mau Mau guerrillas in the insurgency that eventually led to independence in 1963 (M. Carnelly, Harper personal communication, 1985). Satellite images became available from 1973 onwards – the same year as Gaudet's work. He analysed these early images (Gaudet and Falconer 1982) which showed considerable loss of about 50% in area, to around 20 km^2. Harper and Mavuti (2004) showed that there had been two periods of limited papyrus regeneration following clear water-level rises of about one vertical metre, in 1980 and 1988; this germination of seeds and establishment of swamp took place in exactly

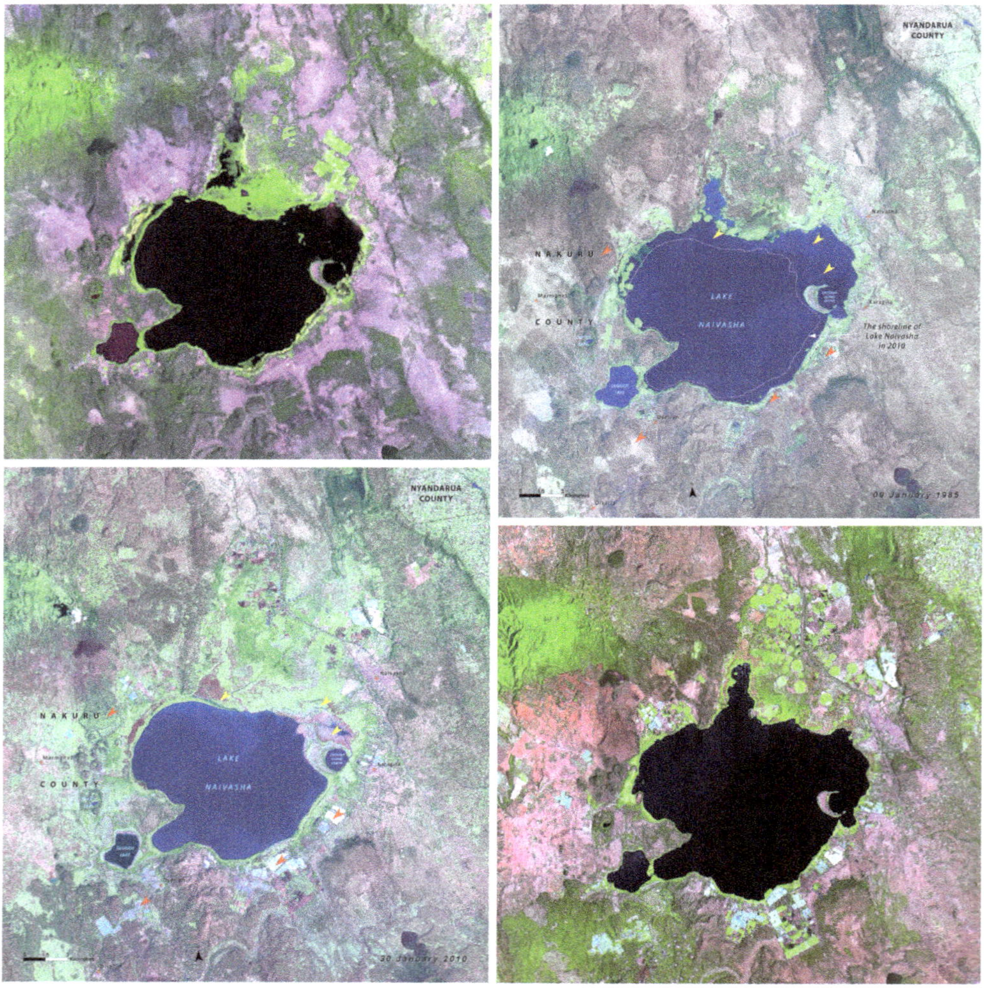

FIGURE 5.2 Four satellite false-colour images taken 50 years apart of Lake Naivasha[7]
SOURCES: LANDSAT IMAGERY COURTESY OF NASA GODDARD SPACE FLIGHT CENTER AND US GEOLOGICAL SURVEY FOR THE IMAGES OF 1973 AND 2023; UNEP (N.D.) FOR THE IMAGES OF 1985 AND 2010

7 Figure 5.2 shows four satellite false-colour images of Lake Naivasha taken at several moments over the past 50 years, showing the dramatic changes of water surface area, together with the equally dramatic surface vegetation changes. The first image, a 1973 Landsat image, shows a very high water level with extensive Papyrus (Cyperus papyrus) stands (the brightest green), particularly in the north, at the southern edge of Crescent Island and between the main lake and Oloidien. All around the shoreline are papyrus clumps, sometimes as in the western side, stranded with lagoons inshore of them; a thin fringe is also shown around most of Oloidien lake. The lighter green in between the papyrus clumps is floating mats of the fern *Salvinia molesta*.

the same ways that Gaudet had described it for four periods between 1924–44 from an analysis of ridges left behind by water-level rises and papyrus rooted in land (Gaudet 1977a).

Harper and Mavuti (2004) show why there was continuous loss of papyrus through the 1980s and 1990s, such that by the turn of this century the total area of papyrus at Lake Naivasha was under 15 km^2 with no functioning North Swamp and very few large islands. It had risen back to 23.3 km^2 by 2013 (Pacini et al. 2018) as a consequence of the water-level rises after 2010. A visual comparison of a 1973 satellite image with one from 2019 (USGS 2022a) shows very superficial similarity of the area of emergent plants at the lake, but this is very deceptive, because 1973 shows extensive papyrus swamps, islands and fringing reefs; 2019 shows extensive floating beds of water hyacinth with little papyrus, clearer under different image interpretation (USGS 2022b).

There have been no large-scale attempts to restore the functionality of papyrus in Lake Naivasha. The small wetland, mentioned above, was constructed at Marula but not replicated. A small experiment growing papyrus on artificial islands was successful (Harper et al. 2013; Pacini et al. 2018; Wanjala et al. 2018) but not upscaled.

2.4.3 Loss of Submerged Macrophytes

Gaudet (1977a) showed the extensive species richness of plants in the submerged shallows (littoral of the lake) in 1973–74 when he carried out his study. Nobody else has studied the littoral of lake Naivasha, but in the mid-1970s two films were made, by Joan and Alan Root and by Cindy Buxton, for UK's Survival, a programme of Anglia Television. These showed extensive beds of yellow water lilies in the shallow lagoons sheltered by papyrus islands and reefs (as Gaudet

The second image, from 1985, shows a smaller lake than twelve years earlier (compared the extent of the northern finger of water) with less papyrus that is more visible as distinct clumps, many stranded in the same place for several years and far less Salvina, which is confined to the shallows. For example, compare Marina Bay, to the immediate SE of Crescent Island, between the two years.

The third image, January 2010, was the time of the lowest lake level since the 1940s. The area can be clearly compared with 1985, by the dotted line in the 1985 lake of 2010 shoreline and also the red and yellow arrows which are in the identical location in both years. Most dramatically, Crescent Island Lagoon is an independent lake and the former island is connected to the N shore, albeit by largely unvegetated soils (pink colour). There is almost no papyrus visible in the image; the edge of the lake is thinly fringed by water hyacinth (Eichhornia crassipes).

The fourth and final image, 2023, shows the lake full again, Crescent Island surrounded by water and the main lake joined with Oloidien. Now, the absence of large papyrus clumps or islands is very clear compared to fifty years earlier, with a similar thin fringe of water hyacinth.

described). On the outside of these lily beds were dense gardens of underwater plants with their flowering heads breaking the surface. In other lakes with a similar transparency to that of Naivasha at that time, such beds would go down to five-meter depth, which would be most of the area of the main lake Naivasha. By 1982 almost all the species found by Gaudet has disappeared; some floating species (*Salvinia, Lemna, Azolla*), survived, but none were submerged and bottom-rooted. Some of them still grew in extensive underwater gardens in Oloidien lake, which at that time was still a freshwater body albeit, with a conductivity (measure of dissolved salts concentration) ten times that of Naivasha. They progressively disappeared from Oloidien, which became more saline and inhospitable to freshwater plants. Some, but not all, of the submerged species reappeared in Lake Naivasha in 1988 and 1989. Gouder de Beauregard et al. (1998) showed a zonation of six different species at depths from 0.4 m to 2.2 m, and beds widely spread along the eastern shorelines of the lake, together with several beds of yellow water lily in sheltered area (such as inside Crescent Island lagoon), but nowhere associated with papyrus islands or reefs. The beds and the water lilies progressively declined and had disappeared by 1996, to reappear briefly in 1999–2002, with the same seven species, but more limited distribution, only along the eastern shores, with yellow water lily quite rare (Ngari et al. 2008). Since that time, submerged plants have been very rare and short-lived.

2.4.4 Nutrients, Transparency, Microscopic Algae and Photosynthesis in the Lake

These components of lake ecology are closely related and widely studied in lakes all over the world, because they give a very precise measure of lake health. The quantity of nutrients – nitrogen, phosphorus and silica – in a lake is an important measure of how enriched, or eutrophic, the lake is. Nutrients determine which species of algae will occur in the lake and which will become dominant. The lake transparency is almost always determined by the density of the algal populations, rarely by sediment (but when levels are very low, shallow lake sediments are easily stirred up into the water by wind). Nutrients, together with the type and the quantity of algae, determine the biomass and then total photosynthesis (primary production) in the lake. Ultimately that determines not only the lake's health, but also the quantity of its provision of fish. All four have been studied at Naivasha by several scientists, but to keep the story simple we will just refer to those few that show the important changes in the lake's state and how useful it can be to people.

Nutrients are usually in short supply and rapidly recycled between living tissues and detritus. In Naivasha, nitrogen is often plentiful because of nitrogen fixation (shown by Gaudet, see above) and by recycling from detritus; the nutrient most limiting is phosphorus. There is often an equilibrium of both elements because nitrogen can be lost back to the atmosphere by anaerobic bacterial breakdown in sediments (denitrification), whilst phosphorus can be locked into sediments, particularly in Naivasha where the sediments are iron-rich (Harper et al. 1993). Sediments will release soluble phosphorus from bacterial action when they are anaerobic, such as at the bottom of Crescent Island Lagoon, where concentrations are often tenfold compared to surface waters, which are mixed by wind action (almost every afternoon) and phosphorus is rapidly taken up by algae. Total Nitrogen has not shown great differences in the ranges of concentrations in the lake for the past fifty years – 0.6–1.5 mg L^{-1} N (Kalff and Watson 1986; Hubble 2000; Ballot et al. 2009; Pacini et al. 2021). But Total Phosphorus has increased three- to fivefold – in 1979–1980 concentrations were between 43–65 µg L^{-1} (Kalff and Watson 1986); in 2014 they were between 115–197 in the Main Lake and 174–217 µg L^{-1} in Crescent Island crater (Pacini et al. 2021). The latter also measured Total Phosphorus concentrations in the Malewa River as 314–319 µg L^{-1} and 420 µg L^{-1} in the Gilgil.

The total number of algal species in the lake has not measurably changed – it just depends how hard one looks and how good a specialist one is. Hubble (2000) found 170 species, but 143 were found in the 1980s (Kalff and Watson 1986), 116 after the turn of this century (Ballot et al. 2009), and 160 in 2013–2014 (Pacini et al. 2021; although a very thorough examination by Olga Lepšová-Skácelová in her lab subsequently found over 500!). The algal composition certainly has changed however, together with transparency, influenced by the size of algae (single-celled earlier, colonial species now) as well as abundance. Until the 1970s the lake water was very clear – transparency was several metres (Milbrink 1977), with a mixture of algae that changed their dominance by season – between green algae, diatoms and blue-green bacteria. By the early 1980s, transparency had reduced to between one and two metres with a dominance by a few species of diatoms and an increase in blue-green bacteria. Through late 1980s and into the 2000s though, transparency was usually lower than one metre; by 2014 it was normally 60–90 centimetres. Transparency has decreased progressively over the past 20 years as the lake has become more and more dominated by blue-green bacteria, often creating floating scums (Harper 2006) and posing a threat to people and livestock because they can produce toxins against their invertebrate grazers to which vertebrates are susceptible (Raffoul et al. 2020); they also can cause oxygen to be depleted in the lake through their decay, which causes fish kills (Njiru et al. 2015). The decrease in turbidity, which was matched by an increase in biomass measured as the quantity of green colour (chlorophyll) in the water, has however been periodically reversed for short periods, whenever lake levels rose rapidly. Thus in 1982, 1999, 2011 and 2014, the lake was clear for several metres with a reduced algal biomass (Pacini et al. 2021).

Photosynthesis has turned out to have doubled, from a mean of 1.8 gm Carbon m^{-2} day^{-1} measured between 1973–1974 (Melack 1979), to a mean of 2.7 between 1997–1998 (Hubble 2000) and then a mean of 3.2 between 2012–2013 (Pacini et al. 2021).

2.4.5 Changes in Abundance of Water Birds

Earlier in this chapter, we used the high number of species of birds – water birds and woodland birds – to show the high biodiversity on the lake and in the woodlands of its riparian zone in the past; now we can look at what 25 years of annual counts show us about the changing numbers of individuals.

The water-bird count species list (the list of all species ever seen) for the lake is around 140 species; 110 have been counted at least once in the census. Some species are intermittent, some migrants; others are seen every year as a handful of individuals; some species are always present in their thousands. A few species might appear in thousands of individuals one year and a few another year. The impression of change in over 100 species is overwhelming, but if one just focusses on one species at a time, it is possible to understand the change. Here are two examples.

The African fish eagle (*Haliaeetus vocifer*) is a majestic predator, said to be "the true voice of Africa" and certainly "the Voice of Naivasha"; birds nest in mature trees, preferably overlooking or close to the lake. The Kenyan ornithologist Leslie Brown studied the Naivasha population in the 1960s and 1970s (Brown and Hopcraft 1973) and wrote a book about them (Brown 1980). Most months he would spend one weekend in a boat on Lake Naivasha, with nothing but binoculars, notebook and a crate of beer, watching them. Since then, they have been counted annually by various scientists since 1986. Fish eagles are so visually and aurally obvious that the numbers recorded on the single-day annual water-bird count are similar to those counted by more intensive surveys over repeated days, and are therefore reliable. In Leslie Brown's days, the population increased from 120 birds in 1968 to over 200 in 1971 as lake level (and shoreline length) rose (Brown 1980). By 1986 there were 160, declining steadily to under 80 in 1998. Between 1999 and 2001 the number increased by 50%, mostly through reproduction, as many of the new birds were juveniles (Harper et al. 2002), up to 110. Numbers declined through the first decade of this century, but then rose progressively until a peak was reached in 2014, of 226 birds in the NK/NMK/KWS census. The one thing which all four population rises have in common is rapid lake-level rise, and population declines mirror lake-level declines. Population rises follow water-level increase, and highs reflect high lake transparency (see above), and it is just this – greater ease of finding food (fish) in the new shallow waters, flushed with released nutrients – that enables fish, and therefore also fish eagles, to breed and increase their respective populations. When the lake levels decline, transparency in the shallows increases through wind stirring of sediments, and fish eagles cannot feed so well.

A second example is red-knobbed coot (*Fulica cristata*). The lake was once renowned for its duck-shooting in colonial days; there were many tens of thousands of yellow-billed duck (*Anas undulata*) and other species, and also tens of thousands of coot. These species are herbivores; they fed on the abundant submerged vegetation. A NMK report showed more than 35,000 duck in the 1970s (Van Someren 1981), a couple of decades after duck-shooting had been banned by the Lake Naivasha Riparian Owners Association (LNROA). Since 1991, however – the first year of the water-bird survey – there have never been more than one-tenth of that number of all species combined from the duck and geese family, *Anatidae*. When there was duck-shooting in the 1950s and 1960s, the hunters wanted to control red-knobbed coot, which they believed damaged their shooting because they were so numerous, but the LNROA did not agree to this after a study showed the coot had no negative impact (Parker 1964); so probably in the 1970s there were many tens of thousands of coot, as well as ducks. In 1991, 16,243 coot were counted. In the following years the numbers declined steadily – to 114 in 1997 and 14 in 1998. Numbers increased rapidly to 7,500 in 2000 and 9,000 in 2001, then dropped again. Another increase started in 2008, rising to 4,275 in 2010. On the face of it, these fluctuations have no pattern, until one realises that coot feed on submerged plants and their numbers are high during the years when plants are growing in the lake, while they disperse elsewhere when the plants disappear.

3 How Do We Explain These Losses and Changes in Natural Resources?

The reasons can be grouped into two major areas – those caused by the impact of alien species and those caused by the impact of more industry and ever-greater human settlement. Sometimes the two have worked together and it is impossible to separate them; at other times they are clearly separate.

3.1 *Alien Species*

Alien species started arriving in Lake Naivasha before the rapid intensification of agriculture and human settlement. It is almost exactly a hundred years since the first known alien – the North American black bass (*Micropterus salmoides*) – was introduced to the lake. The fish in Lake Naivasha was the one form of wildlife which is the exception to the statement made above about high biodiversity arising from unpredictability. Only a few fish can withstand drying of their habitat, which in Naivasha last happened not more than 200 years ago, and their powers of distribution across land are low. The presence of a single species, a small sardine called black lampeye, (full name *Micropanchax* (formerly *Aplocheilichthys*) *antinorii*) as the only native fish in Naivasha when Europeans first recorded life in the lake, is thus perhaps understandable. A plethora of fish introductions – African tilapias – followed, also deliberately, as food for the bass. More recently two predatory species have been "accidentally" introduced: the Eurasian Carp (*Cyprinus carpio*) in 1999 is known to have been washed out of a small dam in the uppermost Gilgil catchment, where it had been deliberately introduced a decade or so earlier; the African Sharptooth Catfish (*Clarias gariepinus*) is the most recent arrival, of unknown origin. For details of the fish and fishery, see Dittmann and Ogolla (this volume) and for a more detailed history of alien introduction/arrival, see Gherardi et al. (2011).

The alien plant species that have impacted the lake all arrived in an unknown fashion, but are widespread invaders of tropical countries on three continents. The first, a South American water fern called *Salvinia molesta*, arrived in the 1960s, either at the same time or shortly followed by another floating fern, *Azolla fillipoides*. *Salvinia molesta* had a serious impact on boat movements and probably also on light penetration to plants in the lake for two decades, *A. fillipoides* less so. The third important plant was also a South American native, the flowering plant Water Hyacinth *Eichhornia crassipes* (Kitunda 2018). This plant has had a serious impact in many lakes and slow-moving rivers, by blocking light for underwater plants, preventing gaseous exchange between water and air, and dramatically impeding boat movements. With hindsight, these floating weeds disrupted boating – especially fishing – but, except when boats were occasionally trapped in solid mats (rare because of the frequent afternoon storms on Lake Naivasha, see USGS (2022b)), they did not of themselves damage lake ecology.

The most damaging, and also deliberately introduced, alien is the Louisiana crayfish, *Procambarus clarkii*. This species, an invertebrate looking like a small lobster, is delicious to eat, and was believed, in the mid-1970s, to be able to diversify the economic value of a commercial fishery, which had been started 20 years earlier. Its impact on the lake ecology, however, was perhaps the most dramatic of any alien before or since. This impact interacted with both the water hyacinth in the 1990s and the carp a decade later.

Put together, the alien introductions help us to understand many of the negative changes to the natural resources of Lake Naivasha, and also the various changes in numbers of individual species described above. The following paragraph explains the combined impact of the aliens (leaving out the fish).

The Louisiana crayfish is a voracious omnivore, which has devasted ecosystems in Mediterranean Europe, India and Africa wherever it has been introduced. It was introduced to Naivasha in 1970, had almost completely eliminated underwater and floating leaved plants by 1974, and totally eliminated them from the main lake by 1982 (Gouder de Beauregard et al. 1998; Smart et al. 2002). Crayfish themselves declined because there was not enough food to sustain the population and, without the shelter provided by a submerged plant forest, they were eaten by bass (Hickley et al. 1994) and piscivorous birds such as cormorant (Childress, Bennun, and Harper 2002) or fish eagles. The crayfish population collapse enabled submerged plants to germinate from seed in the mud, and was aided by the reflooding when the lake level rose, such that submerged plant beds and to a lesser extent water lilies began to recover from 1988. This was limited in area and short-lived, because crayfish populations began to increase again from about 1992. By 1995 almost all plants beds had gone again and crayfish were abundant – but this time they were all found hiding by day in the roots of floating water hyacinth which had recently arrived and rapidly become dominant in shallow waters with no competition for space (papyrus was at a low, lilies were absent). Crayfish would drop from the roots to feed at night on the lake bottom. Between 1999 and 2002 however, water-hyacinth mats progressively disappeared, such that only isolated small clumps were to be seen around the lake. Crayfish population collapse caused by this enabled submerged plants to return. The loss of water hyacinth occurred because the Kenya Agricultural Research Institute had introduced a weevil from its homeland in South America to control hyacinth in lakes Naivasha and Victoria, and over five to six years its population had built up enough to control the plant (Ochiel et al. 2001) (infestation was far, far, greater in Lake Victoria because the higher temperatures at this lake, at lower altitude, had allowed rampant hyacinth growth; control was achieved after two to three years). Since 2006, there have been almost no submerged or floating leaved native plants growing in Naivasha, because crayfish have remained at low numbers, probably short of food (Grey, Jackson, and Carlson 2012), but still enough to eliminate any new green shoots in shallow water. Both carp and catfish are also omnivorous bottom-grubbing fish (Pacini et al. 2021).

It is likely that this new shallow-water (littoral) ecosystem of no submerged plants, scattered floating alien plants, few papyrus islands and plenty of detritus from the remains of floating plants on the bottom, all supporting low densities of zooplankton and high densities of juvenile tilapias and barbus (*Barbus paludinosus*), will continue. It has survived the lowest lake level for decades (2009) and then also the highest (2014) (Pacini et al. 2021).

3.2 *Intensification of Agriculture and Settlement*

Stoof-Leichsenring et al. (2011), analyzing diatoms in sediment cores from the bottom of lake Naivasha in order to interpret the past 200 years of history, divided it into two periods. In the first, the lake's changes were driven by natural climatic fluctuations, but since about 1950 they have been driven by human causes. This is evidenced from the increased sedimentation rate and the increased frequency of diatoms, which indicate slow, progressive enrichment of the water. This confirms with a continuous record the more precise but infrequent analysis of phytoplankton and nutrients described above. Other chapters provide details about the settlement and agriculture, so we do not need to repeat them here, but only interpret the effects, where that is possible. We have explained earlier why papyrus is so important, because it is a biological filter absorbing and slowly releasing chemicals that come into the lake from the catchment. We also explained above how submerged plant beds in the shallow water acted as recyclers of nutrients from detritus; both plant communities retained nutrients by tightly recycling them, releasing them slowly but continuously into the lake water. Without them, nutrients from settlement and erosion, intensive agriculture and pollution can come straight into the lake. That is what is responsible for the three- to five-fold increase in total phosphorus, the change from an abundance of diatom algae to bloom outbreaks of blue-green cyanobacteria, and the doubling of primary productivity (not increased in proportion to phosphorus, because now photosynthesis is limited by the self-shading of the algae/bacteria). The actual increase in nutrients and productivity has been partially mitigated through uptake by the existing, alien, plant community (water hyacinth is well known to take up nutrients and pollutants and is often used in artificial wetlands to clean effluent from domestic and industrial sources; Vymazal (2008)) as well as sedimentation and denitrification, but also by the repeated reversal of its eutrophic state every time heavy rains raise the lake level, bringing dilution.

The lake reached a state in 2022, we suggest, beyond which it cannot survive any more intensive lakeshore settlement and hydrological control of its inflowing waters, because it is retained in an unacceptable state by losing the cycling provided by its "flood pulse". Restoration has been talked about for 20 years but never seriously undertaken, despite all the good words about sustainable management in official plans, so the real benefits that papyrus could bring, are just ignored.

3.3 *The Future of Lake Naivasha*

Kenya Vision 2030 is the national guide to sustainable development of the country. It recognises wetlands and water catchments as key components that can promote economic growth and thus reduce poverty. Lake Naivasha has enormous potential to promote national well-being, but it can only do that if the natural resources which is provides can be understood, restored and sustainably used. Most are declining because the basic principles which underpin them are not understood. They are:

1. Naivasha is biologically unique because its hydrology is unique and unpredictable;
2. that unpredictability is responsible for its still-usable but declining water quality (hence irrigation for industry, raw water for drinking);
3. unpredictability is also responsible for high biodiversity through the habitat diversity of the drawdown zone (hence tourism);
4. sustainable use of the resources must thus maintain the hydrological unpredictability and the drawdown zone, and restore papyrus to maintain water quality.

Only if and when the basic scientific principles of how this lake functions and provides so much natural capital are understood, and social/economic development recognises them, does Lake Naivasha have a future.

Acknowledgements

See the acknowledgements at the end of Chapter 3 and 6 of this volume.

References

Adams, Christopher S., Rosalind R. Boar, David S. Hubble, Mary Gikungu, David M. Harper, Phil Hickley, and Nils Tarras-Wahlberg. 2002. "The Dynamics and Ecology of Exotic Tropical Species in Floating Plant Mats: Lake Naivasha, Kenya." *Hydrobiologia* 488: 115–22. https://doi.org/10.1023/A:1023322430005.

Agassiz, David J.L. 2009. "The Macrolepidoptera Fauna of Acacia in the Kenyan Rift Valley (Part 2 – Description of New Species)." *Tropical Lepidoptera Research* 19 (1): 9–17.

Agassiz, David J.L., and David M. Harper. (2009). "The Macrolepidoptera Fauna of Acacia in the Kenyan Rift Valley (Part 1)." *Tropical Lepidoptera Research* 19 (1): 4–8.

Agutu, Peter Odoyo, Moses Karoki Gachari, and Charles Ndegwa Mundia. 2018. "An Assessment of the Role of Water Hyacinth in the Water Level Changes of Lake Naivasha Using GIS and Remote Sensing." *American Journal of Remote Sensing* 6 (2), 74–88. https://doi.org/10.11648/j.ajrs.20180602.13.

Ballot, Andreas, Kiplagat Kotut, Eberto Novelo, and Lothar Krienitz. 2009. "Changes of Phytoplankton Communities in Lakes Naivasha and Oloidien, Examples of

Degradation and Salinization of Lakes in the Kenyan Rift Valley." *Hydrobiologia* 632: 359–63. https://doi.org/10.1007/s10750-009-9847-0.

Beadle, L. 1932. "Scientific Results of the Cambridge Expedition to the East African Lakes, 1930–1. IV. The Waters of some East African Lakes in Relation to their Fauna and Flora." *Journal of the Linnean Society of London* 38 (258): 135–55.

Brown, Leslie H. 1980. *The African Fish Eagle*. Folkestone: Bailey Bros.

Brown, Leslie H., and J.B.D. Hopcraft. 1973. "Population Structure and Dynamics in the African Fish Eagle *Haliaeetus Vocifer* (Daudin) at Lake Naivasha, Kenya." *East African Wildlife Journal* 11: 255–69. https://doi.org/10.1111/j.1365-2028.1973.tb00091.x.

Buis, Alan. 2020. *Milankovitch (Orbital) Cycles and Their Role in Earth's Climate*. NASA *Climate*: February 27, 2020. https://climate.nasa.gov/news/2948/milankovitch-orbital-cycles-and-their-role-in-earths-climate/.

Butzer, Karl W., Glynn L. Isaac, Jonathan L. Richardson, and Celia Washbourne-Kamau. 1972. "Radiocarbon Dating of East African Lake Levels." *Science* 175: 1069–76.

Childress, R. Brooks, Leon A. Bennun, and David M. Harper. 2002. "Population Changes in Sympatric Great and Long-tailed Cormorants (*Phalacrocorax carbo* and *P. africanus*): The Effects of Niche Overlap or Environmental Change?" *Hydrobiologia* 488: 163–70. https://doi.org/10.1023/A:1023338816801.

Cook, P., B.A. Contor, David M. Harper, and B. Vereecke. 2013. *The Value of Water – An Economic Basis and Mechanism to Assess Voluntary Contributions to a Proposed "Lake Naivasha Basin Sustainable Development Fund"*. Unpublished Report Submitted to LNRA, Imarisha and LNGG.

Cornelissen, M.A.J.M. 2014. "Rehabilitation of the Former Northern Swamp Lake Naivasha – Kenya; On the Modeling of the Sediment Trapping Efficacy for Two Rehabilitation Alternatives." MSc thesis, University of Twente.

Elder, H.Y. 1959. "Report on the Investigation into the Tilapia Population of Lake Naivasha, Kenya." East African Freshwater Fisheries Research Organization. Accessed September 1, https://aquadocs.org/bitstream/handle/1834/32695/Report%2520on%2520the%2520investigation%2520into%2520the%2520Tilapia%2520popn%2520of%2520L.NAI%252CKENYA.pdf?sequence=1&isAllowed=y.

Gaudet, John J. 1977a. "Natural Drawdown on Lake Naivasha, Kenya, and the Formation of Papyrus Swamps." *Aquatic Botany* 3: 1–47. Accessed September 1, 2023. https://vdocuments.net/download/natural-drawdown-on-lake-naivasha-kenya-and-the-formation-of-papyrus-swamps.html.

Gaudet, John J. 1977b. "Uptake, Accumulation, and Loss of Nutrients by Papyrus in Tropical Swamps." *Ecology* 58: 415–422. https://doi.org/10.2307/1935616.

Gaudet, John J. 1979. "Seasonal Changes in Nutrients in a Tropical Swamp: North Swamp, Lake Naivasha, Kenya." *Journal of Ecology* 67: 953–981. https://doi.org/10.2307/2259223.

Gaudet, John J., and Franics M. Muthuri. 1981. "Nutrient Relationships in Shallow Water in an African Lake, Lake Naivasha." *Oecologia* 49: 109–18. https://doi.org/10.1007/BF00376907.

Gaudet, John J., and John M. Melack. 1981. "Major Ion Chemistry in a Tropical African Lake Basin." *Freshwater Biology* 11: 309–31. https://doi.org/10.1111/j.1365-2427.1981.tb01264.x.

Gaudet, John J., and Kenneth A. Falconer. 1982. *Remote Sensing for Tropical Freshwater Bodies. The Problem of Floating Islands on Lake Naivasha.* Nairobi: Regional Remote Sensing Facility.

Gherardi, Francesca, Robert Britton, Kenneth M. Mavuti, Nic Pacini, Jonathan Grey, Elena Tricarico, David M. Harper. 2011. "A Review of Allodiversity in Lake Naivasha, Kenya: Developing Conservation Actions to Protect East African Lakes from Alien Species Impacts." *Biological Conservation* 144: 2585–96. https://doi.org/10.1016/j.biocon.2011.07.020.

Githaiga, J. M. 2008. *River Gilgil Wetlands Rehabilitation Intervention, Environmental Impact Assessment.* Nairobi: School of Biological Sciences, University of Nairobi.

Gouder de Beauregard, Anne-Christine, David M. Harper, François Malaisse, and J.-J. Symoens. 1998. "Dynamique récente et cartographie de la vegetation aquatique (1960–1996) du lac Naivasha (Rift Valley, Kenya)." *Bulletin des Séances, Académie Royale des Sciences d'Outre-Mer* 44: 373–89.

Grey, Jonathan, and David M. Harper. 2002. "Using Stable Isotope Analyses to Identify Allochthonous Inputs to Lake Naivasha Mediated via the Hippopotamus Gut." *Isotopes in Environmental and Health Studies* 38 (4): 245–50. https://doi.org/10.1080/10256010208033269.

Grey, Jonathan, Michelle C. Jackson, and Stephanie M. Carlson. 2012. "Leaves and Eats Shoots': Direct Terrestrial Feeding Can Supplement Invasive Red Swamp Crayfish in Times of Need." *PloS one* 7 (8), p.e42575–e42575. https://doi.org/10.1371/journal.pone.0042575.

Harper, David M. 2006. "The Future of Lake Naivasha." *Swara. Journal of the East African Wildlife Society* 2: 2.

Harper, David M., Charles Adams, and Kenneth M. Mavuti. 1995. "The Aquatic Plant Communities of the Lake Naivasha Wetland, Kenya: Pattern, Dynamics and Conservation." *Wetlands Ecology & Management* 3: 111–23. https://doi.org/10.1007/BF00177693.

Harper, David M., Edward H.J. Morrison, Michael M. Macharia, Kenneth M. Mavuti, and Caroline Upton. 2011. "Lake Naivasha, Kenya: Ecology, Society and Future." *Freshwater Reviews* 4 (2): 89–114. http://www.bioone.org/doi/full/10.1608/FRJ-4.2.149.

Harper, David M., Geoff Phillips, Alison Chilvers, Nzula Kitaka, and Kenneth M. Mavuti. 1993. „Eutrophication Prognosis of Lake Naivasha, Kenya." *Verhandlungen der Internationalen Vereinigung für Limnologie* 25: 861–65. https://doi.org/10.1080/03680770.1992.11900268.

Harper, David M., and Kenneth M. Mavuti. 2004. "Lake Naivasha, Kenya: Ecohydrology to Guide the Management of a Tropical Protected Area." *Ecohydrology & Hydrobiology* 4: 287–305.

Harper, David M., and Maciej Zalewski. 2001. *Science and the Sustainable Management of Tropical Waters; A Summary of the Projects Presented to the Conference.* UNESCO Hydrology. IHP-V Technical Documents in Hydrology 46 2.3/2.4. Naivasha, Kenya, 11–16 April 1999.

Harper, David M., Nic Pacini, Caroline Upton, Edward H.J. Morrison, Richard Fox, and Enock Kiminta. 2013. "Water Cooperation for Sustainable Utilization: Lake Naivasha, Kenya." In *Reaching Water Security through Cooperation. Economic Development and Water,* edited by Jacqui Griffiths, and Rebecca Lambert Free Flow, 256–59. Paris: UNESCO.

Henderson, Ian G., and David M. Harper. 1992. "Bird Distribution and Habitat Structure on Lake Naivasha, Kenya." *African Journal of Ecology* 30: 223–232. https://doi.org/10.1111/j.1365-2028.1992.tb00497.x.

Hickley, Phil, R. North, Mucai Muchiri, and Daivd M. Harper. 1994. "The Diet of Largemouth Bass, Micropterus Salmoides, in Lake Naivasha, Kenya." *Journal of Fish Biology* 44: 607–19. https://doi.org/10.1111/j.1095-8649.1994.tb01237.x.

Hoekstra, Arjen Y., and Ashok K. Chapagain. 2007. "Water Footprints of Nations: Water Use by People as a Function of their Consumption Pattern." *Water Resources Management* 21: 35–48. https://doi.org/10.1007/s11269-006-9039-x.

Hogeboom, Rick H.J., Pieter R. van Oel, Maarten S. Krol, and Martijn J. Booij. 2015. "Modelling the Influence of Groundwater Abstractions on the Water Level of Lake Naivasha, Kenya Under Data-Scarce Conditions." *Water Resourources Managemen*t 29: 4447–63. https://doi.org/10.1007/s11269-015-1069-9.

Holm, LeRoy G., Donald L. Plucknett, and Juan V. Pancho. 1977. *The World's Worst Weeds: Distribution and Biology.* Honolulu: University Press of Hawaii.

Howard-Williams, Clive, and G. M. Lenton. 1975. "The Role of the Littoral Zone in the Functioning of a Shallow Tropical Lake Ecosystem." *Freshwater Biology* 5: 445–59. https://doi.org/10.1111/j.1365-2427.1975.tb00147.x.

Hubble, David. 2000. "Controls on Primary Production in Lake Naivasha, a Shallow Tropical Freshwater." PhD diss., University of Leicester. Accessed September 1, 2023. https://leicester.figshare.com/articles/thesis/Controls_on_primary_production_in_Lake_Naivasha_a_shallow_tropical_freshwater/10180295.

Hutchinson, George E. 1961. "The Paradox of the Plankton." *The American Naturalist* 95 (882): 137–45.

Imarisha. 2012. "Lake Naivasha Sustainable Development Action Plan: First Action Plan (2012–2017) & Priority Actions for Imarisha Naivasha Restoration Programme and Sustainable Development in the Lake Naivasha Basin." Naivasha: Imarisha.

IUCN. 2023. "The IUCN Red List of Endangered Species". Accessed on September 26, 2023. https://www.iucnredlist.org.

Jenkin, Penelope. 1936. "Reports of the Percy Sladen Expedition to Some Rift Valley Lakes in Kenya in 1929. VII. Summary of the Ecological Results with Special Reference to the Alkaline Lakes." *Annals and Magazine of Natural History* 18 (103): 133–81.

Junk, Wolfgang J., and Karl M. Wantzen. 2004. "The Flood Pulse Concept: New Aspects, Approaches and Applications – An Update." In *Proceedings of the Second International Symposium on the Management of Large Rivers for Fisheries*, edited by Robin L. Welcomme, and T. Petr, 117–49. Bangkok: Food and Agriculture Organization and Mekong River Commission, FAO Regional Office for Asia and the Pacific.

Junk, Wolfgang J., Peter B. Bayley, and Richard E. Sparks. 1989. "The Flood Pulse Concept in River-Floodplain Systems." *Canadian Special Publication of Fisheries and Aquatic Sciences* 106 (1): 110–27. Accessed September 1, 2023. https://ftp.cs.ru.nl/toinesmits/Recommended_readings_IWRM_2009/Water_Ecomorphological_principles/1989JunkThe%20flood%20pulse%20concept%20in.pdf.

Kalff, Jacob, and Shawn Watson. 1986. "Phytoplankton and its Dynamics in Two Tropical Lakes: A Tropical and Temperate Zone Comparison." *Hydrobiologia* 138: 161–76. https://doi.org/10.1007/BF00027238.

Khan, Shahbaz, Alain Tchadie, and Carolina Llovera. 2011. Ecohydrology for Sustainability. *UNESCO IHP Library* 1: 1–24.

Kimani, D.K. 2012. 'Avifaunal Diversity in the Acacia Woodlands on the Shoreline of Lake Naivasha." BSc thesis, Africa Nazarene University.

Kitunda, Jeremiah M. 2018. *A History of the Water Hyacinth in Africa: The Flower of Life and Death from 1800 to the Present*. Lanham, MD: Lexington Books.

Klinkhamer, Peter G.L., and Tom J. de Jong. 1993. "*Cirsium vulgare* (Savi) Ten. (*Carduus lanceolatus* L., *Cirsium lanceolatum* (L.) Scop. non Hill)." *Journal of Ecology* 81: 177–91.

KWS. 2012. "Lake Naivasha Basin Integrated Management Plan 2012–2022" (LNBIMP). Naivasha: Kenya Wildlife Services.

Leakey, Louis S. B. (1931) 1971. *The Stone Age Cultures of Kenya Colony*. Cambridge: Cambridge University Press. Reprint, London: Frank Cass.

Marula. 2009. "Lake Naivasha Wetland Rehabilitation Project Northern Swamp Regeneration." Unpublished manuscript, Marula Estate, Naivasha.

Maslin, Mark A., Chris M. Brierley, Alice M. Milner, Susanne Shultz, Martin H. Trauth, and Katy E. Wilson. 2014. "East African Climate Pulses and Early Human Evolution." *Quaternary Science Reviews* 101: 1–17. https://doi.org/10.1016/j.quascirev.2014.06.012.

Mekonnen, Mesfin M., Arjen Y. Hoekstra, and Robert Becht. 2012. "Mitigating the Water Footprint of Export Cut Flowers from the Lake Naivasha Basin, Kenya." *Water Resources Management* 26: 3725–42.

Melack, John M. 1979. "Photosynthetic Rates in Four Tropical African Freshwaters." *Freshwater Biology* 9: 444–571. https://doi.org/10.1111/j.1365-2427.1979.tb01539.x.

Milbrink, Göran 1977. "On the Limnology of Two Alkaline Lakes in Kenya." *Internationale Revue der gesamten Hydrobiologie und Hydrographie* 62: 1–17. https://doi.org/10.1002/iroh.1977.3510620101.

Millennium Ecosystem Assessment (MA). 2005. *Ecosystems and Human Well-Being: Synthesis.* Report. Washington: Island Press.

Misongo, Moses. 2020. "Characterisation of Groundwater Aquifers in a Structurally Complex Region." MSc thesis, University of Nairobi. Accessed September 26, 2023. http://erepository.uonbi.ac.ke/bitstream/handle/11295/154575/Misongo_Characterization%20of%20groundwater%20aquifers%20in%20a%20structurally%20complex%20region%20-%20a%20case%20study%20of%20West%20of%20Lake%20Naivasha%20Area.pdf?sequence=1&isAllowed=y.

Mitich, Larry W. 1998. "Bull Thistle, *Cirsium vulgare*." *Weed Technology* 12: 761–63. https://doi.org/10.1017/S0890037X00044675.

Morrison, Edward H.J., Amy Banzaert, Caroline Upton, Nic Pacini, Jan Pokorny, and David M. Harper. 2013a. "Biomass Briquettes: A Novel Incentive for Managing Papyrus Wetlands Sustainably?" *Wetlands Ecology and Management* 22 (2), 129–42. https://doi.org/10.1007/s11273-013-9310-x.

Morrison, Edward H.J., Caroline Upton, Nic Pacini, Ken Odhiambo-K'oyooh, and David M. Harper. 2013b. "Public Perceptions of Papyrus: Community Appraisal of Wetland Ecosystem Services at Lake Naivasha, Kenya." *Ecohydrology & Hydrobiology* 13 (2): 135–47. https://doi.org/10.1016/j.ecohyd.2013.03.008.

Morrison, Edward H.J., and David M. Harper. 2009. "Ecohydrological Principles to Underpin the Restoration of Cyperus Papyrus at Lake Naivasha, Kenya." *Ecohydrology & Hydrobiology* 9: 83–97. https://doi.org/10.2478/v10104-009-0036-6.

Muchiri, Mucai (1990). "The Feeding Ecology of Tilapia and the Fishery of Lake Naivasha, Kenya." PhD diss., University of Leicester. Accessed September 26, 2023. https://figshare.le.ac.uk/articles/thesis/The_feeding_ecology_of_tilapia_and_the_fishery_of_Lake_Naivasha_Kenya_/10157861/1.

Muchiri, Mucai, Paul J. Hart, and David M. Harper. 1995. "The Persistence of Two Introduced Tilapia Species in Lake Naivasha, Kenya, in the Face of Environmental Variability and Fishing Pressure." In *The Impact of Species Changes in African Lakes,* edited by Paul J. Hart, and Tony J. Pitcher, 299–319. Dordrecht: Springer. https://doi.org/10.1007/978-94-011-0563-7_14.

Mulatu, Dawit W., Pieter R. Oel, and Anne van der Veen. 2015. "Firms' Willingness to Invest in a Water Fund to Improve Water-Related Ecosystem Services in the Lake Naivasha Basin, Kenya." *Water International* 40 (3): 463–482. https://doi.org/10.1080/02508060.2015.1050580.

Muthiri, Francis M., Mike B. Jones, and Simeon K. Imbamba. 1989. "Primary Productivity of Papyrus (Cyperus Papyrus) in a Tropical Swamp; Lake Naivasha, Kenya." *Biomass* 18 (1): 1–14. https://doi.org/10.1016/0144-4565(89)90077-2.

Nasirwa, Oilver, and O. Naliyanya. 2002. "Status of the Lesser Swamp Warbler in the Absence of Migrant Warblers in the Papyrus Habitat Fringing Lake Naivasha, Kenya, July-August 2000." Unpublished Report to Earthwatch Institute.

Negussie, Yohannes Z., Magdalena Urbaniak, and Maciej Zalewski. 2011. "Ecohydrology for a Sustainable Future in Africa – The Cases of Ethiopia, Kenya and Tanzania." *Ecohydrology & Hydrobiology* 11 (3–4): 223–30. https://doi.org/10.2478/v10104-011-0053-0.

Ngari, A.N., J.I. Kinyamario, M.J. Ntiba, and K.M. Mavuti. 2008. "Factors Affecting Abundance and Distribution of Submerged and Floating Macrophytes in Lake Naivasha, Kenya." *African Journal of Ecology* 47: 32–39. https://doi.org/10.1111/j.1365-2028.2007.00911.x.

Nilsson, Erik. 1932. "Quarternary Glaciations and Pluvial Lakes in British East Africa." PhD diss., Centraltryckeriet Stockholm.

Njiru, James, George Morara, Edna Waithaka, and James Mugo. 2015. "Fish Kills in Lake Naivasha, Kenya: What Was the Probable Cause?" *International Journal of Fisheries and Aquatic Studies* 3 (1): 179–84.

Ochiel, G.S., Stephen W. Njoka, A.M. Mailu, and Wycliffe Gitonga. 2001. "Establishment, Spread and Impact of *Neochetina* spp. on Water Hyacinth in Lake Victoria, Kenya." In *Biological and Integrated Control of Water Hyacinth*, Eichhornia crassipes, edited by M. H. Julien, M. P. Hill, T. D. Centre, and Ding Janqing, *ACIAR Proceedings* 102: 89–95.

Ojiambo, S. Bwire, and W. Berry Lyons. 1996. "Residence Times of Major Ions in Lake Naivasha, Kenya, and Their Relationship to Lake Hydrology." In *The Limnology, Climatology and Paleoclimatology of the East African Lakes*, edited by Thomas C. Johnson, and Eric O. Odada, 267–78. Routledge.

Owino, Alfred O., Joseph O. Oyugi, Oliver O. Nasirwa, and Leon A. Bennun. 2001. "Patterns of Variation in Waterbird Numbers on Four Rift Valley Lakes in Kenya, 1991–1999." *Hydrobiologia* 458: 45–53. https://doi.org/10.1023/A:1013115724138.

Owino, Alfred O., Leon Bennun, Oliver Nasirwa, and Jospeh O. Oyugi. 2009. "Trends in Waterbird Numbers in the Southern Rift Valley of Kenya, 1991–2000." *Waterbirds* 25 (2): 191–201. https://doi.org/10.1080/00306525.2000.9639917.

Pacini, Nic, Marek Baxa, M. Kosik, Jonathan Grey, Olga Lepsova-Skacelova, Kamau Mbogo, Timothy Mwinami, Ivo Prikryl, Jan Pokorny, Jo P. E.C. Darlington, Petra Hesslerova, and David M. Harper. 2021. "Ecohydrological Costs and Benefits of Common Carp, the Dominant Species in a 'Novel' Tropical Lake Ecosystem." *Ecohydrology & Hydrobiology* 21: 467–89. https://dx.doi.org/10.1016/j.ecohyd.2021.08.011.

Pacini, Nic, Petra Hesslerová, Jan Pokorný, Timothy Mwinami, Edward H.J. Morrison, Anthony A. Cook, Sikun Zhang, and David M. Harper. 2018. "Papyrus as an Ecohydrological Tool for Restoring Ecosystem Services." *Afrotropical Wetlands* 18 (2): 142–54. https://doi.org/10.1016/j.ecohyd.2018.02.001.

Parker, I. 1964. "The Diet of Duck and Coot at Lake Naivasha." Report commissioned by the East African Natural History Society, Nairobi.

Raffoul, M.H., Eric Enanga, Oscar Senar, Irena F. Creed, and Charles. G. Trick. 2020. "Assessing the Potential Health Risk of Cyanobacteria and Cyanotoxins in Lake Naivasha, Kenya." *Hydrobiologia* 847 (4): 1041–56. https://doi.org/10.1007/s10750-019-04167-z.

Rao, Vakkantham S. 2000. *Principles of Weed Science.* 2nd ed. Enfield: Science Publishers.

Richardson, J.L., and A.E. Richardson. 1972. "History of an African Rift Lake and Its Climatic Implications." *Ecological Monographs* 42 (4): 499–534. https://doi.org/10.2307/1942169.

Sikes, H. 1935. "Notes on the Hydrology of Lake Naivasha." *Journal of the East African Natural History Society* 13 (1–2): 73–84. Accessed September 1, 2023. https://www.biodiversitylibrary.org/content/part/EANHS/Vol.%20XIII_Nos.1___2__57-58__73_1935_Sikes.pdf.

Smart, Andrew C, David M. Harper, François Malaisse, Sophie Schmitz, Stephanie Coley, and Anne-Christine Gouder de Beauregard. 2002. "Feeding of the Exotic Louisiana Red Swamp Crayfish, *Procambarus Clarkii* (Crustacea, Decapoda), in an African Tropical Lake: Lake Naivasha, Kenya." *Hydrobiologia* 488: 129–42.

Stace, Clive 1997. *New Flora of the British Isles.* Cambridge: Cambridge University Press.

Stoof-Leichsenring, Kathleen R., Annett Junginger, Lydia A. Olaka, Ralph Tiedemann, and Martin H. Trauth. 2011. "Environmental Variability in Lake Naivasha, Kenya Over the Last Two Centuries." *Journal of Paleolimnology* 45: 353–67. https://doi.org/10.1007/s10933-011-9502-4.

Terer, Taita, Ludwig Triest, and A. Muthama Muasya. 2012. "Effects of Harvesting *Cyperus Papyrus* in Undisturbed Wetland, Lake Naivasha, Kenya." *Hydrobiologia* 680: 135–48. https://doi.org/10.1007/s10750-011-0910-2.

Thompson, A., and R. Dodson. 1963. "Geology of the Naivasha Area. Geological Survey of Kenya, Rept. 55." Nairobi: Government Printer.

Tyler, Stephanie J. 1991. "Birds of Lake Naivasha. Pt. 2. Foraging Niches and Relationships Between Migrant Resident Warblers in Papyrus Swamps." *Scopus* 14 (2): 117–24.

Upton, C. 2013. "Ecosystem Services and Livelihoods for the African Great Lakes: Overview and Case Studies." Report submitted to Ecosystem Services for Poverty Alleviation, NERC, Swindon, UK.

UNEP. n.d. "Lake Naivasha. Environmental Change Hotspots." Division of Early Warning and Assessment (DEWA). United Nations Environment Programme (UNEP). Accessed on October 31, 2023. https://na.unep.net/atlas/webatlas.php?id=2190.

USGS. 2022a. "Earthshots; Lake Naivasha." Accessed August 1, 2023. https://eros.usgs.gov/image-gallery/earthshot/lake-naivasha-kenya.

USGS. 2022b. "Earthshots; Hyacinth at Lake Naivasha, Kenya Story." Accessed August 1, 2023. https://eros.usgs.gov/image-gallery/earthshot/hyacinth.

Van Someren, G.R.C. 1981. "Report on a Survey of Wildfowl on Lake Naivasha." Unpublished report. Nairobi: NMK, Ornithology Section.

Vereecke, B. 2012 "Knowledge of Papyrus in the Hospitality Sector." Unpublished report submitted to Imarisha, Naivasha.

Verschuren, Dirk. 1994. "Sensitivity of Tropical-African Aquatic Invertebrates to Short-term Trends in Lake Level and Salinity: A Paleolimnological Test for Lake Oloidien, Kenya. *Journal of Paleolimnology* 10: 253–263.

Verschuren, Dirk, Kathleen R. Laird, and Brian F. Cumming. 2000. "Rainfall and Drought in Equatorial East Africa During the Past 1,100 Years." *Nature* 403: 410–13. https://doi.org/10.1038/35000179.

Vymazal, Jan. 2008. "Constructed Wetlands for Wastewater Treatment: A Review." In *Proceedings of Taal 2007: The 12th World Lake Conference*, edited by M. Sengupta and R. Dalwani, 965–80.

Wanjala, Silas, Timothy Mwinami, David M. Harper, Ed H. J. Morrison, and Nic Pacini. 2018. "Ecohydrological Tools for the Preservation and Enhancement of Ecosystem Services in the Naivasha Basin, Kenya." *Ecohydrology & Hydrobiology* 18 (2): 155–73. https://doi.org/10.1016/j.ecohyd.2017.09.004.

Wilkinson, David M. 1999. "The Disturbing History of Intermediate Disturbance." *Oikos.* 84 (1): 145–7. https://doi.org/10.2307/3546874.

WWF. 2012a. "Shared Risk and Opportunity in Water Resources; Seeking a Sustainable Future for Lake Naivasha." WWF Nairobi. Accessed September 1, 2023. https://www.wwf.eu/?206039/Shared-risk-and-opportunity-in-water-resources.

WWF. 2012b. "An Integrated Water Resource Action Plan Programme: Sustainable Land and Water Management for People, Businesses and Nature in the Lake Naivasha Basin, Kenya." Proposal to the Dutch Government. WWF Nairobi.

Zalewski, Maciej. 2002. "Ecohydrology – The Use of Ecological and Hydrologîcal Processes for Sustainable Management of Water Resources." *Hydrologîcal Sciences Journal* 47 (5): 823–32. https://doi.org/10.1080/02626660209492986.

CHAPTER 6

Linking Hydrology to Biodiversity, Ecosystem Services and Climate within the Lake Naivasha Catchment

Nic Pacini, Giulia Silvia Giberti and David Harper

1 Introduction[1]

Freshwater is a rare resource in the Rift Valley and a primary factor regulating the development and distribution of all forms of life. The availability of moisture is particularly critical in this highland region subject to high evaporation rates. The high daily solar radiation due to the sub-equatorial location of Lake Naivasha and the basin's "muted seasonality" (Melack 1979), with nearly steady temperature conditions all year round, imply that seasonal changes are controlled primarily by the spatial and temporal distribution of rains and river discharge rather than by temperature. Two annual rainy periods alternating with two dry periods promote a seasonal succession of plant and animal life forms, and patterns of animal behaviour, that develop across the catchment and in the lake, by following a more or less predictable temporal yearly pattern.

While animal communities react actively to moisture availability by performing diurnal and/or seasonal migrations, plants have developed mechanisms to withstand, but then also to regulate, the conditions of their surrounding habitat. As described by the concept of *ecohydrology*, of which the Naivasha catchment is a UNESCO demonstration site (Harper et al., this volume), forests and wetlands interact with the annual precipitation cycle by absorbing water within soils and plant biomass. This controls peak flows and short-circuits seasonal rains through a "Small Water Cycle", based on daily evapotranspiration and local condensation. Water retention and gradual release by upland and stream-corridor forests moderate discharge releases to the stream drainage network, ensuring water availability in the dry lowland portions of the catchment even in the absence of precipitation.

This natural regulatory mechanism has been greatly affected by human encroachment through deforestation, grazing, and through the establishment of farming plots in the highlands and within stream corridor forests. Such

1 First submission date: March 2, 2021. Acceptance date: March 13, 2023.

 DOI:10.1163/9789004695429_007

impacts have decreased the capacity of natural ecosystems to oppose climatic variability and are over-exposing the basin to risk of drought or flooding, depending on the intensity and timing of seasonal rains.

This chapter describes plant biodiversity and its relation to the vulnerability of the catchment's hydrology and local climate, which are linked to the distribution of soils and vegetation types. Emphasis is given to the role and to the condition of catchment streams, which reflects on their potential to provide essential ecosystem services to the local population. The Gilgil sub-catchment is described in detail as a case-study representing Naivasha catchment streams. It was selected because, apart for the Aberdares' equatorial alpine zone (not inhabited as it is entirely within a National Park), it includes the main characteristic landscape physiographic units that are relevant to human settlement in the Naivasha Basin (highland wetlands, portions of pristine forest, farming plots, tree plantations, small dams, lowland pasture, semi-arid savannah, stream corridors and vegetated lake fringe). The distribution of plant species within the Gilgil stream corridor is analysed in detail to highlight mechanisms that relate hydrology to ecology and to biodiversity, and that document the progressive degradation of the catchment under the increasing anthropogenic impact on upland and stream corridor forests.

The thesis proposed in this chapter underlines the relevance of the hydrological control exerted by upland and stream-corridor forests on the catchment stream network, on water quality and quantity, on key ecosystem services sustaining the resident population, and on catchment biodiversity. Arguments are brought forward to emphasise that loosening these ecological linkages through forest degradation is increasing the exposure of the Naivasha catchment and its people to the impact of global climate changes.

2 Orography and Climate of the Naivasha Basin

The Rift Valley is a tectonic depression developed along a series of superimposed fault lines that runs across the whole of East Africa from Mozambique to the Red Sea. Being a "valley without a river", the hydrology of the semi-arid Rift Valley relies on rains that are captured primarily in the surrounding forested highlands and that flow towards the valley floor draining the valley's steep ridges. Along the valley floor, most water is stored in ancient craters and in tectonic basins from where it progressively evaporates generating shallow saline lakes (such as Natron, Elmenteita, Bogoria); a smaller portion of it reaches underground aquifers and can be transferred for long distances along deep conduits connected through the fault system. In few cases, this deep-water percolation reaches sufficient flow magnitude, and an inflow-outflow balance can be achieved preventing salt accumulation at the surface. This explains the freshness of lakes Naivasha and Baringo (see Harper et al., this volume).

Situated just South of the Equator, the 3,500 km^2 broad Lake Naivasha catchment, lying between 1980 and 3990 m a.s.l., experiences a daily average temperature ranging annually from 8 to 30 °C with a minor seasonal variation (Odongo et al. 2015). The eastern portion of the basin is surrounded by the Nyandarua Mountains (Aberdares NP), while the Mau Massif borders the western slopes. The northern end comprises the stream corridors of the two main tributaries: the Malewa and the Gilgil; the southern and South-western portion of the basin grade towards semi-arid lowland savannahs. Average annual precipitation is strongly influenced by altitude, with 610 mm year^{-1} in the southern drylands, while the eastern highlands enjoy up to 1,550 mm year^{-1} (Odongo et al. 2015).

Hydrological and biogeochemical processes taking place in upstream catchments dictate the balance between salts, nutrients and sediments in lowland streams and wetlands. In the case of Lake Naivasha, this is even more so, given its prevailing endorheic nature and its slow and limited outflow at the bottom of the Rift Valley. Along the Rift Valley basement, potential evaporation driven by high temperatures exceeds precipitation during most of the year, apart from the few days that correspond to the peak of the rainy seasons (Odongo et al. 2015). Less than 30% of the water budget of Lake Naivasha is derived from rainfall generated from local evaporation and evapotranspiration (Becht and Harper 2002). Most of the water budget of the semi-arid Rift Valley relies on long-distance moisture transfer operated by monsoonal rains that are captured by highland forests covering the valley ridges. The longitudinal N-S orientation of the Rift Valley determines that a more significant portion of the monsoon rains, coming from the Indian Ocean, tends to fall on the windward side of the mountain massifs that enclose it (Figure 6.1). In particular, moisture condenses along the eastern Rift Valley ridges represented by Mount Kenya and by the Nyandarua Range. On their eastern flank, these "Water Towers" supply the sources of the >800 km-long Tana River and its tributaries, the largest river system in the country. A lesser portion of the moisture reaches the leeward side of the mountains and the Rift Valley with the Naivasha catchment that finds itself in a lee rain shadow (Everard and Harper 2002; Odongo et al. 2015).

The two annual rainy seasons, associated with the solar heating maxima, during the equinoxes (approximately on September 23 and March 21), are connected to Indian and West African monsoons and under the influence of sea surface temperatures. Interannual rainfall patterns tend to be relatively unstable and subject to the variable influence of global-scale climate teleconnections

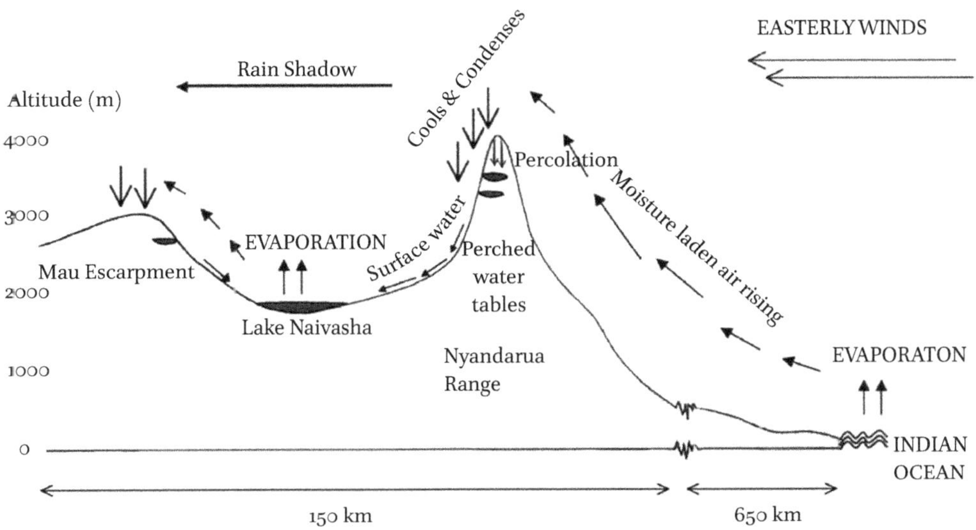

FIGURE 6.1 Schematic East-West section of the Rift Valley passing through the Naivasha Basin with an idealised hydrological model
EVERARD ET AL. (2002, 17). REPRODUCED WITH MINOR MODIFICATION WITH PERMISSION FROM KLUWER ACADEMIC PUBLISHERS

such as the El Niño Southern Oscillation (ENSO). Predictions of the effects of anthropogenic climate forcing (Cook and Vizy 2013) indicate a likely extension of the Short Rains (boreal fall period), to the detriment of the Long Rains (boreal spring period). However, global circulation models failed to explain the succession of events that caused East African climatic changes during the past millennium (Klein et al. 2016); climate predictions for this part of the continent are generally inaccurate possibly because of insufficient meteorological data recording stations, the complex orography, and the naturally high rainfall spatial and temporal variability (Herrnegger et al. 2021). Analyses of Lake Naivasha Basin rainfall trends, conducted after a careful scrutiny of available local field data for the period 1960–2010, were unable to detect changes in annual rainfall averages (Odongo et al. 2015) despite significant lake-level declines experienced up to 2010, and lake-level increase thereafter. Last decade's lake-level surge was recently attributed to a 25% increase in precipitation above the long-term average on the basis of rainfall estimates derived from remotely sensed satellite data (Herrnegger et al. 2021).

East African climate seems to have always had a high intrinsic variability (Verschuren 2001; deMenocal and Rind 2019). The dependence on long-distance moisture transport exposes the Rift Valley ecosystems to a high range of seasonal and interannual variability because precipitation patterns and annual river flows are ultimately under the influence of distant processes, subject to global climatic equilibria, such as sea-surface temperatures in the western Indian Ocean (Nicholson 2017, 2019), which drive wind direction and strength and determine the magnitude of precipitation over the region. These inherently unstable conditions highlight the relevance of local evapotranspiration-condensation cycles that naturally regulate climatic extremes by ensuring more regular low-intensity rains (Kravčík et al. 2008). Locally generated precipitations are essential to maintain acceptable temperatures and available water resources that ensure the survival of the basin's broad diversity of life forms, despite the variability of global forcing mechanisms. The preservation of mountain forests, densely vegetated stream corridors and wetlands is key for the maintenance of essential climate-regulating ecosystem services that contribute to the well-being of local communities. Their relevance is emphasised by the threat of global climate change, whose impacts still remain unclear.

3 The Connection between Hydrology, Soils, and Human Settlement

At 1,988 m a.s.l., Lake Naivasha is the highest wetland along the bottom of the Gregory Rift (the eastern branch of the Great Rift Valley) running across the whole of East Africa. The largest portion of the basin is drained by the Malewa River (1730 km^2), which receives several permanently flowing tributaries from the western slopes of the Nyandarua (highest peak at 3,990 m a.s.l.), and notably from the Turasha sub-catchment, contributing the greatest portion of overland flow reaching the lake.

The western and southern portions of the basin are instead drained by short ephemeral streams active only during intense rains. The Karati (temporary) stream rises at 2,648 m in the South-East and drains 149 km^2; its runoff reaches the lake only during the rainy seasons. Apart from the Marmanet, which feeds directly into the lake aquifer without reaching the lake's edge, there are few other highly seasonal streams draining the southern and western edges of the catchment. These belong to a portion of the Lake Naivasha Basin that is hydrologically inactive during the greatest part of the year, when it receives virtually no rain and produces no drainage; it corresponds to nearly half of the basin surface (Odongo et al. 2014).

The Gilgil River (527 km^2) comes straight southwards from the Bahati Highlands at the far northern edge of the basin. The Bahati Escarpment consists of Tertiary to Quaternary (Pliocene) basalt and other igneous rocks (Saemundsson 2010), overlain by deep and fertile volcanic nitisols. Here average annual rainfall reaches 1,300 mm and the river flows in a southerly direction between narrow ridges. It begins as two streams: the Morendat and the Kiriũndũ, which merge to become the Gigil River and are then joined by the Little Gilgil at Gilgil Town. Its discharge tends to become very low during the dry season and once in a few years the main river may dry out completely during particularly severe droughts, such as in 2009 and more recently in early 2017. The lower portion of its sub-catchment reaches the valley floor underlain by the Pre-Cambrian Basement Complex consisting of weathered gneisses, schists and granulites. A large area to the North of Lake Naivasha is covered in lacustrine sediments deposited during the Pleistocene, when present-day lakes Naivasha, Elmenteita and Nakuru used to form a single basin (Richardson and Richardson 1972). Today, in proximity of the lake, the lowland Basement Complex is overlain by poorly drained, infertile and partially alkaline sandy soils that originated from the evaporation of these ancient shallow wetlands.

The distribution of water resources and soils in the Naivasha catchment closely matches the distribution of human activities and settlements. As emphasised in the series "Farm Management Handbook of Kenya" (see Jaetzold et al. 2006), while the lowlands are infertile, the well-watered slopes of the Nyandarua Mountains, above 2100 m a.s.l., represent some of the most productive farmland in the country. These ideal conditions derive from the combination of moderate and relatively stable mid-mountain temperatures, a dense drainage network ensuring water availability, and fertile volcanic soils that until recently were covered by pristine forests. Until the beginning of the last century, until the European settlers arrived, the forests above the 2,000 m altitude line remained nearly intact apart for few clearings on ridge tops and level areas visited by migrant pastoralists, and for scattered crop patches (Church 2020). Most of the early colonial farmers that first established in these areas did so by taking advantage of British veteran-soldier-settlement schemes promoted by the British government. By the 1930s, they cut down extensive portions of forest and farmed introduced food and cash crops. The great majority of them fled during the 1950s state of emergency during the Mau Mau uprising. Only a few remained, took Kenyan citizenship and continued developing the land (Church 2020).

Today, the basin slopes above 2,100 m are occupied by small-scale subsistence farmers, mostly belonging to the Kikuyu community. In the upper portion of the Gilgil sub-catchment, most of the forest and bamboo thicket has been replaced by extensively cultivated potato, cabbage and maize plots stretching up the riverbank. The number of smallholder farmers has risen significantly in recent years under sustained population-growth rates. The drier lowland sub-catchment, covered by semi-arid savannah, is characterised by a much lower population density apart from the trading centres of Gilgil and

Naivasha, and is used mainly for extensive cattle ranching. As mentioned by Jaetzold et al. (2006), this lowland portion of the sub-catchment is not considered fit for horticultural crops.

Despite adverse soil conditions, large farms producing crops for export developed around the lake edge. Here temperatures are higher, soils are alkaline and nutrient poor, groundwater tends to be strongly alkaline, and the catchment is covered by semi-arid savannah vegetation. Stark interannual change in lake levels (either too much or too little water depending on the year) frustrate traditional farming attempts. These unfavourable conditions are addressed by large-scale agroindustrial solutions, which make use of greenhouses, hydroponic farming, and constant irrigation by means of high-capacity pumps that extract water from the lake or from the lake aquifer through boreholes (see Müller-Mahn and Gemählich, this volume).

The difference in soil quality determined strongly contrasting patterns of human activity and the different final target to which the Basin's resources are put. Traditional farming by small-scale settlers occupies the well-watered fertile highland slopes of the basin and produces mainly for the farmers' own subsistence (see Rouillé-Kielo, Calas, and Racaud, this volume). Lowland farming run by multinational agricultural corporations has developed large greenhouses for lake-water-fed crops (often grown on artificial substrates) to produce high-value horticultural vegetables and flowers targeted for the export market. These operations employ large numbers of labourers settled around the lake edge. A relatively limited area, situated to the North of Olkaria, is dedicated to large-scale wheat farming. Migrant pastoralists, mostly belonging to the Maasai community, take advantage of marginal lands throughout the lowlands and in the south-western dry portion of the catchment.

Beside farming, the semi-arid western portion of the basin stretching into Hell's Gate NP is occupied by Africa's largest geothermal power plant, Olkaria, run by the parastatal company KenGen (see Nweke-Eze and Adongo, this volume).

4 The Key Role of Ecosystem Services Provided by Streams

Flowing off the Water Towers – i.e.: the mountain forests that capture precipitations and store water resources, over 5 million streams and rivers, covering an estimated length of 12.8 million km (Welcomme 1976) – run towards the lowland valleys and represent the "life-blood of the African continent" (ibid.). Their permanent flow distributes moisture across the basin and to lowland wetlands, which moderates excess heat in lowland ecosystems (Pokorný and Huryna 2016a); this contributes to generating multiple ecosystem services including the maintenance of high biodiversity (Dobson et al. 2022). By distributing water across the Basin, streams connect highland and lowland ecosystems, while the multitude of smaller tributaries that constitute the drainage network ensures lateral connectivity across the landscape. Streamflow provides a wide range of direct and indirect regulatory services – not just support to biodiversity, but climate regulation, carbon sequestration, water storage and purification, soil formation and erosion prevention.

Streams bring life-bearing moisture and create a cool vegetated streamside habitat able to shelter a wide range of life forms. Outside the corridor, a relatively restricted number of species that are heat- and drought-resistant predominate; these are typical of heat-stressed African lowland savannahs, such as the lower portion of the Lake Naivasha catchment, facing moisture deficit during most of the year. Water vapour (latent heat) constantly produced by stream corridors, forests and wetlands is conveyed through lateral advection towards the open landscape and farming plots on the valley sides (Figure 6.2). In this function, stream corridors act as the catchment's steam valves. The release of water vapour contributes to solar-energy dissipation by moderating the generation of sensible heat across whole catchments (Pokorný and Huryna 2016a, 2016b). This flux produces a moderate cooling effect on the areas adjacent to the riparian forest. However, deforestation, wetland drainage and conversion to cropland imply that most of this vapour is very quickly transported towards the higher troposphere by the turbulent ascending flux of hot air that is generated by solar heating of dehydrated exposed soils (Figure 6.2). Through this process, evaporation-produced vapour is transported far higher than it would be normally, and its chances of re-condensing over the Basin are reduced.

Stream-corridor vegetation controls water temperature as well as water quality; the two are closely connected. Firstly, because low temperature facilitated by the presence of riverine vegetation increases oxygen concentration in water supporting the degradation of organic matter and stream self-purification by aquatic bacterial communities (Allan 1995); secondly, stream-corridor forests increase the water infiltration capacity of surrounding soils, reducing runoff intensity and discharge velocity. This enhanced permeability reduces soil erosion and the production of suspended sediments that otherwise accumulate in the drainage network and that tend to fill dams and natural wetlands. Slowing down water discharge enhances the capacity to retain and process nutrients within streams and reduces eutrophication impacts on downstream standing-water ecosystems.

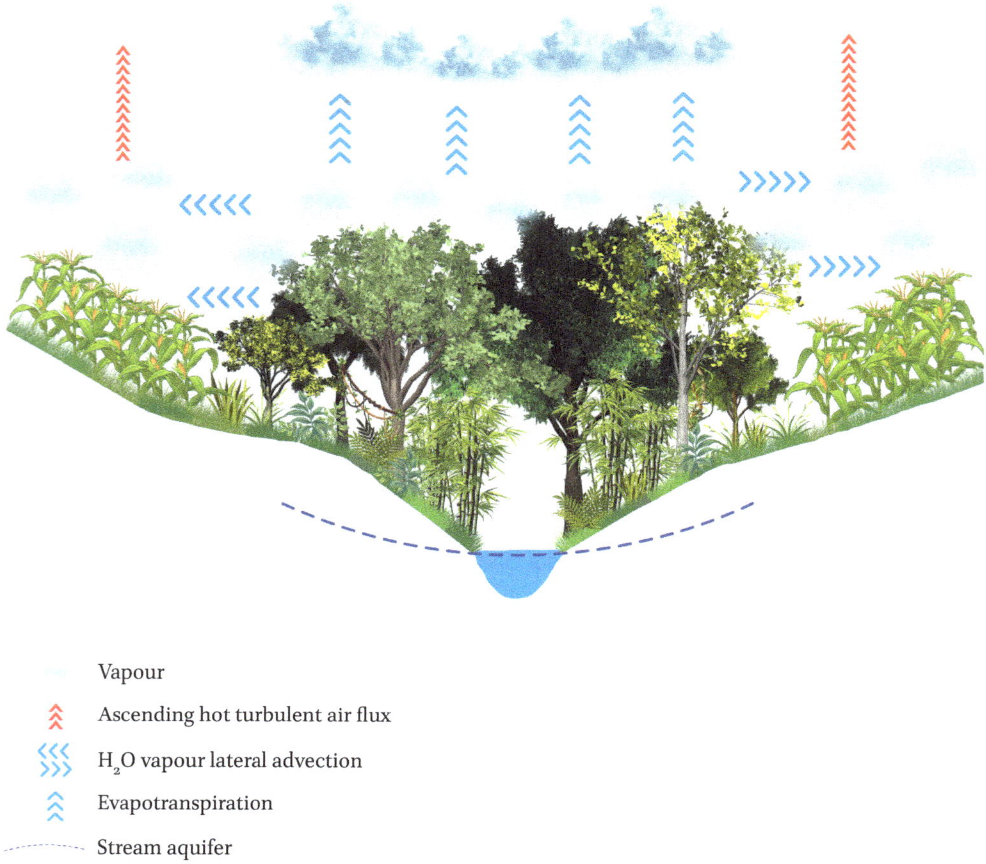

FIGURE 6.2 Idealised river section illustrating a shallow and wide riverbed. Evapotranspiration by riparian trees causes lateral advection of water vapour towards deforested and cropped valley sides. Here, a rapidly ascending flux of hot air (red arrows) transports moisture towards the higher troposphere. The deforested valley sides "steal" water from riparian corridors
FIGURE BY NIC PACINI AND GIULIA SILVIA GIBERTI

Human settlements are critically dependent on the availability of permanent-stream runoff for their direct water needs, for domestic animals and for watering smallholdings and gardens. By connecting highland forests to lower floodplains, streams distribute the benefits of the Water Towers throughout the catchment and supply communities with a continuous flux of direct ecosystem services that are not provided by the neighbouring dry savannah. These include mitigation of temperature extremes (both hot and cold), maintenance of riparian vegetation during the dry season, provision of wild edible and medicinal herbs/fruits, provision of cattle fodder, source of water for daily use, creation of a comfortable place where locals and their visitors relax, sense of identity, spiritual/existence values, and provision of sites for carrying out social/political gatherings and customary rites. At the lake edge, pastoralists, small- and medium-scale farmers dependent on rainfed crops, fishermen, large-scale horticultural and flower farms and their workers, and lakeside communities, all depend on services provided by stream-corridor and lake-fringing wetlands. All of these services associated with riparian ecosystems take place within the Lake Naivasha Basin and are threatened by deforestation and increasing water abstraction; several of them are in need of support/restoration by means of ecohydrological tools (as illustrated by Wanjala et al. 2018).

During the colonial period, water-related ecosystem services were enhanced through light infrastructure by early settlers, who developed over 900 small dams scattered throughout the catchment (Wanjala et al. 2018). Today the great majority have silted up and fallen into disrepair, and became unmanageable because of uncertain/disputed ownership. Stream power from early colonial infrastructure – such as water mills, stream-powered sawmills and ram pumps, as well as wells – has almost entirely disappeared, leaving the local rural population dependent on the availability of unmanaged natural water resources.

5 Local Impact on the Naivasha Basin Streams

From the high ridges (2,700 m and above), vegetated stream corridors descend towards the Rift Valley basement and break the monotony of its sun-baked lowland savannahs dominated by xeric woodlands and heat adapted grasses and bushes. Their status is changing under rapid demographic growth and intensifying land-use practices (fuelwood harvesting, charcoal burning, grazing within the stream-corridor forest). In large portions of the basin, logging and farming have been so severe that in most of the catchment outside protected areas, stream-corridor vegetation constitutes the only remaining forested ecosystem within a landscape whose natural habitats have become highly fragmented by farming plots, tree plantations and infrastructure.

Stream corridors are not only vulnerable to these localised impacts, but also and foremost to damage to upland forests that would naturally regulate stream discharge by preventing flash floods and thus supporting a gradual release of water during drought periods. Since the beginning of this century, hydrological processes within the Lake Naivasha Basin have become more vulnerable because of the deforestation of the mountain massifs that border the Rift Valley, in particular the eastern slopes of the Nyandarua. According to Global Forest Watch, Nakuru County came second only to the neighbouring Narok County as the most intensively deforested region in Kenya, and the two counties together (<1/20 of Kenya) saw the loss of >100 kha of tree cover during the period 2001–2021, equivalent to 1/3 of the entire deforestation experienced by the country as a whole (Global Forest Watch 2024). At the local scale, declines in forest (5.4%) and bushland (26.4%) have been recorded from the Naivasha Basin between 1973 and 2011 using remote-sensing techniques (Odongo et al. 2014). This has significantly affected the discharge network. The combination of deforestation, excessive sediment accumulation in dams and wetlands, and stream-water abstraction during the dry season has produced a highly variable runoff characterised by sudden river spates amidst long periods of drought. This has contributed to enhanced riverbank erosion, stream-corridor forest degradation, dam siltation and loss of ecosystem services. The significance of forest loss and of the desiccation and shrinking of stream-corridor forests for the maintenance of local and regional biodiversity, as well as climatic stability, are poorly understood. The causes of catchment degradation tend to be attributed to external sources, in particular to greenhouse gases-fuelled global climate change (USAID 2018).

6 Stream-Corridor Biodiversity

At the scale of Lake Naivasha Basin, given the significant recent local changes in deforestation, increasing encroachment by farming plots, rapid population increase, urbanisation and water abstraction from rivers and wetlands, it seems very hard to argue that local changes may reflect global climate-change effects linked to an increase in greenhouse gases in the atmosphere. We show here how mapping of plant biodiversity gives an indicator of a potential biological impact caused by changes in hydrology. We analysed the altitudinal succession of riparian plant diversity in a mountain stream that recently showed signs of increasingly irregular water levels and frequent droughts, which up until recently were unusual.

We considered that across the catchment, plant distribution changes primarily according to a species succession that responds to the altitudinal gradient, from the cool and well-watered highlands towards the warm and dry lowland savannah. In contrast to this general pattern, within stream corridors, where moisture is provided by the stream itself, the distribution of plant species is strongly influenced by stream hydrology and the microclimate is controlled by river dynamics rather than altitude. For this reason, within a stream corridor, we should expect that changes in species distribution, from the sources towards the valley floor, are likely to reflect the impact of recent hydrological changes in the catchment.

Throughout the catchment, plant-species assemblages follow an altitudinal pattern, from sources to mouth, but vary also in a lateral direction, i.e.: across the stream, from the moist edge of the riverbank towards the drier valley sides. At a finer scale, within the stream corridor, variable morphoedaphic conditions that are continuously reworked by river dynamics tend to create a fine-scale mosaic whereby, every few metres, local plant species may include strictly aquatic species, close to terrestrial and/or drought-adapted ones, depending on the frequency of inundation of the local microhabitats and on the presence/absence of a shallow stream aquifer. In addition to this naturally diverse assemblage, a well-established stream-corridor forest sustained by permanent river discharge tends to include a characteristic stable riverine plant community, specifically adapted to infrequent river spates and to periodically inundated alluvial soils deposited by floods. This riverine plant community is typically absent from all other habitat types within the catchment. Studying the diversity and the spatial distribution of riverine vegetation, it can be postulated that the plant-species assemblage is shaped by the catchment's "hydrological history".

To investigate the relationship between hydrology and stream-corridor vegetation in a typical stream draining the slopes of the Rift Valley, vegetation quadrats and tree transects were collected at 11 sites along the River Gilgil including its main tributaries: the Kiriũndũ and the Morendat (Figure 6.3). The objective was to highlight direct impacts caused to the natural riverine forest by deforestation, cattle ranching, tree planting and water abstraction in a river characterised by an increasingly unstable hydrology. The study area was selected because of its moderate length, its simple drainage pattern, and its relative ease of access.

After an initial survey of the longitudinal species succession, it became apparent that the difference in species composition between the stream-corridor vegetation and the outer plant community increased longitudinally from the well-watered cool highlands to the dryland lower basement. In the upstream (little difference), riparian vegetation merged with the remains of the surrounding forest that used to be dominated by Pencil Cedar (*Juniperus procera*), with remnants of bamboo thickets (*Arundinaria alpina*) in wetter patches. In contrast, the stream-corridor vegetation in the lower section included taxa that differed entirely (large difference) from the vegetation of the nearby open savannah, and that were closely dependent on the presence of the stream microclimate.

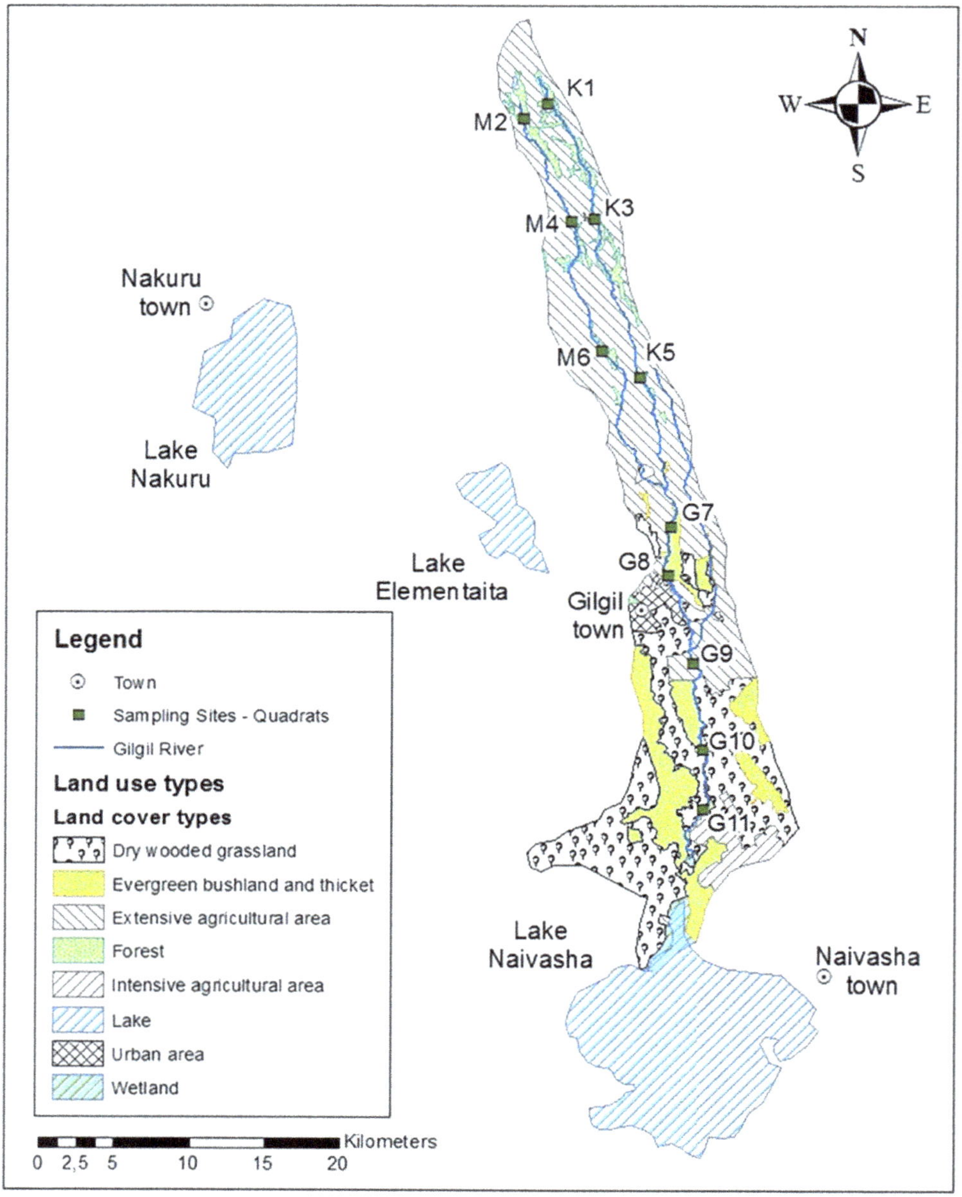

FIGURE 6.3 Location of vegetation quadrat sites along the Gilgil River and tributaries (Morendat and Kiriundu) and main land use

REPRODUCED FROM GIBERTI ET AL. (2021, 32) WITH PERMISSION BY WILEY

Description of the methodology employed to run the vegetation survey (Giberti et al. 2021)

The composition of the vegetation within the stream corridor was assessed in detail in 2018-2019 with a two-pronged strategy: a detailed survey conducted in a series of 22 (10 x 15 m) quadrats dislocated in 11 stations and dedicated to the mapping of the entire plant community, and 52 (3 x 100 m) belt transects for the recording of trees and bushes taller than 1 m. The survey identified 365 taxa, believed to represent about 2/3 of the actual extant plant diversity; this estimate was derived by applying β diversity distribution models based on bootstrapping statistical techniques (Fortin et al. 2020). The survey described the presence of contrasting upstream and downstream vegetation groups, whose distribution correlated with changes in lithology and soils. The consultation of maps contained within the Nakuru County Atlas of Agro-Ecological Zones (Jaetzold et al. 2006) provided a relatively detailed insight of the distribution of soils within the basin. In the upstream, two categories were included: 1) UBP1 well-drained deep luvisols developed over volcanic ashes overlying basalt and nepheline tuffs, with inclusions of gleyic soils; 2) RB3 deep friable eutric nitisols developed from basalt, nepheline and phonolite tuffs. Two categories were included in the downstream: 3) P1PC – imperfectly drained sandy cambisols mostly developed from ancient lake sediments laying over pyroclastic rocks, 4) PlU1 – very deep poorly drained moderately sodic solonetz developed over lacustrine plains.

When these soil categories were fitted as vectors in Non-metric Multi-Dimensional Scaling ordination (Figures 6.4 and 6.5 for quadrats and transects respectively; vector significance was assessed by envfit R "vegan" package Oksanen et al. 2016; Strohbach, Audorff and Beierkuhnlein 2009), it became apparent that the grouping of sites according to assemblage composition was significantly influenced by soil type, as indicated by significant statistical correlation (For details see Giberti et al. 2022). Quadrat samples partitioned into two groups, while transect samples that were more numerous than quadrats and more spatially continuous indicated that three groups could be identified. In both quadrats and transects, particularly high significance in site grouping was achieved by the sandy cambisols comprised under category P1PC that supported a distinctive plant assemblage in relation to the other groups.

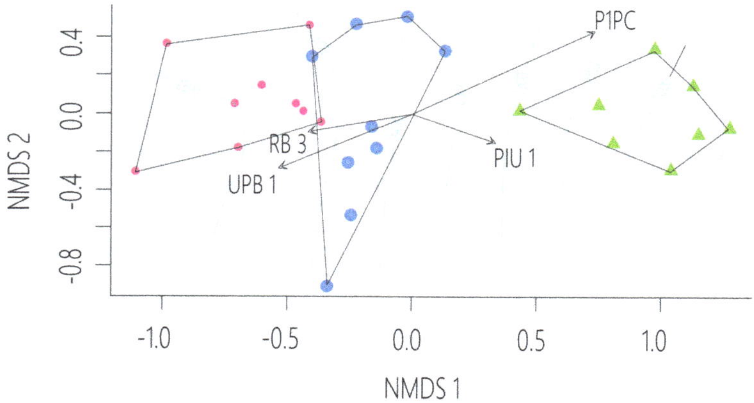

FIGURE 6.4 Non-metric Multi-Dimensional Scaling (NMDS) of transect data (Stress: 0.133); sites are clearly divided into three groups (Upstream diamonds on the left, Midstream circles at centre, and downstream triangles on the right; Permanova 999 permutations P<0.001). Physiographic/lithological UBP1 and P1PC soil category (see main text) vectors correlated significantly with NMDS axes, but not the other two. Altitude contours (m a.s.l.) were fitted using R ordisurf

REPRODUCED FROM GIBERTI ET AL. (2021, 34) WITH PERMISSION BY WILEY

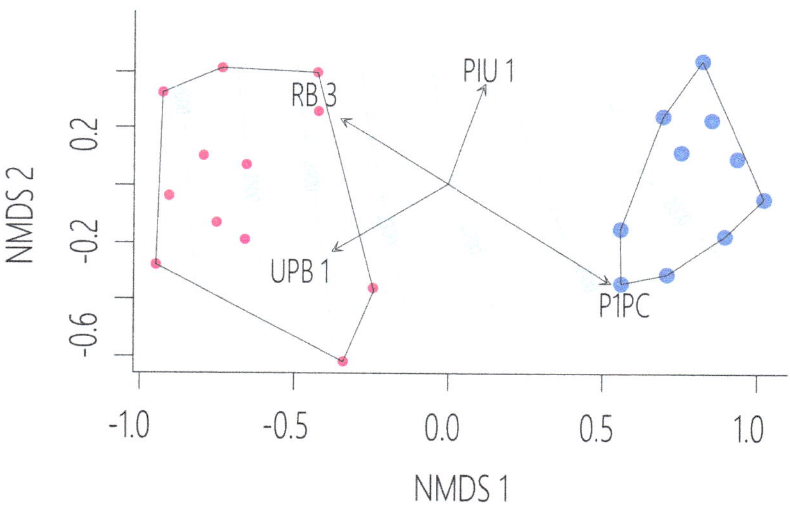

FIGURE 6.5 Non-metric Multi-Dimensional Scaling of quadrat data (Stress: 0.095); sites are clearly divided into two groups (Upstream diamonds on the left, Downstream dark circles on the right, Permanova 999 permutations P<0.001). Physiographic/lithological UBP1 and P1PC (see main text) vectors correlated significantly with NMDS axes. Altitude contours (m a.s.l.) were fitted using R ordisurf

REPRODUCED FROM GIBERTI ET AL. (2021, 34) WITH PERMISSION BY WILEY

A closer look at the species distribution indicated that the number of riverine and native tree species tended to increase in a downstream direction, while forest taxa preferred sites at higher altitude. Species distribution was then assessed according to selected indicator species and to changes in β diversity. Exotic tree plantations had a significant influence on the composition of upstream transects where farmers plant trees around agricultural plots to prevent erosion and to provide for timber. Some of these tree plantations typically extend up to the water's edge and interrupt the continuity of the natural stream-forest corridor by reducing natural habitats for local plant and animal species. The farmers' choice tends to fall on a small number of fast-growing alien species, such as *Acacia mearnsii* and *Eucalyptus globulus*. The expansion of these plantations represents a potential threat to the biotic integrity of the stream-corridor ecosystem. Similar tree plantations were not found in the dry lowlands, where tree planting would require irrigation.

By taking into account the prevailing conditions dictated by altitude, geology, climate and the biogeographical distribution of East African plant species, the *Potential Natural Vegetation of Eastern Africa* project (VECEA) elaborated theoretical vegetation maps for most of East Africa, including our study area (Kindt et al. 2015). We compared the distribution of the plant community in the stream corridor we surveyed to these theoretical maps and we realised that several taxa belonging to the "Afromontane vegetation community" plant-species assemblage, described by the VECEA project, found refuge in the stream-corridor forest. The presence of these typically montane species indicated that local microclimatic conditions created by the stream corridor, and the substrate accumulated as a result of river spates, allowed mountain taxa to extend their natural distribution and to migrate towards lower altitudes. Inside the shaded stream corridor, the stream contributed to local micro-habitats that induced the establishment of upland species normally requiring cool and well-watered conditions (Figure 6.6). This allowed them to expand towards the lowlands, where they would not be generally expected, at sites where the catchment vegetation found just outside the corridor was dominated by dryland-adapted bushland. This colonisation by mountain taxa was not observed along the entire stream length, however. At several sites in the lower portion of the basin, the stream-corridor forest appeared disturbed and discontinuous; several transects indicated the presence of drought-adapted woodland taxa that in places invaded the stream-corridor forest, interrupting the continuity of the stream-driven microhabitat.

FIGURE 6.6 River Gilgil close to station G7 (Mawingo). River terraces are deforested, grazed and few farming plots are visible. Stream-corridor trees include a mixture of endemic species, some of which are adapted to high-altitude habitats, as well as alien species (cypress in the foreground) established from tree plantations

Finally, we noticed several lowland sites characterised by larger trees (*Ficus sur*, *Syzygium* spp., *Sesbania sesban, etc.*) belonging to an entirely distinctive riverine community, not found elsewhere, either in the lowlands or in the upland forests. These species, characteristic of floodplain forests, tend to withstand prolonged inundation and to grow on alluvial soils generated by river spates and are highly specific of riverine habitats; they are hardly seen anywhere else. The fact that this riverine community comprised only large individual trees and that our quadrats did not contain signs of any small tree saplings belonging to this community, indicated that these riverine tree species are not regenerating and are likely to represent the last surviving components of a previous larger stream-corridor forest that may have progressively faded away, while the floodplain was desiccating and shrinking.

In the lower river section, characterised by these large riverine trees, the morphology of the channel showed signs of significant erosion, with deepening of the channel bed and presence of loose sediments. Consequently, the stream-corridor vegetation tended to find itself high above the channel bed, often exposing its roots through the channel banks. Most lowland sites showed evidence of this progressive deepening of the river channel and disconnection of the riparian vegetation from the stream aquifer (Figure 6.7). Increasing the distance between the riparian vegetation and the riverbed is likely to

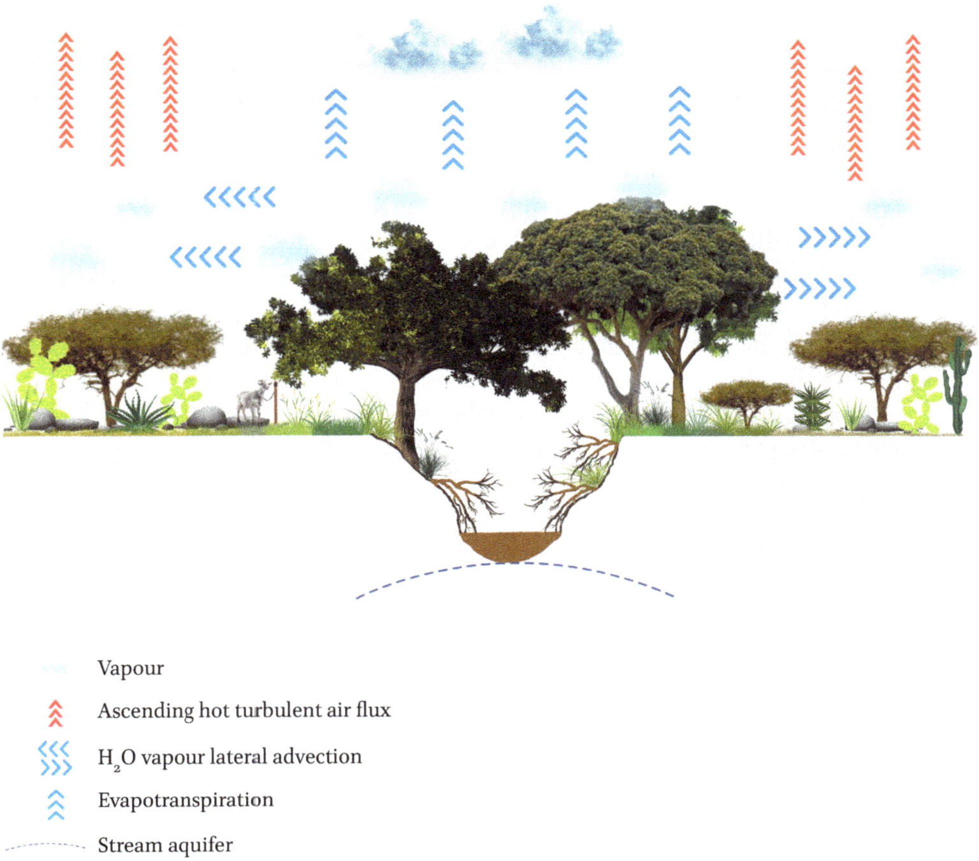

FIGURE 6.7 Shrinking of the stream corridor caused by overgrazing of the valley sides reduces evapotranspiration in conjunction with the deepening of the stream bed due to scouring and riverbed erosion during seasonal flood events. Only a few large trees are able to reach close to the stream aquifer with their deep roots, while grasses and forbs are disconnected and depend on the moisture provided by the few remaining trees. The dry floodplain emits intense fluxes of sensible heat causing ascending turbulent air
FIGURE BY NIC PACINI AND GIULIA SILVIA GIBERTI

promote further desiccation and narrowing of the stream corridor. Intuitively, it could be hypothesised that because of their shallower roots, grasses, herbs and saplings adapted to wet conditions are likely to bear the greatest impact, while large riparian trees provided with ancient and deeper roots may withstand channel deepening for a longer period. Desiccation and narrowing of the stream corridor prevent riparian-tree regrowth because of a reduction in seedling regeneration, as the tree propagules tend to find themselves too far from the water table to become established.

Rather than being indicative of local (i.e. site-scale) effects, poor tree regeneration and channel deepening are in agreement with the recent changes in stream hydrology related to upland deforestation, water abstraction, and degradation of the upper catchment. Studies conducted by Hesslerová and Pokorný (2010a, 2010b) in the nearby Mau Forest indicate that recent highland deforestation has reduced potential evapotranspiration and increased local temperature in the highlands surrounding the Naivasha Basin. The observation of changes in local temperature indicate that the upper catchment has likely partially lost its Water Tower properties and the "sponge effect" through which mountain forests reduce runoff response to precipitation. Consequently, runoff generated by tropical thunderstorms tends to build up faster than before. It produces violent spates that descend the stream valley causing bank erosion and severe riverbed scouring, as reported by several other authors describing the ongoing degradation of catchment processes along the Rift Valley slopes (Shivoga et al. 2007; Guzha et al. 2018). The repeated scouring and deepening of the riverbed reduce the connection between the riparian vegetation roots and the stream-recharged groundwater, accentuating the shrinking and desiccation of the riparian vegetation. The progressive degradation of the stream corridor is likely to cause a reduction in evapotranspiration and an increase in sensible heat across the catchment. This will tend to fuel an amplifying feedback loop leading to higher average temperatures and increasingly severe droughts.

7 Recent Hypotheses Underline the Relevance of Forests for Local Climate

While it is often emphasised that tropical forests are the "planet's lungs", as they produce a significant portion of the oxygen we breathe, it should be remembered that forests also act as "fridges", by dissipating solar energy through evapotranspiration (i.e. water-vapour release through plant leaves). This controlled cooling effect, produced by turning liquid water into water vapour (latent heat), contributes to creating stable climatic conditions that are ideal for life, even in tropical regions, despite constantly high insolation rates. It is not so long ago that scientists realised that forests are active components of the water cycle that circulates through them (Salati et al. 1979), and that most of the rainfall that maintains life at tropical latitudes is recycled through living forests connected to large rivers, such as the Amazon and in the Congo basins (Lovejoy and Nobre 2018). These regions receive a large proportion of the solar radiation and host the largest fluxes of H_2O on the planet, both in terms of runoff (blue water) and fluxes of water vapour (green water or "flying rivers"), generated by the forest evapotranspiration. Blue and green water fluxes are closely coupled by local evapotranspiration-condensation cycles that regulate the extant temperature avoiding excessive overheating of the earth surface and ensuring near-constant moisture availability.

Worldwide, forests tend to stabilise the water cycle; their effects are relevant to the stability of local ecosystems but are also globally significant. Forests release vapour through evapotranspiration, promote cloud condensation and rainfall, and, when it rains, moderate runoff by retaining moisture within upland vegetation and forest soils (the so-called "sponge effect"). Excess water that is not consumed through evapotranspiration gradually flows downstream, contributing to the maintenance of baseflow during droughts. Forest control over the local water cycle is essential to ensure permanent water sources that generate streams, vital for the persistence of lowland ecosystems.

The importance of local water cycling is reflected in indigenous knowledge that understands the close relationship between forests and water availability – worldwide (FAO and FILAC 2021) and in rural Sub-Saharan Africa (Bhatasara 2017). Within Kenya, it was reported by elders living within the Mau Massif (Silale 2021). For this same reason, a small Maasai community living on the western flank of the Naivasha Basin bordering the eastern Mau strictly protects 650 acres of pristine pencil cedar-dominated forest (*Juniperus procera*). "Cutting the forest would run dry the last permanent spring feeding the Marmanet, our sole source of water during the dry season. The other springs that used to flow around our village are now dry because of deforestation" states community elder Sonkoi Kiminta (Forest & Water 2013[2]). In conformity with the Kenyan *Forest Conservation and Management Act* (GOK 2016), gazetted in 2008, which provided for community involvement in joint forest management through Community Forest Associations (CFAs), the Maasai from the Enaiboro Ajijik community, comprising about 2,000 people, today hold the communal title deed to Olosho Rongai (Narrow Ridge) Forest. The forest is situated between the Mau and the Eburru Massifs and represents one of the few remaining intact portions of the formerly extensive Mau Forest complex, known up to the 1980s as the largest single forest block within East Africa. This is a rare example of forest conservation managed by local communities in partnership with the government forestry authority (KFS).

Lower downstream, at the Lake's edge, former farmer, conservationist and LNRA-secretary Sarah Higgins used to underline how weather patterns

2 Documentary filmed in 2013 in the Naivasha Basin as part of a local filming workshop conducted by the Naivasha Basin Sustainability Initiative.

changed alongside damage to the Mau Forest. Rains used to be sufficient to grow wheat during the 1970s and 1980s but thereafter deforestation and overgrazing destroyed the soil cover and rains started to fail (Schwartz 2013).

Despite the intuitively persuasive linkage between forests and water availability, only a minority of scientists shares Sonkoi Kiminta's and Sarah Higgins' opinion. The relationship between forest and water lies at the centre of a long-standing dispute with hydrologists on the one side claiming that by reducing evapotranspiration, deforestation actually increases surface runoff, and ecologists on the other side arguing that forest disruption inevitably leads to the loss of hydrological regulatory services, a change in local climate, and a consequent reduction in water availability in the catchment downstream. Both parties provide evidence supporting their arguments. This fundamental and apparently simple issue is unlikely to be easily resolved because field observations are catchment-specific, subject to high interannual variability, and strongly influenced by medium and long-term effects that are difficult to measure in human time scales. A taste of these arguments can be gained in a recent review of catchment studies by Guzha et al. (2018) highlighting East African forest-streamflow relationships. Empirical science seems unable to decide whether the protection of forest reserves in the Naivasha Catchment will support the persistence of sufficient water supplies to its growing population. Many people find it more convenient to argue that all will depend on external rain inputs, controlled by global climate change, regardless of local land-management choices.

Towards the end of the last century, it was noticed that the "recycling of moisture" by vegetation must have an impact on precipitation, and therefore that climate and land use are closely connected (Savenije 1995). However, recycling cannot suffice to maintain local forests; they must be supported by a mechanism that enables moisture convergence. A more comprehensive theory, called the *Biotic Pump*, was formulated for the first time only some 15 years ago by two Russian physicists from the Saint Petersburg Nuclear Physics Institute (Makarieva and Gorshkov 2007). Their analysis started from the consideration of the thermodynamics ruling the transfer of moisture from the surface of the oceans (the Earth's water reservoir) towards the interior of the continents. They realised that the cool temperatures generated by forest evapotranspiration reduce the altitude of the atmospheric boundary layer and generate preferential condensation sites (moisture acceptors) towards which clouds tend to converge by overcoming the atmosphere's frictional drag, in which condensation occurs (Makarieva and Gorshkov 2010; van der Ent et al. 2010; Makarieva et al. 2022). In parallel to this, recent evidence highlighted that forests are also able to promote precipitation from clouds above them through volatile "rainfall catalysts" (organic compounds as well as bacterial spores released from tree leaves, Morris et al. 2014). According to these views, forests are able to "seed" and harvest rains (Ellison et al. 2017). When precipitation occurs, the condensation of cloud vapour causes a depression in the lower atmosphere that attracts distant air masses and the moisture they carry, multiplying rain yields. It is therefore not surprising that forest persistence could be related to permanent mountain springs, as in the case of the Marmanet, through a process that is dependent on local catchment conditions rather than solely on pressure gradients generated by temperature differences measured in distant oceans and air masses, as purported by mainstream science. The Naivasha Basin has been an important case study highlighting the relevance of the role of mountain forests, stream corridors and lake-fringing wetlands in regulating the water cycle and moderating climate extremes (Hesslerová et al. 2019).

Although the Biotic Pump theory is disputed (Meesters et al. 2009), it remains the best-articulated and most coherent set of arguments supporting the functional role of forests in moderating catchment hydrology and climate. It also provides a warning to land managers and large-scale farmers who for too long have considered that their main role has been to increase food production in Sub-Saharan Africa through extensive deforestation, establishment of mechanised agriculture and irrigation, and intensive use of pesticides and mineral fertilisers. Naivasha's horticultural and flower industry is an expression of the same. In fact, most UN Agencies (such as FAO, World Agroforestry Centre, the UN Environmental Programme) stopped sharing this view long ago and tend to advocate for a more sustainable approach; however, none of them makes a clear reference to Makarieva and Gorshkov's Biotic Pump. The model proposed by the two Russians is highly controversial because it leads towards a very different interpretation of the causes and impacts of global climate changes from what is portrayed by the official views expressed by the Intergovernmental Panel on Climate Change (IPCC) obsessed with CO_2 emissions.

8 Conclusion

The attention given to global climate change, attributed to the rise in the concentration of greenhouse gases, indirectly implies neglecting locally induced changes that do impact local climate and that go largely unreported. Our brief analysis of the climate and hydrology of the Naivasha catchment suggests that perturbations caused by deforestation, agricultural encroachment and water abstraction are likely to cause severe repercussions. These impacts are not systematically monitored and go unrecorded. Changes to stream hydrology are reflected in the distribution of plant species within stream-corridor forests and in changes of stream morphology. These indicators suggest the progressive loss of an endemic riparian forest community caused by the shrinking and desiccation of stream corridor forests.

Linkages between upland forests, catchment drainage and local climate evidenced by concepts such as the Water Towers and the Biotic Pump Theory indicate that greater attention should be given to land-use changes within the catchment, and that the degradation of the natural infrastructure consisting of upland forests, stream corridors and catchment wetlands, and their interconnections, should be prevented. This infrastructure is essential to withstand the forcing imposed by outer climate drivers, such as the increase in atmospheric anthropogenic greenhouse gases. Similarly to what has been suggested for the Mediterranean coastline by Millán (2014), local desiccation and reduced evapotranspiration tend to increase the height of the atmospheric boundary layer (where condensation has a chance to start), contribute to increases in air pressure, and reduce the Small Water Cycle related to convective precipitation. The contribution of stream corridors' evapotranspiration to the moisture budget of the local Water Towers similarly decreases and the Biotic Pump weakens. These degenerative alterations enhance the exposure of catchment ecosystems to the variability of long-distance rain-generating processes, associated with the migration of the ITCZ (see Harper et al., this volume) and with changes in sea-surface temperatures. The cascading effects described imply that impacts of the deforestation of mountain forests and of riparian corridors contribute to surface warming far beyond what could be expected if only carbon sequestration is taken into account (see Butt et al. 2023 and references therein).

At present, it could be argued that given the scale of local catchment degradation, the impact of global mechanisms of climate change cited by USAID (2018) and by Kenya government reports such as *The National Climate Change Action Plan* 2018–2022 (Government of Kenya 2018) could be overestimated in relation to local causes, and in any case very hard to evaluate. Given the extent of the dynamic alterations that the basin is undergoing, linked to land-use change and to rapid population growth, it is likely too soon to come to conclusions about the real impact of greenhouse gas-driven global climate changes for this catchment.

At present, the reasons behind the recent lake-level rise remain debated, while recent plans ventilated by the government to significantly increase abstractions from the lake aquifer and to develop a large artificial reservoir above Gilgil Town are a matter of great concern (see Kioko and Wanjala, this volume).

Acknowledgements

The authors whole-heartedly acknowledge the commitment shown by Silas Wanjala, Timothy Mwinami, Alfred Koech Kiprotich, and Steven Wanjiku Methu who accompanied us as colleagues and friends during numerous field surveys sharing our passion for scientific enquiry, and the endless encouragement and shelter offered by Diane and Sandy McLeish.

References

Allan, David J. 1995. *Stream Ecology: Structure and Function of Running Waters.* Dordrecht: Springer.

Becht, Robert, and David M. Harper. 2002. "Towards an Understanding of Human Impact upon the Hydrology of Lake Naivasha, Kenya." *Hydrobiologia* 488: 1–11. https://doi.org/10.1023/A:1023318007715.

Bhatasara, Sandra. 2017. "Towards a Sociology of Adaptation to Rainfall Variability in Rural Zimbabwe: The Case of Charewa in Mutoko." *Fudan Journal of the Humanities and Social Sciences* 10 (4): 547–68. https://doi.org/10.1007/s40647-017-0177-8.

Butt, Edward W., Jessica C.A. Baker, Francisco G. Silva Bezerra, Celso von Randow Celso., Ana P.D. Aguiar, and Dominick V. Spracklen. 2023. "Amazon Deforestation Causes Strong Regional Warming." *Proceedings of the National Academy of Sciences* 120 (45) e2309123120. https://doi.org/10.1073/pnas.2309123120.

Church, Colin. 2020. "*Mel-el-Lek's Mountain: Aberdares – Place of Mystery, Drama and Compelling Events in Kenya's Unfolding Story*". Independently published. ISBN-13: 979-8639198533.

Cook, Kerry H., and Edward K. Vizy. 2013. "Projected Changes in East African Rainy Seasons." *Journal of Climate* 26 (16): 5931–48. https://doi.org/10.1175/JCLI-D-12-00455.1.

deMenocal, Peter B., and David Rind. 2019. "Sensitivity of Subtropical African and Asian Climate to Prescribed Boundary Condition Changes: Model Implications for the Plio-Pleistocene Evolution of Low-Latitude Climate." In *The Limnology, Climatology and Paleoclimatology of the East African Lakes,* edited by Thomas C. Johnson, and Eric O. Odada, 57–77. London: Routledge. https://doi.org/10.1201/9780203748978-3.

Dobson, Barnaby, Saoirse Barry, Robin Maes-Prior, Ana Mijic, Guy Woodward, and William D. Pearse. 2022. "Predicting Catchment Suitability for Biodiversity at National Scales." *Water Research* 221: 118764. https://doi.org/10.1016/j.watres.2022.118764.

Ellison, David, Cindy E. Morris, Bruno Locatelli, Douglas Sheil, Jane Cohen, Daniel Murdiyarso, Victoria Gutierrez, et al.. 2017. "Trees, Forests and Water: Cool Insights for a Hot World." *Global Environmental Change* 43: 51–61. https://doi.org/10.1016/j.gloenvcha.2017.01.002.

Everard, Mark, Vale A. Jacqueline, David M. Harper, and Håkan Tarras Wahlberg. 2002. "The physical attributes of the Lake Naivasha catchment rivers." *Hydrobiologia* 488: 13–25. https://doi.org/10.1023/A:1023349724553.

FAO and FILAC. 2021. *Forest Governance by Indigenous and Tribal People. An Opportunity for Climate Action in Latin America and the Caribbean.* Santiago. https://doi.org/10.4060/cb2953en.

Fortin, Mathieu, Anna Kondratyeva, and Rosalinde Van Couwenberghe. 2020. "Improved β-diversity Estimators Based on Multiple-site Dissimilarity: Distinguishing the Sample from the Population." *Global Ecology and Biogeography* 29 (6): 1073–84. https://doi.org/10.1111/geb.13080.

Giberti, Giulia S., Bruna Gumiero, Alfred K. Kiprotich, Stephen Wanjiku Methu, David M. Harper, and Nic Pacini. 2021. "Longitudinal Vegetation Turnover in an Eastern Rift Valley Riparian Corridor." *African Journal of Ecology* 60 (1): 27–42. https://doi.org/10.1111/aje.12920.

Global Forest Watch. 2024. "Kenya." Accessed May 9, 2024. https://www.globalforestwatch.org/dashboards/country/KEN/.

Government of Kenya. 2016. *Forest Conservation and Management Act (No. 34 of 2016).* This Act entered into Force on 31 August 2016 and repealed the *Forests Act* of 2007. www.kenyalaw.org.

Government of Kenya. 2018. *The National Climate Change Action Plan (Kenya) 2018–2022.* Nairobi: Ministry of Environment and Forestry of Kenya. Accessed September 11, 2023. http://faolex.fao.org/docs/pdf/ken190169.pdf.

Guzha, Alphonce C., Mariana C. Rufino, Sheila Okoth, Suzanne Jacobs, and Rodolfo L. B. Nóbrega. 2018. "Impacts of Land Use and Land Cover Change on Surface Runoff, Discharge and Low Flows: Evidence From East Africa." *Journal of Hydrology: Regional Studies* 15: 49–67. https://doi.org/10.1016/j.ejrh.2017.11.005.

Herrnegger Mathew, Stecher Gabriel, Schwatke Christian, and Olang Luke. 2021. "Hydroclimatic Analysis of Rising Water levels in the Great Rift Valley Lakes of Kenya". *Journal of Hydrology: Regional Studies* 36: 100857. doi.org/10.1016/j.ejrh.2021.100857.

Hesslerová, Petra, and Jan Pokorný. 2010a. "Effect of Mau Forest Clear Cut on Temperature Distribution and Hydrology of Catchment of Lakes Nakuru and Naivasha: Preliminary Study." In *Water and Nutrient Management in Natural and Constructed Wetlands*, edited by Jan Vymazal, 263–73. Dordrecht: Springer. https://doi.org/10.1007/978-90-481-9585-5_19.

Hesslerová, Petra, and Jan Pokorný. 2010b. "Forest Clearing, Water Loss, and Land Surface Heating as Development Costs." *International Journal of Water* 5 (4): 401–18.

Hesslerová, Petra, Jan Pokorný, Hanna Huryna, and David Harper. 2019. "Chapter 4: Wetlands and Forests Regulate Climate via Evapotranspiration". In *Wetlands:*

Ecosystem Services, Restoration and Wise Use, Ecological Studies 238, edited by Shuqin An, and Jos T. A. Verhoeven, 63–93. Switzerland: Springer Nature.

Jaetzold, Ralph, Helmut Schmidt, Berthold Hornetz, and Chris Shisanya. 2006. *Farm Management Handbook of Kenya. Volume II: Natural Conditions and Farm Management Information.* Annex: Atlas of Agro-ecological Zones, Soils and Fertilising by Group of Districts. Subpart B1a: Southern Rift Valley Province Nakuru County. Nairobi: Ministry of Agriculture.

Kindt, Roelandt, Paulo van Breugel, Caleb Orwa, Jens-Peter B. Lillesø, Ramni Jamnadass, and Lars Graudal. 2015. "Useful Tree Species for Eastern Africa: A Species Selection Tool Based on the VECEA Map." Version 2.0. World Agroforestry Centre (ICRAF) and Forest and Landscape Denmark. Accessed September 11, 2023. https://vegetationmap4africa.org.

Klein François, Hugues Goosse, Nicholas E. Graham, and Dirk Verschuren. 2016. Comparison of simulated and reconstructed variations in East African hydroclimate over the last millennium. *Climate of the Past* 12: 1499–1518. https://doi:10.5194/cp-12-1499-2016.

Kravčík, Michal, Jan Pokorný, Juraj Kohutiar, Martin Kováč, and Eugen Tóth. 2008. *Water for the Recovery of the Climate – A New Water Paradigm*. Accessed September 11, 2023. www.waterparadigm.org.

Lovejoy, Thomas, and Carlos A. Nobre. 2018. "Amazon Tipping Point." *Science Advances* 4: eaat2340. https://doi.org/10.1126/sciadv.aat2340.

Makarieva, Anastassia M., Andrei V. Nefiodov, Antonio Donato Nobre, Douglas Sheil, Paulo Nobre, Jan Pokorný, Petra Hesslerová, and Bai-Lian Li. 2022. "Vegetation Impact on Atmospheric Moisture Transport Under Increasing Land-Ocean Temperature Contrasts." *Heliyon* 8, e11173. https://doi.org/10.1016/j.heliyon.2022.e11173.

Makarieva, Anastassia M., and Victor G. Gorshkov. 2007. "Biotic Pump of Atmospheric Moisture as Driver of the Hydrological Cycle on Land." *Hydrology Earth System Sciences* 11: 1013–33. https://doi.org/10.5194/hess-11-1013-2007.

Makarieva, Anastassia M., and Victor G. Gorshkov. 2010. "The Biotic Pump: Condensation, Atmospheric Dynamics and Climate." *International Journal of Water* 5 (4): 365–85. https://doi.org/10.1504/IJW.2010.038729.

Meesters, Antoon A. J. Dolman, and L. A. Bruijnzeel. 2009. "Comment on 'Biotic Pump of Atmospheric Moisture as Driver of the Hydrological Cycle on Land' by Anastassia Makarieva and Victor G. Gorshkov, Hydrol. Earth Syst. Sci., 11: 1013–33, 2007." *Hydrology and Earth System Sciences Discussions* 13 (7): 401–16. https://doi.org/10.5194/hess-13-1299-2009.

Melack, John M. 1979. "Temporal Variability of Phytoplankton in Tropical Lakes." *Oecologia* 44: 1–7 https://doi.org/10.1007/BF00346388.

Millán, Millán M. 2014. "Extreme Hydrometeorological Events and Climate Change Oredictions in Europe." *Journal of Hydrology* 518: 206–24. https://doi.org/10.1016/j.jhydrol.2013.12.041.

Morris, Cindy E., Franz Conen, J. Alex Huffman, Vaughan Phillips, Ulrich Pöschl, and David C. Sands. 2014. "Bioprecipitation: A Feedback Cycle Linking Earth History, Ecosystem Dynamics and Land Use through Biological Ice Nucleators in the Atmosphere." *Global Change Biology* 20 (2): 341–51. https://doi.org/10.1111/gcb.12447.

NBSI. 2013. *Forest and Water.* Documentary filmed in the Naivasha Basin as part of a local filming workshop conducted by the Naivasha Basin Sustainability Initiative.

Nicholson, Sharon E. 2017. "Climate and Climatic Variability of Rainfall Over Eastern Africa." *Reviews of Geophysics* 55 (3): 590–635. https://doi.org/10.1002/2016RG000544.

Nicholson, Sharon E. 2019. "A Review of Climate Dynamics and Climate Variability in Eastern Africa." In *The Limnology, Climatology and Paleoclimatology of the East African Lakes,* edited by Thomas C. Johnson, and Eric O. Odada, 25–56. London: Routledge. https://doi.org/10.1201/9780203748978-2.

Odongo, Vincent O., Christiaan van der Tol, Pieter R. van Oel, Frank M. Meins, Robert Becht, Japheth O. Onyando, and Zhongbo Su. 2015. "Characterisation of Hydroclimatological Trends and Variability in the Lake Naivasha Basin, Kenya". *Hydrological Processes* 29 (15): 3276–93. https://doi.org/10.1002/hyp.10443.

Odongo, Vincent O., Dawit W. Mulatu, Francis K. Muthoni, Pieter R. van Oel, Frank M. Meins, Christiaan van der Tol, Andrew K. Skidmore, et al. 2014. "Coupling Socio-economic Factors and Eco-hydrological Processes Using a Cascade-modeling Approach." *Journal of Hydrology* 518 (October): 49–59. https://doi.org/10.1016/j.jhydrol.2014.01.012.

Oksanen, Jari F., Guillaume Blanchet, Michael Friendly, Roeland Kindt, Pierre Legendre, Dan McGlinn, Peter R. Minchin, et al. 2016. Vegan: Community Ecology Package. *R Package Version* 2 (5–6). Accessed September 11, 2023. https://CRAN.R-project.org/package=vegan.

Pokorný, Jan, and Hanna Huryna. 2016a. "The Role of Water and Vegetation in the Distribution of Solar Energy and Local Climate: A Review." *Folia Geobotanica* 51: 191–208. https://doi.org/10.1007/s12224-016-9261-0.

Pokorný, Jan, and Hanna Huryna. 2016b. "Weather, Climate and Wetlands: Understanding the Terms and Definitions." In *The Wetland Book,* edited by C. Max Finlayson, Mark Everard, Kenneth Irvine, Robert J. McInnes, Beth A. Middleton, Anne A. van Dam, and Nick C. Davidson, 1–5. Dordrecht: Springer. https://doi.org/10.1007/978-94-007-6172-8_355-1.

Richardson, Jonathan L., and A. E. Richardson. 1972. "History of an African Rift Lake and Its Climatic Implications." *Ecological Monographs* 42 (4): 499–536.

Saemundsson, Kristjan. 2010. "East African Rift System – an Overview." Presented at Short Course V on Exploration for Geothermal Resources, organised by UNU-GTP, GDC and KenGen, at Lake Naivasha, Kenya. Kenya, Oct. 29–Nov. 19. Accessed September 11, 2023. https://orkustofnun.is/gogn/unu-gtp-sc/UNU-GTP-SC-11-18.pdf.

Salati, Eneas, A. Dall 'Ollio, Eiichi Matsui, and Joel R. Gat. 1979. "Recycling of Water in the Amazon, Brazil: An Isotopic Study." *Water Resources Research* 15: 1250–58.

Savenije, Hubert H. G. 1995. "New Definitions for Moisture Recycling and the Relationship with Land-use Change in the Sahel." *Journal of Hydrology* 167: 57–78.

Schwartz, Judith D. 2013. "Clearing Forests May Transform Local—and Global—Climate". *Scientific American* (website), March 4. https://www.scientificamerican.com/article.cfm?id=clearing-forests-may-transform-local-and-global-climate.

Shivoga, William A., Mucai Muchiri, Samuel Kibichi, Jethro Odanga, Scott N. Miller, Tracy J. Baldyga, Eric M. Enanga, and Maina C. Gichaba. 2007. "Influences of Land Use/Cover on Water Quality in the Upper and Middle Reaches of River Njoro, Kenya." *Lakes and Reservoirs: Research and Management* 12 (2): 97–105. https://doi.org/10.1111/j.1440-1770.2007.00325.

Silale, Keit. 2022. "Poverty-Fueled Deforestation Threatens Kenya's Largest Water Catchment". *Mongabay* (website). August 22. https://news.mongabay.com/2022/08/poverty-fueled-deforestation-threatens-kenyas-largest-water-catchment/.

Strohbach, Michael, Volker Audorff, and Carl Beierkuhnlein. 2009. "Drivers of Plant Species Composition in Siliceous Spring Ecosystems: Groundwater Chemistry, Catchment Traits or Spatial Factors?" *Journal of Limnology* 68 (2): 375–84. https://doi.org/10.3274/JL09-68-2-20.

USAID. 2018. "Climate Risk Profile: Kenya." July 27, 2018. www.climatelinks.org/sites/default/files/asset/document/2018_USAID-ATLAS-Project_Climate-Risk-Profile-Kenya.pdf.

Van der Ent, Rudi J , Hubert H.G. Savenije, Bettina Schaefli, and Susan C. Steele-Dunne. 2010. "Origin and Fate of Atmospheric Moisture Over Continents." *Water Resources Research* 46 (9): 1–12. https://doi.org/10.1029/2010WR009127.

Verschuren, Dirk. 2001. "Reconstructing Fluctuations of a Shallow East African Lake during the Past 1800 Years from Sediment Stratigraphy in a Submerged Crater Basin." *Journal of Paleolimnology* 25: 297–311. https://doi.org/10.1023/A:1011150300252.

Wanjala, Silas W., Timothy Mwinami, David M. Harper, Ed H.J. Morrison, and Nic Pacini. 2018. "Ecohydrological Tools for the Preservation and Enhancement of Ecosystem Services in the Naivasha Basin, Kenya." *Ecohydrology and Hydrobiology* 18 (2): 155–73. https://doi.org/10.1016/j.ecohyd.2017.09.004.

Welcomme, Robin L. 1976. "Some General and Theoretical Considerations on the Fish Yield of African Rivers". *Journal of Fish Biology* 8: 351–64.

PART 3

Agricultural Intensification in and around Naivasha

CHAPTER 7

The Cut-Flower Industry in the Social-Ecological System of Lake Naivasha: Production Networks and Marketisation

Detlef Müller-Mahn and Andreas Gemählich

The Lake Naivasha area has often been portrayed as an extremely dynamic social-ecological system, an epitome of complex societal relations with nature, and an example of environmental problems caused by economic globalisation.[1] Numerous studies, including some chapters in this book, attribute the detrimental changes in the environment of the lake and its riparian land to the concentration of huge cut-flower farms (see e.g. Harper et al., 2011; Becht 2007; Calas 2013; Kimani, Mwangi, and Gichuki 2012; Mekonnen, Hoekstra, and Becht 2012). Roses play a conspicuous role in the social-ecological transformation of the Lake Naivasha area, as tons of mostly short-stemmed flowers are shipped from here to the distribution points in the Netherlands and neighbouring countries every day (Styles 2019; Kuiper 2019; Gemählich 2022). Since the beginning of the new millennium concerns have increased that the growing demand for cheap cut flowers in Europe may cause further degradation or even the collapse of the sensitive lake ecosystem in the Kenyan Rift Valley. Western media aggravated these concerns through reports about the presumably disastrous consequences of flower production, culminating in an apocalyptic narrative of exploitation and local misery that caught a lot of attention among European consumers (Gemählich 2022, 176ff.). In contrast to this popular narrative, we argue in this chapter that the social ecology of Naivasha requires a more comprehensive view of cross-scalar connections and their impact on the flower industry. We propose to complement the analysis of social-ecological transformation with a study of the economic linkages and interaction between Naivasha and the wider world, or in other words, between horizontal and vertical relations of the flower industry.

The empirical fieldwork for this chapter was part of a PhD project in the years 2014 to 2017 (Gemählich 2022), together with repeated visits of both authors to the area in subsequent years, the last one in September 2022.

1 First submission date: September 30, 2022. Acceptance date: March 13, 2023.

 | DOI:10.1163/9789004695429_008

Fieldwork employed a mixed-methods approach, building primarily on expert interviews with stakeholders in Naivasha, Nairobi, and the Netherlands.

The chapter consists of three parts. First, we present Naivasha as a distinct geographical *place* and discuss how *"place-making"* happens at the intersection of horizontal and vertical relations. "Horizontal" relations refer to connections or mutual effects within the local social-ecological system, and "vertical" relations to connections across scales or hierarchies between the local system and the wider world. Secondly, we scrutinise the ongoing change in vertical relations by referring to economisation and marketisation processes. Thirdly, we focus on water governance and corporate social responsibility as examples of the effects of marketisation and growing consumer awareness.

1 Naivasha's Place at the Intersection of Horizontal and Vertical Relations

The geographical place of Naivasha may be approached from different angles. On the one hand, the place may be seen as a territory within distinct boundaries, for example a hydrological catchment area, an ecosystem, or an administrative unit. On the other hand, the place of Naivasha may also be conceived as a knot in a network, or a hub that is constituted primarily through its outward relations. While the first view relates to connections and coupling mechanisms within the local social-ecological system of the lake area, the second one highlights the vertical cross-scalar relations of Naivasha as a production site embedded in globally organised economic networks. Important for our argument is the proposition that horizontal as well as vertical relations contribute to the specific "place-making" of Naivasha. This raises the question of how horizontal and vertical relations intersect and influence each other.

Some of the processes that contribute to shaping Naivasha's place already have a long history. Over the past five decades the riparian land around the lake developed into the most intensively cultivated area in the Kenyan Rift Valley. The spatial concentration of agro-industrial enterprises is not coincidental, as Naivasha Town lies only a two-hour drive away from Nairobi and its international airport, making the place particularly attractive for export-oriented high-value horticultural production. Since the early 1970s, the lake area developed into an agro-industrial hub with a specialisation in cut flowers, primarily short-stem roses (Kuiper 2019; Gemählich 2022). The unique location of the lake area not only allows for a quick road connection to the airport, but also provides excellent conditions for a highly specialised form of floriculture due to a maximum amount of sunshine, moderate temperatures around the AQ 1year, access to fresh water, and cheap labour. Under these conditions, export-oriented flower farming instigated an extremely rapid land-use change – with problematic consequences for the environment, however. Many studies have investigated the problematic environmental impact of flower farming, as documented inter alia by the publications of Lowery and Mendes (1977), Åse (1987), Verschuren et al. (2000), Becht and Harper (2002), Everard and Harper (2002), Everard et al. (2002), Otiang'a-Owiti and Oswe (2010), and Awange et al. (2013). With a background in environmental studies and ecosystem approaches, these publications shed light on the local impact of flower farming and help us to understand the intricate relationship between the massive growth of the agro-industrial sector and changes in the local ecosystem (see also Harper et al., this volume, and Pacini, Giberti, and Harper, this volume).

Not coincidentally, the understanding of Naivasha as a defined territorial unit is also characteristic of the agencies responsible for resource management and environmental governance. Most stakeholders in the local flower industry and the concerned national administration align with ecological and hydrological conceptualisations of Naivasha as a distinct place. For example, the government organisation for water regulation, WRMA, builds upon the *Lake Naivasha Catchment Area Protection Order* of 2011, which is confined in its authority to the boundary of the catchment area. However, the spatial delineation of Naivasha becomes problematic when the boundaries of various definitions do not exactly overlap. For example, ecocentric and hydrological definitions do not fully coincide with administrative boundaries. This may lead to confusion over responsibilities, as the following quote from an expert interview with the coordinator of an environmental project indicates.

> Lake Naivasha is in the county of Nakuru, but the source of water, the surface water of Lake Naivasha, comes from the county of Nyandarua. So, if Nyandarua decides to do some things, it will affect Nakuru and if Nakuru and Nyandarua are not talking to each other, (…) those are the kinds of things, those are the complexities.
>
> ENVIRONMENTAL PROJECT COORDINATOR, 2015

In contrast to the ecology-centred view described above, economic geography is more interested in the vertical relations that link the particular locality to the wider world. In that sense, Naivasha can be seen as an agro-industrial "hub", a "node" (Hopkins and Wallerstein 1986, 160) or a "box" (Gereffi, Korzeniewicz, and Korzeniewicz 1994, 2) within commodity or value chains, including also the institutional setting of production centres and trade relations (Dannenberg and Revilla Diez 2016, 169). In the case of Naivasha, the importance of these structural arrangements is obvious. The flower industry with its roughly 2,000 hectares of greenhouses is by far the biggest employer in the region, the major income generator, and the biggest water abstractor (Gemählich 2022).

The point to be made here is that horizontal as well as vertical relations are relevant for the place-making of Naivasha. Ecological and economic perspectives, together with the study of horizontal and vertical relations, complement each other in the explanation of the extremely rapid land-use change in the area. To embrace this complementarity, one may refer to the concept of a "global sense of place" proposed by Massey (1991) and Amin (2002). According to this concept, Naivasha is positioned within a relational network produced by linkages between economic and ecological spheres across scales. Such an integrative perspective embraces the horizontal relations of coupled social-ecological systems, and their vertical cross-scalar linkages. In this regard, the cut-flower industry presents a focal point within the specific actor-network of Lake Naivasha, the node that connects the horizontal and vertical relations (Murdoch 1998, 361; Law and Mol 2001; Bosco 2006).

In the following section we describe how the vertical relations between Naivasha and the European market have changed in recent years and continue to drive the transformation of the flower industry in the lake area.

2 Economic-Geography Perspectives: Marketisation and Networks

Economic geography addresses the question of how things become economic, or more precisely, how markets come into being. To this end, it investigates marketisation processes and the emergence of new market orders, focussing on new ways of economising things and creating relationships between producers and consumers. The producer-consumer relationship is crucial for the understanding of cut-flower marketisation and the reconfiguration of the industry in Naivasha, as will be shown below.

Two broad conceptual approaches provide explanations for the economisation of things and the emergence of markets, one focussing on chains, and the other (and more recent one) on networks Revilla Diez, Hulke, and Kavelage 2023). As for the first, the concept of Global Commodity Chains (GCC) addresses multi-scalar linkages in a globalised economy by reconstructing the path or "chain" along which a specific good passes on its way from the producer to the consumer. According to this concept, the analysis focusses on the input-output structure, the territorial dimension, and most prominently the power structure that keeps the chain together. Depending on the end of the chain that is considered to be more powerful, one distinguishes between producer- and buyer-driven commodity chains (Gereffi 1994).

The second concept is that of Global Production Networks (GPN), which frames economic relations more broadly by viewing them in connection with other factors and networks. GPN are defined as "an organisational arrangement, comprising interconnected economic and non-economic actors coordinated by a global lead firm, and producing goods or services across multiple geographical locations for worldwide markets" (Coe and Yeung 2022, 1–2). In the following, we concentrate on network approaches, without discarding the relevance of commodity chains. We assume that the focus on networks is more appropriate to explain the ongoing processes of marketisation in the flower industry.

Marketisation refers to the creation of markets for a particular product such as flowers. Regarding the cut-flower industry of Naivasha, economic geography and its specific perspectives on GCC and GPN formation help to explain the dynamics of the local agro-industrial hub and the organisation of flower production in relation to the global cut-flower market. These perspectives focus on marketisation, economisation processes, and in a very general sense, the inclusion of all aspects of life in the economy. In this context it is important to note that marketisation in the cut-flower industry is not confined to the dynamics of economic relations, but relates also to the spheres of ecology and society. Capturing these phenomena requires us "to connect the processes of production, distribution and consumption to the natural environment in which they are fundamentally grounded" (Coe, Dicken, and Hess 2008, 278).

Marketisation in the flower industry of Naivasha consists of three dimensions, namely farm expansion, technical innovation, and the regulation of cut-flower production through standards. Among these three dimensions, the spatial expansion of flower farms is the immediately visible result of international trade and the new market order (Gemählich 2022). It goes along with competition over land ownership, rising land prices, and an increased economisation of land. The control over land has been a constant source of conflict in Naivasha since colonial times. This is due to increased pressure, but also to contested land titles, coerced leases and a fuzzy delineation of the riparian zone because of lake-level fluctuations (Lang and Sakdapolrak 2014, 189; Harper et al., this volume). Flower farms play a key role in these disputes, because they require large areas, are enclosed by fences, and guarded by security services. Furthermore, they have a direct effect on all other land-use groups in the area, because they deny pastoralists, smallholders and fishermen access to the lake. Land and water conflicts are further exacerbated by other recent developments. A growing number of hotels along South Lake Road, newly established conservancies, and the creation of Hell's Gate NP just south of the lake occupy territories that are no longer available for the previous forms of land use like pastoralism and small-scale cultivation. New markets for cheap flowers established by European retail chains encouraged many farms to expand their greenhouses, resulting in severe competition over land and water. While the flower farms are struggling with the shrinking availability of affordable land, especially at South Lake, other stakeholders and especially the pastoralists are seriously affected by the inaccessibility of the lake and riparian land. As a consequence, the riparian zone has become a hotspot of land conflicts. Legislation alone does not really help to curb the conflicts, because the law has relatively little to do with everyday practice:

> Cultivation is illegal, building permanent structures is illegal, logging is illegal. Also, under the Fisheries Act, poaching is illegal, under the Wildlife Act, poaching is illegal.
>
> MONITORING OFFICER, ENVIRONMENTAL NGO, 2015

As the quote indicates, many activities that contribute to the multitude of conflicts around Lake Naivasha are forbidden by law, but this is often not enforced. To make things worse, various user groups have competing claims to the use of this zone, which mostly remain unresolved. Pastoralists, for example, demand access to the lake via public corridors. The Kenya Wildlife Service demands free access to the lake via wildlife corridors, while at the same time it welcomes the protective effect of fences around the riparian zone. At the same time, the expansion of flower farms intensifies this competition, as the riparian zone is the only option for further growth, especially at the south side of the lake. The expansion of farmland has already led to direct confrontations with local pastoralists, when farms have demanded the closure of public corridors and the eviction of pastoralists.

It is interesting to note that these problematic social consequences of flower farming were closely observed and reported by international media and NGOs. In recent years, critical media reports informed European consumers about the conditions under which the "cheap" flowers were produced in Kenya, including land conflicts, the marginalisation of pastoralists, and other social consequences of the flower industry in Naivasha (Gemählich 2022). The rising awareness of these problems among European consumers was in turn observed by marketing agents and farm managers, who were anxious to maintain the reputation of their industry. They therefore took action to mitigate some of the immediate negative impacts, for example by establishing social standards and a new code of conduct to prohibit cultivation in the riparian zone of the lake. This response of the flower industry to the growing consumer awareness induced a number of positive changes, as the following statement indicates:

> [S]ome of these issues have been fading away, because you find some flower farms agreeing with a group of pastoralists to graze in their farms. Some have been allowing them to use their roads to access the lake; some have been giving water to their livestock.
>
> REPRESENTATIVE, PASTORALIST OUTREACH SERVICE, 2015

3 The Effect of Marketisation on Water Governance and Corporate Social Responsibility

Marketisation is a key driver of change in the relations between producers and consumers. In the following, we examine how this affects the local conditions in Naivasha in regard to water governance and corporate social responsibility.

The careful management of scarce water resources is a prerequisite not only for the sustainability of large-scale horticultural production, but also for the reputation of the Kenyan flower industry in the Global North. This issue became critical in more recent years, because the expansion of flower farms in Naivasha went along with an increase of water abstraction from the lake. Roses from Naivasha have a water footprint of seven to 13 litres per stem (Mekonnen, Hoekstra, and Becht 2012), making the flower industry the single biggest abstractor of fresh water in the area. When the lake's water table declined in several consecutive years shortly after the millennium, the situation was widely reported in international media, to the extent that consumers in Europe became increasingly concerned. In response to the problems of water scarcity, and also to the scrutiny of critical consumers, many farms introduced new technologies like drip irrigation, hydroponics, water recycling, and other water-saving methods. More recently, the water table started to rise again. In 2022, water abstraction from the lake ceased to be a problem, but as technological innovation and responsive action had already been well established, the technological innovations remained in place and became a pillar of the new market order.

The interplay of marketisation and economisation of resources becomes visible in the institutions regulating water abstraction and discharge. Due to their role as commercial water abstractors, flower farms are automatically part of the Lake Naivasha Water Resource Users Association (LANAWRUA), an organisation for stakeholder representation and inclusion in water-resource management. Flower farms are active members in these organisations and mostly well respected by the other actors, not least because of their financial contributions (Gemählich 2022). In contrast to their officially proclaimed critical position towards flower farms (see e.g. Food & Water Watch 2008), environmental NGOs and other stakeholders acknowledge that farms do also act as facilitators for sustainable development at the local level. For example, flower farms support local organisations informally by allowing them access to their premises, providing boats for ecosystem monitoring, facilitating educational activities, or just providing food and equipment for community events. Thus, the flower industry is not only the biggest employer, but also the biggest single donor in the region for all sorts of projects, often with substantial contributions and professional support. Quite remarkable in this context is the fact that the biggest donations do not stem from local farms, but from European retail chains, once again a sign of the enormous importance of cross-scalar or "vertical" relations.

Governmental regulations of water use are generally based on the *Kenya Water Act*, which introduced water permits and block-rate water pricing in 2002 (Kuhn et al. 2016, 377). The regulations make the regional WRMA office the key authority of water-use regulation. Its tasks comprise the issuing and control of permits, the ensuring of proper water allocation, quantity and quality, and the collection of fees. Regulations of this sort are particularly important in the context of GCC and GPN formation, because they "shape the environment for the embedded companies and are key determinants for economic success" (Dannenberg and Revilla Diez 2016, 170). However, GCC and GPN perspectives have so far paid little attention to the relationship between the emergence and transformation of the institutional framework in the context of global economic linkages. Concerning the flower industry at Lake Naivasha, water is not only a vital resource of the global market, but its regulation has also become entangled in the new market order.

The rules and regulations of water use do not exactly reflect the reality of water abstraction, which explains why the specific linkages between institutions and marketisation processes are so important. This is partly due to the fact that the official water-management framework remains highly inefficient, and its implementation is lagging behind. Another reason is the WRMA's incapability to provide accurate data, which weakens its authority among water users (Verstoep 2015).

> [I]f it is a flower farm or another industry or whatever; there is need for proper water allocation and regulation. If WRMA would function correctly, then there would not be any issue because it is clear and they

> follow up correctly on how much water we have and who is taking water out according to a legal permit.
>
> PROJECT COORDINATOR, DEVELOPMENT AGENCY, 2015

These institutional weaknesses make it easy for water users to evade restrictions and fines. For instance, data from a WRMA abstraction survey indicate that water users avoid the restriction of water allocation in the case of shortages. The restrictions refer to the official permit issued to water abstractors. In the case of a low water table in the lake, abstractors are only allowed to use part of the allotted amount. Many farms cope with these regulations by acquiring permits well above the actually required amount of water. Therefore, more than one third of water users abstract less than their permit allows them. In 2016 the abstraction at Lake Naivasha was 44% lower than the authorised amount. This is also a consequence of the low financial impact of costs for permits, abstractions and fines (cf. Kuhn et al. 2016). The fee for commercial irrigation water is KSh 0.75[2] per cubic metre and typically accounts for less than 0.1% of the production cost of a flower farm.

Besides regulating water abstraction and management, the Kenya Water Act of 2002 also instigated the creation of new organisations for water distribution, conflict resolution and service provision – the Water Resource Users Associations (WRUAs). The WRUAs have a double role in the institutional framework of water governance by providing a platform for stakeholder participation and at the same time controlling compliance with the rules set by WRMA. The Lake Naivasha WRUA (LANAWRUA) takes a pioneering role in this double purpose of merging stakeholder participation and rule enforcement. As its internal organisation is not clearly regulated, the LANAWRUA and its members agreed upon a model of stakeholder representation, as its chairman points out:

> Actually, we are the pioneers; we are the guys who were constructed first by WRMA (…). So, (…) we thought of how to move this forward. So, we came up with the idea of categories; pastoralists, individual people with private shareholders, flower farms and large-scale farmers, we have tour operators that are the hotels, and we have (…) the water companies. (…) We categorised ourselves in those groups and invited each category to a wider meeting. There were some trainings on the roles of LANAWRUA, benefits to members and all that. After everybody knew all that, we asked each category to go back to their people and tell what they heard

2 At January 1, 2014, KSh 1 was equivalent to $0,0114 (source for the exchange rate: https://www.oanda.com/currency-converter/en/).

> and after 30 days, they brought along two representatives who sit on the LANAWRUA committee.
>
> CHAIRMAN, LANAWRUA, 2015

Quite remarkably, the WRUA is not only enforcing regulation, but also lobbying in the interest of its members:

> The main interest is that we are able to influence some of the issues affecting us as an association. We are able to lobby some decisions on how we want things done here as an association. We are able to advocate when it comes to issues of what the law is saying. We advocate and force WRMA to enforce on issues of the rules of water resources.
>
> CHAIRMAN, LANAWRUA, 2015

The organisational structure of the WRUAs is clearly to the advantage of the flower industry, since it is better networked and organised than the other stakeholder groups. This allows the farm managers to play a dominant role in the LANAWRUA. The cut-flower industry actively influences the governance of water resources through the relevant institutions. In this way it responds to a problem that many scholars and development agencies see as one of the root causes of environmental degradation in the Lake Naivasha area, namely the inefficiency of government institutions to control the situation (see e.g. Verstoep 2015). The example of the Water Allocation Plan (WAP) Naivasha Basin (2010) underlines this point. The WAP limits water abstraction for different user groups depending on the availability of water, which is measured according to the lake level. This plan itself states that:

> [the] process of developing the Water Allocation Plan was started in 2005 by the Lake Naivasha Growers Group (LNGG) who engaged consultants and participated in the development of the initial WAP which was submitted to WRMA in 2007. This draft has been exposed to stakeholders and revised to form the current version.

The WAP reflects the cut-flower industry's interests in several ways. Flower growers greatly depend on reliable water supply and sufficient water quality to reduce treatment costs. Due to fluctuating lake levels, farms had to make substantial investments in their water-abstraction systems in recent years. Many drilled boreholes to diversify their water sources, which is a costly investment. In the face of an imminent water crisis, the farms endorsed an integrated approach to which all stakeholders are committed to securing the availability of irrigation water. Furthermore, the flower industry is highly supportive of all measures that help to improve the image of a sustainable environment at Lake Naivasha. The practice of water use and abstraction is explicitly justified in the WAP by pointing at the

> [b]ad publicity for the horticultural and floricultural export industry which may be seen to be over exploiting [*sic*] the water resources to the detriment of other water users; Loss of export markets (for agricultural products) and tourism potential due to the bad publicity associated with economic activities around the lake (…).

Exerting an influence on institutions of water governance has an additional positive side effect for growers. It helps them to facilitate market access, since certification schemes regulating access to European markets demand compliance with regional legislation regarding the use of resources. Not surprisingly, the WAP aims to "make compliance to [*sic*] WAP conditional to market access" and "establish compliance to WAP as a criteria [*sic*] for certification."

The institutional framework for water abstraction including the co-development of institutions such as WAP links up to the new market order in several ways. The increased public attention given to the Lake Naivasha cut-flower industry due to direct sales urges farms to engage in projects propagating a positive image of the industry and its surroundings. The capital-intensive, technically advanced production technology gives the industry an edge compared to other stakeholders such as smallholders or horticultural farms. Hence, the cut-flower industry often appears to be the most compliant stakeholder group at Lake Naivasha, as for instance the LANAWRUA-chairman confirms:

> They are guys who are very cooperative, and I really enjoy working with them because (…) they are very compliant. No issues in their operational areas, no issues with abstraction and pollution. Actually, they are my study case; for any errant abstractor, I try to encourage them to go and see what these guys are doing. So, for me, they are benchmarks since they are really cooperative and make my work easier.
>
> CHAIRMAN, LANAWRUA, 2015

The above examples show that economisation of nature and its transformation into a "resource" goes hand in hand with marketisation. As a consequence, the new market order does not simply lead to an over-exploitation of resources and labour, but it also shapes new institutions and helps to implement improved technologies for more efficient water use, a control of pollution and better labour conditions.

The second area where the influence of the new relations between producers and consumers becomes visible is the growing importance of *corporate social responsibility* (CSR). Social welfare in Naivasha is closely linked to the labour conditions, employment and wages in the agro-industrial hub. So far, the cut-flower industry has contributed to the transformation of the local society in Naivasha in multiple ways even beyond the farms, for example through the provision of basic medical, educational and transport infrastructure. These activities can all be interpreted as expressions of the CSR of the flower industry. In general, corporate social-welfare projects are considered as a form of private governance, either serving as a strategy to enhance the performance of a firm, or simply responding to external pressures exerted by critical media reports or NGO activities (Bair and Palpacuer 2015). Since around 2010, the growing influence of the new market order convinced many farm managers to adopt CSR. Social welfare measures are now increasingly implemented as a response to external pressures. A farm manager describes this new attitude:

> I am here to produce and make a business, not only to help people. (...) I built half the secondary school, but I can't take the whole burden of development on my shoulders.
>
> FARM MANAGER, GROWER, 2014

CSR activities contribute to improvements of labour conditions and social welfare, even though investments are limited by the industry's weak financial capacities. A community worker described the volatility of the engagement of the farms in the CSR projects of his organisation as follows:

> The financial support that can come in is, for instance, if we have an activity and the flower farm will chip in once or twice but neither is it consistent support nor is it reliable support as we may sit, discuss and agree; then later their finance department declines the request.
>
> COMMUNITY WORKER, DEVELOPMENT AGENCY, 2015

The examples of social activities around the flower farms give evidence of the fact that CSR is not simply a philanthropic engagement, but may also be considered as a tool to maintain the hegemonic position of firms within economic relations by absorbing the criticism of external actors, such as NGOs and the media. That means that CSR activities serve also economic purposes by stabilising the workforce, keeping the semi-skilled labourers in their employments, and promoting the social reputation of particular farms. Corporate retailers require farms to fund CSR projects, as a farm manager describes:

> (…) they do require us to have a demonstrable project ongoing which is socially related. So you are putting a school up, or a social or a bus stop. We got big plans, but it depends on the money.
>
> HEAD OF PRODUCTION, GROWER, 2014

Certification schemes help to make economisation even more explicit. Similar to what Brown (2013) observed in a study on Fairtrade banana production in Colombia, certificates and labels fit well into the already existing framework of philanthropic measures and development practices. The labels allow the cut flower industry to "sell" these measures and the positive image created by them to buyers and customers. In this regard, Fairtrade stands out with the payment of a Premium, which is primarily used for social-welfare projects of communities adjacent to the farms. Social-welfare projects are beneficial for the improvement of the local health and educational system:

> I see quite a number of farms doing a lot for the local communities in Naivasha (…). For example, on the issue of water supply, sanitation, health facilities, education sector; they are doing a remarkable job.
>
> REPRESENTATIVE, PASTORALIST OUTREACH SERVICE, 2015

Farm workers generally appreciate projects funded by the Fairtrade Premium and other CSR activities (see also Kuiper and Gemählich 2017, 43), and they do benefit from these new developments. However, one also has to acknowledge that CSR engagement is on the one hand "framed in terms of benevolence and voluntarism rather than as a fundamental responsibility" (Brown 2013, 2581), but on the other hand it is also part of the industry's competitive strategy. To put it in the words of a local board committee of a Humanitarian NGO:

> Social responsibility: to cover their faces and throw a carrot and tell you, 'you can eat on that while we are busy having buffet of all sorts of meals'.
>
> BOARD COMMITTEE MEMBER, HUMANITARIAN NGO, 2015

4 Conclusion

The chapter took as its point of departure debates about Naivasha's place at the intersection between horizontal and vertical network relations. It focussed on the cross-scalar ("vertical") economic relations that connect the cut-flower industry of Lake Naivasha to global markets, especially in Europe. The article addressed the question how cross-scalar economic relations influence the transformation of the local social-ecological system of the Lake Naivasha area through value chains, production networks, and marketisation.

The chapter argued that the local effects of cross-scalar relations have so far been rather ambiguous, but should not only be reduced to the apocalyptic narratives presented by some media reports. On the one hand, the growing demand for short-stemmed roses in Europe led to a massive expansion of flower farming in the Kenyan Rift Valley, with ensuing harmful effects on the local environment and renewed conflicts between different user groups. On the other hand, European consumers are becoming increasingly aware of the problems caused by flower production in Naivasha, with negative effects for the reputation of cut flowers. These customer concerns alarmed farm managers and trade organisations, who then made major efforts to mitigate potentially harmful effects of flower farming. In a similar vein the intricate processes of marketisation contributed to the creation of new agencies and governance structures in Naivasha.

The results of these recent endeavours of the flower industry to improve the conditions in Naivasha become visible in two fields. First, sustainable water governance is generally supported by all stakeholders including the flower farms. Water abstraction and the regulation of water use are now centrally controlled to guarantee a sustainable and fair distribution of water. These measures are admittedly not sufficient to improve the situation of pastoralists and smallholder farmers, but they do help to curb environmental harm. Secondly, the new market order and efforts toward greater transparency and traceability in cut-flower value chains turn the public attention on the conditions of flower production into a business risk for flower growers. The cut-flower industry therefore actively engages in gaining control over these discourses through new engagements in corporate social responsibility and the implementation of social-welfare projects.

The chapter concludes that a comprehensive understanding of the dynamic changes in the flower industry as well as in the social-ecological system of Lake Naivasha requires a broader, cross-scalar perspective that addresses the role of this very peculiar agro-industrial hub in relation to global marketisation processes.

Acknowledgements

The chapter presents findings of a project that was part of the research group *Resilience, Collapse and Reorganisation of Social-Ecological Systems in African Savannahs*. The project investigated the relevance of cross-scalar linkages for local social-ecological systems and their transformation. It was funded by the German Research Foundation DFG. The authors wish to express their gratitude for the financial support.

References

Amin, Ash. 2002. "'Spatialities of Globalisation." *Environment and Planning A* 34 (3): 385–99. https://doi.org/10.1068/a3439.

Åse, Lars-Erik. 1987. "A Note on the Water Budget of Lake Naivasha, Kenya. Especially the Role of Salvinia Molesta Mitch and Cyperus Papyrus L." *Geografiska Annaler. Series A, Physical Geography* 69 (3/4): 415–29. https://doi.org/10.2307/521355.

Awange, Joseph, Ehsan Forootan, Jürgen Kusche, John Kiema, P. Omondiernhard Heck, Kevin Fleming, S.O. Ohanya, and Rodrigo Gonçalves. 2013. "Understanding the Decline of Water Storage Across the Ramser-Lake Naivasha using Satellite-Based Methods." *Advances in Water Resources* 60: 7–23. https://doi.org/10.1016/j.advwatres.2013.07.002.

Bair, Jennifer, and Florence Palpacuer. 2015. "CSR Beyond the Corporation: Contested Governance in Global Value Chains." *Global Networks* 15: 1–19. https://doi.org/10.1111/glob.12085.

Becht, Robert. 2007. "Environmental Effects of the Floricultural Industry on the Lake Naivasha Basin." Enschede: ITC International Institute for Geo-Information Science and Earth Observation. Accessed July 27, 2023. https://ftp.itc.nl/pub/naivasha/ITC/Becht2007.pdf.

Becht, Robert, and David Harper. 2002. "Towards an Understanding of Human Impact Upon the Hydrology of Lake Naivasha, Kenya." *Hydrobiologia* 488: 1–11. https://doi.org/10.1023/A:1023318007715.

Bosco, Fernando. 2006. "Actor-Network Theory, Networks, and Relational Approaches in Human Geography." In *Approaches to Human Geography*, edited by Stuart Aitken and Gill Valentine, 136–46. SAGE. https://doi.org/10.4135/9781446215432.

Brown, Sandy. 2013. "One Hundred Years of Labor Control: Violence, Militancy, and the Fairtrade Banana Commodity Chain in Colombia." *Environment and Planning A: Economy and Space* 4 (11): 2572–91. https://doi.org/10.1068/a45691.

Calas, Bernard. 2013. "Mondialisation, clusterisation et recyclage colonial. Naivasha, laboratoire d'une Afrique émergente." *EchoGéo* 26: 1–25. https://doi.org/10.4000/echogeo.13586.

Coe, Neil. M., and Henry Wai-Chung Yeung. 2022. *"Global Production Networks: Theorizing Economic Development in an Interconnected World."* Oxford: Oxford University Press. https://doi.org/10.1093/acprof:oso/9780198703907.001.0001.

Coe, Neil, Peter Dicken and Martin Hess. 2008. “Global Production Networks: Realizing the Potential.” *Journal of Economic Geography* 8 (3): 271–95. https://doi.org/10.1093/jeg/lbn002.

Dannenberg, Peter, and Javier Revilla Diez. 2016. “Editorial: Socio-Economic Networks and Value Chains in the Global South — An Institutional Perspective.” *DIE ERDE–Journal of the Geographical Society of Berlin* 147: 169–72. https://doi.org/10.12854/erde-147-12.

Everard, Mark, and David Harper. 2002. “Towards the Sustainability of the Lake Naivasha Ramsar Site and its Catchment.” *Hydrobiologia* 488: 191–203. https://doi.org/10.1023/A:1023390430571.

Everard, Mark, Jacqueline Vale, David Harper, and Håkan Tarras-Wahlberg. 2002. “The Physical Attributes of the Lake Naivasha Catchment Rivers.” *Hydrobiologia* 488: 13–25. https://doi.org/10.1023/A:1023349724553.

Food & Water Watch. 2023. “Lake Naivasha. Withering under the Assault of International Flower Vendors.” Washington, Ottawa: The Council of Canadians. Accessed July 27, 2023. http://www.foodandwatereurope.org/wp-content/uploads/2009/11/FoodandWaterEuropeLakeNavaisha.pdf.

Gereffi, Gary. 1994. “The Organization of Buyer-Driven Global Commodity Chains: How U.S. Retailers Shape Overseas Production Networks.” In *Commodity Chains and Global Capitalism*, edited by Gary Gereffi and Miguel Korzeniewicz, 95–122. Westport: Praeger Publishers.

Gereffi, Gary, Miguel Korzeniewicz, and Roberto Korzeniewicz. 1994. „Introduction: Global Commodity Chains.” In *Commodity Chains and Global Capitalism*, edited by Gary Gereffi and Miguel Korzeniewicz, 1–14. Westport: Praeger Publishers.

Gemählich, Andreas. 2022. *The Kenyan Cut Flower Industry and Global Market Dynamics.* Woodbridge: Boydell & Brewer.

Harper, David M., Edward H.J. Morrison, Michael M. Macharia, Kenneth M. Mavuti, and Caroline Upton. 2011. “Lake Naivasha, Kenya: Ecology, Society and Future.” *Freshwater Reviews* 4 (2): 89–114. http://www.bioone.org/doi/full/10.1608/FRJ-4.2.149.

Hopkins, Terence K. and Immanuel Wallerstein. 1986. “Commodity Chains in the World-Economy Prior to 1800.” *Review: A Journal of the Fernand Braudel Center for the Study of Economies, Historical Systems and Civilizations* 10 (1): 157–70.

Kimani, Rosemary W., Benson M. Mwangi, and Cecilia M. Gichuki. 2012. “Treatment of Flower Farm Wastewater Effluents Using Constructed Wetlands in Lake Naivasha, Kenya.” *Indian Journal of Science and Technology* 5 (1): 1870–88. https://doi.org/10.17485/ijst/2012/v5i1.8.

Kuhn, Arnim, Wolfgang Britz, Daniel Kyalo Willy, and Pieter van Oel. 2016. “Simulating the Viability of Water Institutions under Volatile Rainfall Conditions – The Case of the Lake Naivasha Basin.” *Environmental Modelling & Software* 75 (2016): 373–87. https://doi.org/10.1016/j.envsoft.2014.08.021.

Kuiper, Gerda. 2019. *"Agro-Industrial Labour in Kenya. Cut Flower Farms and Migrant Workers' Settlements."* New York: Palgrave Macmillan.

Kuiper, Gerda, and Andreas Gemählich. 2017. "Sustainability and Depoliticisation. Certifications in the Cut-Flower Industry at Lake Naivasha, Kenya." *Africa Spectrum* 52 (3): 31–53. https://doi.org/10.1177/000203971705200302.

Lang, Britta, and Patrick Sakdapolrak. 2014. "Belonging and Recognition after the Post-Election Violence: A Case Study on Labour Migrants in Naivasha, Kenya." *Erdkunde* 68 (3): 185–96. http://www.jstor.org/stable/24365231.

Law, John, and Annemarie Mol. 2001. "Situating Technoscience: An Inquiry into Spatialities." *Environment and Planning D* 19 (5): 609–21. https://doi.org/10.1068/d243t.

Lowery, R. S., and A. J. Mendes. 1977. "Procambarus Clarkii in Lake Naivasha, Kenya, and its Effects on Established and Potential Fisheries." *Aquaculture* 1 (2): 111–21. https://doi.org/10.1016/0044-8486(77)90069-2.

Massey, Doreen. 1991. "A Global Sense of Place." *Marxism Today* (June): 24–29.

Mekonnen, Mesfin Mergia, Arjen Hoekstra, and Robert Becht. 2012. "Mitigating the Water Footprint of Export Cut Flowers from the Lake Naivasha Basin, Kenya." *Water Resources Management* 26 (13): 3725–42. https://doi.org/10.1007/s11269-012-0099-9.

Murdoch, Jonathan. 1998. "The Spaces of Actor-Network Theory." *Geoforum* 29: 357–74.

Otiang'a-Owiti, Oswe Georg, and Abiya Ignatius. 2010. "Human Impact on Lake Ecosystems: The Case of Lake Naivasha, Kenya." *African Journal of Aquatic Science* 32 (1): 79–88. https://doi.org/10.2989/AJAS.2007.32.1.11.148.

Revilla Diez, Javier, Carloin Hulke, and Linus Kalvelage. 2023. "Value Chains and Global Production Networks: Conceptual Considerations and Economic Development in the 'Wild'." In *Conservation, Markets, and the Environment in Southern and Eastern Africa. Commodifying the 'Wild',* edited by Michael Bollig, Selma Lendelvo, Alfons Mosimane, and Romie Nghitevelekwa, 56–78. Woodbridge, Suffolk: James Currey.

Styles, Megan. 2019. *Roses from Kenya. Labor, Environment, and the Global Trade in Cut Flowers.* Seattle: University of Washington Press.

Verschuren, Dirk, Kathleen Laird, and Brian Cumming. 2000. "Rainfall and Drought in Equatorial East Africa during the Past 1,100 Years." *Nature* 403: 410–14. https://doi.org/10.1038/35000179.

Verstoep, J. 2015. "Challenges in Water Management in the Lake Naivasha Basin. Analysis on the Effects and Performance of IWRAP for Different Irrigation Water User Groups in the Lake Naivasha Basin, Kenya." Thesis, Wageningen University. Accessed July 27, 2023. http://edepot.wur.nl/345816.

CHAPTER 8

Labour Relations in the Naivasha Cut-Flower Industry: A Moral Economy Approach

Gerda Kuiper

1 Introduction[1]

Naivasha as a "nerve center" (Styles 2019) is to a large extent defined by the presence of the cut-flower industry. Work opportunities on the mostly foreign-owned flower farms, ranging from unskilled to managerial jobs, attract people from across Kenya and even from abroad. The industry is in constant flux. The flower farms of today look different from the flower farms that were established at the shores of the lake in the mid-1970s. The type of crop has changed, from seasonal flowers that are produced outdoors to the year-round cultivation of roses, predominantly taking place in greenhouses. In more recent years, farms have increasingly diversified their activities, venturing into for instance vegetable farming and tourism (see Kioko et al., this volume). The makeshift camps where workers were housed have mostly been closed, with the workers moving to private rental housing in mushrooming settlements, while a remaining handful of compounds have been transformed into relatively orderly housing estates. In addition to these profound transformations, there is an imminent threat that single farms or even the industry as a whole will move away from Naivasha, either when natural resources such as water and land appear increasingly scarce, or because conditions for production seem more advantageous elsewhere (Becht, Odada, and Higgins 2005; Gemählich 2022; Kuiper 2019; Riisgaard and Gibbon 2014).

Despite the industry's volatility and a highly uncertain future, it has existed for almost five decades now and has unquestionably become the most important pillar of Naivasha's economy. Many of the farms that I encountered during my fieldwork in the mid-2010s had been there for years or even decades, and continue to be present up until today, albeit sometimes under a different name or different ownership. A substantial minority of the managers and workers I met had a decades-long history of staying in Naivasha, even though virtually all

1 First submission date: February 22, 2021. Acceptance date: March 13, 2023.

 | DOI:10.1163/9789004695429_009

of them considered their stay there as temporary. In short, despite its ultimate impermanence, the industry has become one of Naivasha's most stable defining features since its establishment in the 1970s.

In this chapter, I interrogate the particular moral economy that has developed within the Naivasha flower industry. I argue that the moral economy was able to come into existence because of the industry's long-term, overall stable presence in Naivasha. Moreover, this moral economy in its turn contributed to the relatively harmonious and stable labour relations there.

The use of the term "moral economy" might seem odd to anyone familiar with the critical attention the industry has received within global discussions on ethical trade (Dolan, Opondo, and Smith 2002; Kuiper and Gemählich 2017). The term "moral economy" is sometimes used to refer to an "alternative", non-capitalistic production system, based on norms such as fairness and justice (Weiler, Otero, and Wittman 2016). The Naivasha flower industry, with its neoliberal characteristics (Dolan 2007) and stark hierarchies (Kuiper 2019), does not qualify as such. However, I do not aim to evaluate the industry according to certain norms. Instead, I seek to ascertain the content of the different norms and values that motivate the main actors in this industry. In global contestations over labour conditions within the flower industry, the particular values that flower-farm workers – including higher-ranked employees such as supervisors and managers – themselves attach to their work have received little attention (Kuiper and Gemählich 2017). In this chapter, I interrogate these values and the particular moral expectations that workers and managers of the Naivasha flower farms have towards each other (cf. Carrier 2018). I thus use "moral economy" as an analytic lens, in order to gain a more profound understanding of (and not to justify or to criticise) the unequal labour relations within the Naivasha flower industry (cf. Berman and Larin 2016, 8).

After an introduction to the conceptual and methodological background to this chapter, the empirical analysis shows that Naivasha's moral economy is primarily constituted between actors who are considered and consider themselves to be non-locals, and who enter the industry with their own moral values. I focus on three main groups, which are fairly well delineated within farm hierarchies: foreign top managers, Kenyan middle-level managers, and the large base of farm workers, commonly referred to as "general workers" in the industry. Only a minority of farm employees from any of these groups were born in Naivasha. Moreover, only a few envision themselves to be still there at old age. I interrogate the mutual expectations between these three groups of employees and analyse how their expectations are shaped by their different intersectional positions, shaped by for instance class, ethnicity and gender. This chapter therewith, first, contributes to scholarship on the concept of moral economy, which tends to focus on class alone rather than on intersectional positionalities. Second, it highlights the enduring relevance of employees' backgrounds for the remarkable stability of labour relations in the Naivasha flower industry.

2 Moral Economy: An Interactional and Intersectional Approach

Moral economy presents itself as a useful concept for analysing unequal labour relations. The concept has been most influentially used by E.P. Thompson and James Scott in their analysis of class relations and inequality. They described the moral reactions among subaltern groups after economic arrangements have changed. Thompson (1971) used the concept to signal a transition in the English political economic system in the eighteenth century. He showed that food riots occurring at the time were motivated by more than deprivation alone: they provided a moral commentary on the transformation of the economy and on resulting changes in the relation between the elites and the "English crowd". Scott (1976) similarly highlighted that exploitation in itself does not necessarily lead to moral outrage. The Southeast Asian peasants he studied had a "subsistence ethic": they would only resist when technical and social arrangements changed in such a way that they could no longer subsist themselves. As stated by John Lonsdale, "injustice is as much moral as material fact. Observers may measure economic inequality, but cannot assume that this is mirrored in felt moral inequity". More specifically, people "judge any new class inequality by the social value which past experience has warned them to put on their labour, the price of their sweat" (Berman and Lonsdale 1992, 351). Crucially, such moral judgements and interpretations both reflect and (re)constitute unequal relations (Neumark 2019).

Whereas Thompson and Scott focused on the morals of subaltern groups, Palomera and Vetta (2016) pointed out that subalterns' moral notions are formed within a broader social field. That is, even when subaltern groups and elites do not share the same values, they are part of one moral economy.[2] In line with this insight, Carrier (2018, 18) described moral-economic activity as those activities defined by "the mutual obligations that arise when people transact with each other over the course of time". Long-term interactions are "the basis of the expectations that each party to the relationship has about the other party, about the obligations each has toward the other and about the

2 Despite taking on a holistic approach here, I do believe that it is crucial to bring specifically subaltern views to the fore. As pointed out by Neumark (2019, 43), the possibilities to *debate* economic inequality are also unequally distributed.

relationship itself" (Carrier 2018, 25). Specific moral economies are, according to such an understanding, "historical bargains between informed people" (Berman and Lonsdale 1992, 351), implying they are based on interaction and are continuously undergoing change.

Salverda (2020) has pointed out that moral-economic activities do not take place in a vacuum: they are shaped by "global" ideas and norms. Moreover, specific actors who are not located where the transactions take place and who are not directly involved in them exert an influence on the shape of (moral) economic activities. For contemporary agro-industries, regulations as laid down in international codes of conduct are an example of this "extra-local" influence. Although I primarily highlight the interactions and obligations between actors located in Naivasha in this chapter, I also touch upon the role of such "extra-local" actors in shaping moral expectations locally.

In addition to an *interactional* understanding of moral economy, I suggest adopting an *intersectional* approach. The concept of moral economy has often been used to gain a better understanding of class relations. However, class is not the only factor shaping people's moral judgments or their economic possibilities. People have diverse commitments and are part of a "nexus of moral spheres", which extends beyond the workplace, for instance into the household (Kofti 2016, 438). Thus, apart from class, factors such as nationality, gender, religion, and ethnicity further shape mutual expectations within a moral economy.

Although studies on moral economies classically focus on peasant agriculture, the concept has more recently also been fruitfully applied to the analysis of labour relations on plantations and in industries (Besky 2014; Kofti 2016). In such a wage-labour context, expectations are influenced by specific ethics of work – historically-formed judgements about what constitutes proper or "good" work. The different values attached to work can be capitalistic and focused on immediate economic gain, but can also be about, for instance, reputation, providing for dependents, self-fulfilment, and friendly work relations (Monteith and Giesbert 2017; Fisher 2018). An intersectional understanding is also crucial in this regard. Fisher (2018) for instance pointed out that gender influences workers' understanding of what constitutes "good work" as well as their opportunities to shape labour conditions to their advantage.

3 Methodology

This chapter is based on data collected during ethnographic fieldwork on labour arrangements within the Naivasha flower industry. The fieldwork was conducted for a total period of 13 months between 2013 and 2016. I make use of data obtained through interviews with farm managers, farm workers, and other industry stakeholders;[3] from comments made during listing and ranking exercises with low-income Naivasha residents on the different types of jobs that are available to them; and from observations made during extensive visits to flower farms and to workers' settlements and compounds. Particularly fruitful were the regular visits I paid to one specific flower farm ("Karibu Farm") over a period of several months. During these visits, I had the opportunity to conduct interviews with general workers, supervisors, middle-level managers, and the managing director, who was a member of the Dutch family owning the farm.[4] Finally, I searched the digitalised archive of the newspapers that are part of the Nation Media Group in Nairobi, supplemented by a recent search of the website of its main newspaper, the *Daily Nation*.

A common approach in analysing moral economies is to look at moments of conflict and resistance after existing economic arrangements change or break down (Besky 2014, 38). However, I argue that breakdown is not the only moment in which a moral economy shows itself. The contours of an (inherently dynamic) moral economy can also surface from an analysis of periods of relative stability. As noted above, deprivation or hardship in itself does not lead to conflict or dissent. Indeed, whereas the labour relations in the Naivasha flower industry are contested on a global scale, they are remarkably stable and harmonious locally. The analytic question then becomes why certain conditions are considered tolerable, even if they seem exploitative from an outsider perspective. This question can be answered by investigating how the conditions fit into existing moral expectations about reciprocal obligations (Berman and Larin 2016; Palomera and Vetta 2016; Carrier 2018). In this chapter, I therefore interrogate the diverse moral expectations among flower-farm workers and managers, in relation to their understandings of the existing labour conditions and hierarchies within the Naivasha flower industry.

3 Some of these interviews were carried out in cooperation with Andreas Gemählich, and some of the data used for this chapter is also reported on in his monograph published in 2022.

4 I was mostly unaccompanied during the interviews that I carried out on and outside this flower farm. The interview language was either Swahili, English, or Dutch, depending on the interviewee's preferences. See Kuiper (2019, 6–16) for a more elaborate account of the methodology.

4 Moral Expectations and Mutual Obligations in the Naivasha Flower Industry

In the only article on moral economies in the Kenyan flower industry that has appeared so far, Dolan (2007) primarily focused on the motivations of the flowers' consumers, paying less attention to the values of managers and workers. In addition, critical global campaigns and reports on labour conditions in foreign-owned agro-industries in African countries tend to evaluate the conditions based on universal standards. General workers' own understandings of the value of their work and their expectations towards their employers are largely glossed over in such discussions on "ethical work" (McEwan, Hughes, and Bek 2017; Smith and Barrientos 2005). But the expectations and motivations of more powerful actors, such as the foreign top managers, have also received little attention. They have generally been expected to be only profit-driven, and are "villainised" (Styles 2019, 10).

I therefore explicitly discuss managers' views and their expectations of their workers here. That is, I look at how economic decisions in the Naivasha flower industry, for instance around the recruitment of workers, are shaped by (perceptions of) mutual obligations that formed over time between foreign top managers, Kenyan middle-level managers, and general workers. I also briefly sketch the backgrounds of these three groups of workers, as these respective backgrounds have shaped the expectations with which they arrived in Naivasha.

Some overlap and mobility between these groups exists, especially via the position of supervisor, as I explain below. However, levels of income are for instance clearly distinct. These three groups are also socially segregated and reside in different locations. Some general workers are provided with one or two rooms on farm compounds, while most of them reside in private rental rooms, often of poor quality, in dense settlements (see Map 1.1). Middle-level managers tend to reside in rental apartments in Naivasha Town, while top managers usually live in a house on the farm or in gated communities. These three groups of farm employees can thus clearly be distinguished from each other, despite some overlap and socio-economic mobility between them.

4.1 *Top Managers: Work Ethics*

Top management of the Naivasha flower farms consists in majority of expatriates and Kenyan nationals of European or Asian descent. These top managers are moreover invariably male. It is important to note that the following description of the moral considerations and expectations of these managers is largely limited to the viewpoint of European, especially Dutch, farm managers. They are considered leading in the industry (Gemählich 2022), but another reason for this bias is that I have become more familiar with their viewpoints than with those of, for instance, Kenyan managers from Indian or European descent. This partial access I had is a reflection of the segregated bubbles these managers live in, with limited other-than-professional interaction between top managers from different backgrounds, let alone with lower-ranked Kenyans of African descent (McIntosh 2016; Styles 2019, 153). The Dutch managers based in Naivasha, for instance, "form an outpost of the Dutch cut flower network", and these connections "are the backbone of recruitment systems for farm managers" (Gemählich 2022, 95).

Interviews with European farm managers indicated that they aim to run a viable and prosperous business in Naivasha. They also enjoy the dynamic trade of cultivating flowers, especially those who grew up in flower-growing families. Moreover, they take pride in providing employment to hundreds to thousands of workers. A Dutch manager said he considered flower-farm employment an honest way of providing people with a living. He mentioned proudly that labour turnover rates of flower farms in Kenya were much lower than in neighbouring Ethiopia. Such managers look for loyal employees who ascribe value to their work, and they emphasise the importance of "proper" work ethics.

Several farm managers stated that they find the presence of skilled and disciplined labour more important than the cost of labour. As Richard Fox, top farm manager and then president of the Kenya Flower Council, stated during a workshop in Naivasha in May 2015: "labour is not cheap here, but Kenyans have an aptitude for work and for agriculture".[5] The value of a disciplined, stable, and loyal workforce also surfaces from farms' changing practices. Conditions in the Kenyan flower industry are increasingly geared towards retaining labour, such as the increased prevalence of permanent contracts, at least in unionised farms (Gibbon and Riisgaard 2014). The managers' attachment to loyalty and to a "proper" work ethic is partly informed by the material process of cultivating flowers and the requirements of the market, that desires spotless flowers. As Besky (2014) argued for tea, perennial and vulnerable crops need workers with long-term commitment, who get a "feeling" for the crop. Disciplined and time-conscious labour is similarly indispensable for the production of roses

5 Nevertheless, from a global perspective, wages of Kenyan flower-farm workers are comparatively low. Gemählich (2022, 125) estimated the labour costs for firms in the Netherlands to be thirty times higher than for firms located in Kenya. Moreover, within the Naivasha industry, differences in income and wealth between general workers and top managers are immense (see Kuiper, forthcoming, for a discussion of wages in the industry).

(Kuiper 2019). In addition, the Dutch managers' emphasis on proper work ethics, expressed through orderliness, timeliness, and the adherence to procedures, seems to be related to their cultural background (Faludi 2005).

The increased formalisation of recruitment procedures, also noted by Gibbon and Riisgaard (2014), fits into this emphasis on proper labour procedures. The managing director of Karibu Farm explained that he did not like to take on people through recommendations. Instead, the farm would put out an advertisement at the gate. The human-resources manager would make a first selection of applicants and the managing director would do a final interview round (even though it is rare for foreign managers to be actively involved in recruitment). When I asked him whether applicants with poor English-language skills were not then in a disadvantaged position, he replied that he looked for people with the right "attitude". An example of what he meant by this is that he was adamant that he would not let ethnicity play a role in managerial decisions. He emphasised the need for a good Kenyan production manager, who avoids the impression that he is favouring anyone. The managing director explained that in the first years he was there, he had trouble with people who did not work according to his "work ethics". For instance, he received anonymous messages containing allegations that certain farm workers had made mistakes, and employees tried to tell him "how things go in Kenya". The situation had calmed down after some employees were fired and after he had replaced his production manager.

4.2 *Middle-Level Managers: Professional Ethics*

As the previous example shows, foreign managers tend to depend on their Kenyan middle-level managers, especially human-resource and welfare managers, for direct daily interactions with the general workers and for "managing" workers' expectations (see also Styles 2019, 96). Clayton and Savage (1974, XIV) pointed out that in the colonial period, European settlers and African farm workers only knew each other "in the relationship of employee or employer". The situation is no different in the present-day flower industry. Kenyan middle managers function as intermediaries, a role they find stressful.

Whereas Karibu Farm's managing director was happy with his production manager, this manager himself complained to me about the complexity of his work. He said that he had to work with a highly diverse group of people who resort to flower-farm work because they have nothing else they can do. The social part of the work is therefore difficult and requires skills. He said workers might try to "trick" him, and he had to be tough because people sometimes "tested" him. Now and then he fired workers who were still on their probation period, or even those on a permanent contract when they made gross mistakes or were considered unruly. On the other hand, the manager told me that some workers are very bright and they only work on the farm because their "time has not come yet". In other words, coming from a similar class and ethnic background as some of the general workers made it stressful for this Kenyan manager to work within the stark farm hierarchies. Similarly, another Kenyan manager told me he advised his children not to look for employment in the horticultural industry. He considered it strenuous to balance the requests of the workers, the directors, and the markets.

The general workers also expect top managers to leave the direct supervision of labourers to a Kenyan manager, whom they usually feel more comfortable around. One Dutch manager was for instance described to me as "a beast" by one of the workers of the same farm, as he tended to bypass his HR manager in taking the decision to fire people. The worker moreover held it against the top manager that he addressed general workers directly instead of having supervisors correct their mistakes. Similarly, one of the supervisors of Karibu Farm was rather disgruntled about the behaviour of a young Dutch management intern. She had shown the intern the work in the greenhouse she was overseeing, until he had learned enough to notice mistakes made in "her" greenhouse after a few weeks. What vexed her was that he directly told the managing director about these mistakes, instead of informing the Kenyan production manager.

Perhaps because pressure is high, Kenyan managers are highly mobile. They tend to move farms every few years, the reason for which several of them described as "to look for greener pastures". Most middle managers, such as human-resources officers and technical experts, have higher-education diplomas from Kenyan universities (Styles 2019, 89). Despite their education, many of them had to start in modest – such as supervisory – positions. By moving farms every few years, they slowly make their way up the managerial ladder – and in some cases fall down again. I interviewed a former (lower-level) quality-assurance manager, who according to his own account lost his job on a vegetable farm due to ethnic discrimination. He then applied for a job as a supervisor on a Naivasha flower farm, but first had to work as a general worker for at least a year before being allowed to enter that position. This job was especially degrading to him as he considered the harvesting work in the greenhouses to be unfit for men (I come back to this gendered aspect below).

Styles (2019) has elaborated on the aspirations of these Kenyan managers. They share some aspirations with general workers, such as the wish to invest in land or small-scale business. However, some managers also see their work as an opportunity to establish themselves as patrons, redistributing wealth and employment through their networks. Others perceive their job as an opportunity for upward mobility, and aim for "professionalism". Middle-level managers take on ambiguous positions in the process of "moral boundary drawing on class and race lines" (Sayer 2010, 174). They are clearly not equal to foreign managers, but also aim to distinguish themselves from general workers. It seems that it is exactly this ambiguity with regard to whose interests they are expected to look after that makes them effective as intermediaries. Professionalism is a way to conceal this double moral allegiance and to deal with the stress that comes with the job.

It is worthwhile to elaborate briefly on the position of supervisors here, which is even more ambiguous. They roughly oversee a dozen to three dozen workers, but their exact role varies greatly between farms. In many cases, supervisors initially started as general workers, and their superior position is then justified by referring to skill and experience. Whereas most managers spend their time going around the farm or sitting in the office, supervisors have a fixed working place and they usually work along with the general workers part of the time. They have slightly higher salaries than general workers but earn well below what managers earn (see Kuiper, forthcoming). The supervisors in Karibu Farm were aware of the work ethics their managing director looked for. One of them told me they train their workers to do their work responsibly. I also noticed when she trained two new workers that she explained specific procedures, such as the removal of so-called "blind shoots" (non-productive stems), by stating that no resources such as fertiliser, which were brought by "the white man" (meaning the director), should be wasted, as this would mean a loss for the company.

Despite this loyalty to the management, I nevertheless noticed on multiple occasions that she allowed small transgressions of rules whenever there were no managers in sight, such as measuring several stems at once to gain speed (rather than measuring them one by one for accuracy). In fact, when a new supervisor who came in from another farm was stricter than the workers she was overseeing were used to, they went on an unannounced "go-slow", harvesting far fewer flowers than usual with remaining stems going to waste. Resistance by workers mostly takes such subtle forms and remains small-scale, as in this case. The difficult position this supervisor found herself in when the perception she had of her role did not fit workers' expectations illustrates the highly ambiguous but crucial role of supervisors in the everyday operation of the flower farms.

4.3 *General Workers: "Future Ethics"*

General workers' expectations towards their employers are shaped by established practices of agricultural wage labour in Naivasha that predate the arrival of the flower industry. Naivasha has attracted agricultural wage labour since the beginning of the twentieth century. Colonial land and labour policies created a system of labour migration in which Kenyans residing elsewhere were induced or even forced to come to Naivasha to work on European settlers' expansive ranches, either as temporary workers or as squatters.[6] After Kenya gained independence, the established system of temporary chain labour migration remained in place, along with a new influx of migrants who acquired a small piece of land and settled in Naivasha permanently (see Kuiper and Gravesen, this volume). Just as with many of these predecessors, the majority of the workers employed by the present-day flower farms do not intend to reside in Naivasha for the rest of their lives. Their stay there is a means to an end: they hope to be able to save money to invest in a future outside Naivasha.

Workers' life goals centre on being or becoming an independent and responsible adult, which implies for instance owning a plot of land and financially supporting dependent family members, such as children and parents (Kuiper 2020, 241–42; Styles 2019, 76–77). When one of the largest Naivasha flower farms went bankrupt and did not pay out salaries, one of the employees explained to me that he and his colleagues got into trouble with their families back home, who were by then used to receiving remittances. Workers therefore consider the provision of stable employment to be the main moral obligation of the farms – their expectations thus being quite similar to those of top managers. Workers at Karibu Farm for instance appreciated the trustworthiness of the farm management. Quite a number of workers I met in 2014 and 2015 had been working on the farm since it was taken over by the Dutch management around the year 2003. One of them told me that she thought the Dutch manager was a good employer: "he knows his workers, and he knows how to stick to a budget". Moreover, their salaries – however modest – were always paid on time, before the third of the month.

When conflicts do occur in the industry, they usually revolve around wage levels. Despite top managers downplaying the importance of wage levels, there were limits to what increase in labour costs they, and especially farm investors, still found acceptable, and such decisions were also largely taken outside Naivasha (cf. Giger et al. 2020). The managing director of Karibu Farm mentioned in 2014 that the considerable wage increases in prior years had in his view been necessary, but he feared people would expect similar increments in the years to come, while the Dutch farm directors would not agree to that. This fear was not unfounded, as the workers' union KPAWU sometimes resorts to (a threat to) strike when it cannot reach a new agreement with the employers on wage

6 For a more comprehensive description of this history of labour migration to Naivasha, see Kuiper (2019, 35–84).

levels. Especially in the early days, when the farms paid little more than the statutory wages for agricultural labour, which were fully inadequate for Naivasha's urbanising environment, there were a number of strikes (Mwathi and Njuguna 1996; Styles 2019, 102).

Even though nominal wages in the industry have increased continuously, real wages have declined due to inflation and general workers do not commonly earn a living wage (Kuiper, forthcoming). Nevertheless, as union officials and one of Naivasha's labour officers told me, strikes have become rare, and conflicts over wages are usually solved through negotiations or court cases. There are moreover strict procedures for strikes, whenever they do occur. It is rare for workers to mobilise and to openly protest outside of such official frameworks. Uncontrolled collective protests in the form of wild strikes, riots, and the blocking of roads occur only in exceptional cases, such as when the above-mentioned bankrupt farm left workers jobless without paying their due salaries and other payments (*Daily Nation* 2014).

It is important to stress that although the flower industry is the main employer in Naivasha, it is not the only one. That is, migrant workers have other options for earning an income. A labour officer of the Naivasha Sub-County for instance mentioned vegetable farms, construction companies, and tourism as other important sectors providing low-skilled employment. However, few of these other sectors provide as much security and stability as the flower industry, and there are always more people who hope to get a permanent flower farm position than vacancies to be filled (Kuiper 2019).[7] The differences in payment and job stability between the flower and vegetable industries, and the consequences for the opportunities available to workers to make investments in a future outside Naivasha, are particularly noteworthy (Kuiper and Greiner 2021).

The high level of discipline in the flower industry is not something all workers could cope with. It was a reason for some migrant workers not to apply for a flower-farm job in the first place, and I met a few former workers who had left their jobs for this reason. However, most former workers I spoke to had either left due to external circumstances, such as losing their job or sickness in the family, or simply because of their wish to return "home". Flower-farm workers accept the high demands placed on them with regard to work ethics in return for the stability of their employment. The workers have more than what Scott (1976) called a "subsistence ethic" – they have a "future ethic": their work ideally not only provides for their subsistence but also enables them to

7 Giger et al. (2020) similarly observed for the Mount Kenya region, where wage levels are even lower, that flower farm jobs are sought after because of the reliability of payment.

make material investments in a future after wage labour. Even if by no means all the general workers eventually succeed in reaching their goals, flower-farm work provides more opportunities for saving and making investments than other jobs do, at least for those who are lucky enough to land a permanent job (Kuiper 2019; Kuiper and Greiner 2021). This aspect can partly explain the absence of massive dissent among general workers.

5 Class, Ethnicity, and Gender in the Flower Industry's Moral Economy

Class differences in the Naivasha flower industry, in the sense of income and wealth differences, are pronounced. Local journalists have noted the marked inequality immediately visible in Naivasha Town – the town where "poverty and affluence live side-by-side", as Macharia Mwangi titled a 2007 article in the *Daily Nation* – and the discrepancy between the meagre wages of general workers and the profits the farms make (Omondi 2007). When driving on the main road around the lake, one immediately notices lush gated communities and dense, poorly constructed workers' settlements located side by side. As argued elsewhere, socio-economic disparities in the area increased after the establishment of the flower industry, and these disparities have become inscribed in the built environment (Kuiper and Greiner 2021).

Whereas these – partly racialised – wealth inequalities have attracted criticism from outsiders, they are not a new phenomenon locally. Even when workers challenge their own low wage levels, the great wealth of foreign managers is not questioned. Inequality among Kenyan employees is more susceptible to debate. These varying evaluations of wealth inequalities indicate that – at least in the Kenyan context – the (social and moral) meaning of class depends on the intersection with other markers of inequality. I discuss the role of two of these markers, ethnicity and gender, in shaping mutual expectations in the Naivasha flower industry in what follows.

5.1 *Ethnicity and Work Ethics*

"Ethnic identities are socially constructed historic and moral communities that are subject to ongoing processes of negotiation and contestation" (Lynch 2011, 8). Such negotiations and (moral) contestations take place internally, but can also be shaped by external actors and can become more about relations *between* than *within* ethnic groups. Ethnicity became an explicitly political category in Kenya during the colonial period. Resources such as access to education and wage labour were distributed unevenly to the elites of different ethnic groups, who distributed the resources further. Ethnicity therewith became intertwined with class (Berman and Larin 2016, 13). Land and labour policies also created a specific system of chain labour migration, in which different ethnic groups were ascribed different economic roles. After Kenya gained independence, land policies that were, at least in their effects, discriminatory intensified these ethnicised class disparities (Kanyinga 2009).

In Naivasha, the flower industry became an additional factor shaping class relations. At first sight, having (ethnic) patron-client relations play a role in the distribution of resources such as jobs seems unacceptable within the industry. An allocation of resources based on personal ties is defined as "corruption" in the moral economies the European managers stem from (Berman and Larin 2016, 9).[8] The Dutch manager of Karibu Farm explicitly mentioned his dislike of preferential treatment because of ethnic background. He wanted his employees to know that they are to be rewarded for the work they do, not for who they are. Despite such resolutions on the part of foreign managers, I found that members of ethnic groups originating in Western Kenya were strongly overrepresented among general workers in specific farms, even though the contemporary workforce is far from being as starkly divided along ethnic lines as during the colonial period.[9]

These imbalances in the ethnic composition of the workforce are partly due to the hiring practices of the farms, which despite their formalisation depend for a large part on "word of mouth", tapping into workers' own networks for accessing new labour (Kuiper 2019, 110–16). Such imbalances are also the product of certain implicit ideas among Kenyan HR managers on the relation between ethnicity and work ethics, ideas which have roots in the colonial period. Workers on certain farms moreover complained to me that some Kenyan managers used their position to establish themselves as patrons, favouring members of their own ethnic group.[10] Finally, even when Kenyan managers do not (consciously or unconsciously) take ethnicity into account in their decisions, the farms' reliance on educated Kenyan managers for their daily

8 Remarkably, Dutch managers seem to be sceptical of basing recruitment decisions on personal ties only when it comes to recruiting Kenyan employees. Personal networks are of great importance in landing a top management job in Dutch-led companies (Gemählich 2022).

9 For example, among 50 residents of the compounds of one of the farms, 26 were born in former Western Province while 17 were born in former Nyanza Province.

10 However, other employees would consider such distribution of resources within one's network acceptable or even a virtue, not a sign of corruption. Importantly, patron-client relations are not purely "economic". They are affective relations based on certain values (Droz 2006; Lynch 2011).

operations further institutionalises the ethnicised class differences. These managers mostly come from a different class background than the general workers. Justified by level of education, the higher status of certain (ethnic/class) groups is thus confirmed in farm hierarchies.

Ethnicity's enduring political relevance in Kenya was tragically exemplified by the ethnicised post-election violence that also rocked Naivasha Town and its surroundings in early 2008. Many farm workers were affected and fled, some never to return. Although some flower farms functioned as refuges, and they even managed to continue most of their production at the time, they ultimately depended on the army (and thus on the state) to restore order (Styles 2019, 56–62). Since then, farm managers have– understandably – been apprehensive about the potentially divisive role of ethnicity, yet their policy of not letting ethnicity play a role on the farms contradicts their reliance on the established system of chain labour migration and the implicit use they make of ethnicised discourses on belonging in securing their workers' consent to work. These discourses, in which there is no place for flower-farm workers in Naivasha after they retire, shape the workers' "future ethic" of looking for a job that allows them to save money to "go home".[11]

Ethnicity shapes workers' "work ethic" in yet another way. As stated by John Lonsdale in his work on the Kikuyu, ethnic groups "are changing moral arenas of political debate" (Berman and Lonsdale 1992, 267). That is, they are constituted through debates on what makes – for instance – a "proper Kikuyu". The value of work has been an important part of these debates since pre-colonial times, and is an integral part of how relations within an ethnic group are constituted. Droz (2006) described the Kikuyu ethos of working hard in order to obtain land and to establish oneself as a patron. According to Lonsdale (2008, 333), the Kikuyu developed a "labour theory of (moral and material) value", as it was labour that turned soils into wealth. Cohen and Odhiambo (1989, 5) described similar debates concerning ethnic and class identity among the Luo. They for instance referred to the "pressures placed on Luo men to maintain good homes (*dala*) in the Siaya countryside, as places to be at home and to be buried, and as concrete acknowledgements of links to the past". One way in which Luo men dealt with this pressure was to engage in labour migration. In both instances, moral discourses within an ethnic group encouraged people to work hard. These ethnic moral ideas continue to shape visions on work in the present.

11 Labour migrants' wish to "go home" is often taken for granted. Elsewhere, I have questioned the meaning of the notion of "going home" and pointed at its partly involuntary nature (Kuiper 2019, 239–64; Kuiper and Greiner 2021).

In sum, the continuing relevance of ethnicity can threaten the unity of the flower industry's workforce. On the other hand, their ethnic backgrounds also shape workers' values, including their valuation of work, and their visions for the future. Such ethnic moral debates therewith foster employees' consent to work and "sweat" on the farms.

5.2 *Gender and Ethical Work*

Gender is another important factor shaping mutual expectations within the flower industry, and one which extra-local actors (Salverda 2020), namely certification schemes and global labour-rights campaigns, have a considerable influence on.[12] Due to the relatively large proportion of female workers in the industry, the position of women has been an important topic in the above-mentioned discussions on "ethical work" and in the codes of conduct that were an outcome of these discussions. An example is the code of conduct of the Kenya Flower Council, which was established in the early 2000s. A newspaper article by Riungu (2004) described that the "code stipulates that females will not be employed in handling, mixing or application of pesticides or work in areas that are hazardous to their role as mothers. At the same time expectant and nursing mothers will be assigned duties appropriate to their physical conditions". Through a reference to their reproductive roles, women are thus portrayed as vulnerable and in need of protection. Another example is the emphasis on the prevention of sexual harassment in these campaigns.

Farm managers also have pronounced ideas about gender roles within the farms. Among foreign managers, these ideas intersected with certain racial stereotypes. Several managers portrayed Kenyan women as more efficient in the precise work that harvesting is, and also to be more trustworthy and responsible than Kenyan men. One manager stated he preferred to hire only women, who he considered to be more "loyal", but he needed men for the more technical tasks. Another manager recognised explicitly that especially single women with children are good workers, as they depend on their jobs. The

12 Salverda (2020) mentions government officials as another group of important "extra-local" actors shaping moral economies in specific industries. The role of the (national) government in the development of the Kenyan flower industry is contested. The industry has been portrayed as a neoliberal success story based on "self-regulation", also by government officials (*Daily Nation* 1999). In contrast, Tyce (2020) recently analysed the decisive role of Kenya's political economy in the industry's early success. In addition, Styles (2019, 115–45) highlighted the mutual dependency of government officials and industry players for realising their respective goals. The fuzzy role of government officials in shaping labour conditions and workers' expectations in the Naivasha flower industry is too complex to discuss comprehensively in this chapter.

consequence is that a majority of general workers are women, as both managers and workers consider these jobs simply more fit for them. Men on the other hand are over-represented in managerial positions, and, partly due to regulations in codes of conduct, women are completely excluded from other better-paying jobs that are considered hazardous, such as working night shifts or in spraying pesticides (Kuiper 2019, 163–66). Both extra-local and local ideas about "proper" work for men and women thus have concrete effects, shaping the composition of the workforce.

In other contexts, (agro-)industrial jobs have enabled women to slightly improve their positions within their households through the income they earn. However, when women depend on their jobs for this better bargaining position within their households, this at the same time reduces their space for making demands in the work place (Kofti 2016; Friedemann-Sánchez 2009). In other words, women's moral obligations outside the workplace can ensure their consent to work. In Naivasha, flower-farm jobs make women less dependent on their husbands. At the same time, the farm jobs contribute considerably to the women's heavy work burden while responsibilities between men and women outside the workplace seemingly have not shifted much. Styles (2019, 76) for instance found women felt more responsibility for paying children's schools fees than men, and would prioritise that over investing in property or a small business. As highlighted by Kioko et al. (this volume) and Obiria (2020), this double workload became especially poignant during the recent COVID-19 pandemic, as men more easily relinquish their domestic responsibilities. Notably, Berman and Lonsdale (1992) repeatedly pointed at the inseparableness of ethnic thought on civic virtue and on gender, and these gendered divisions of labour and responsibilities outside the farms can be expected to be influenced by workers' respective ethnic backgrounds.

Both extra-local critics' portrayals of female workers in need of protection and farm managers' emphasis on women's "responsibility" does not encompass a profound recognition of women's double workload. However, the double workload was not a topic of debate among workers either. Mutual expectations regarding gender divisions of labour within and outside the farms seem to be largely in concordance with each other. Gender is thus not as "divisive" as ethnicity can be, but gendered roles similarly ensure the consent to work of, at least, female workers.

6 Conclusion

Naivasha was the first flower-growing hub in Kenya, but several others have developed since then. A distinctive feature of the moral economy of the

Naivasha flower industry is the migratory background of the majority of the employees, influencing the ethnic composition of the workforce (Kabiru, Mbatia, and Mburugu 2018). For instance, farms around Mount Kenya, another major rose-producing area, are set in a rural environment and partly draw their unskilled labour from surrounding smallholder households (Giger et al. 2020). Whereas the farms there have the same gendered division of labour as in Naivasha, with a relatively large proportion of female employees, the implications of this division of labour for households are different in such a setting. Women are for instance possibly even more overburdened, as they also have tasks on their family farm in addition to their jobs (Ulrich 2014). Workers' expectations of their jobs will also be different in a context where they already possess what most Naivasha farm workers aspire to: their own plot of land. Turnover rates around Mount Kenya are much higher than in Naivasha, which Kabiru, Mbatia, and Mburugu (2018, 8) ascribe to the even lower wages paid there, but which can also be related to the fact that those who come from surrounding households do not have to pay rent.

In this chapter, I have shown that the particular moral economy of the Naivasha flower industry is constituted by diverse actors, many of whom do not consider themselves to be "locals". However, the moral economy is highly specific and tied to Naivasha as a locality, and should therefore be distinguished from translocal moral economies stretching across a distance through migrant's networks, as described by for instance Petrou and Connel (2017). The actors in the Naivasha moral economy not only have a shared history in Naivasha but also had highly disparate histories before arriving there. In other words, the actors – most of whom only arrive in Naivasha in their adult years – have been part of other moral economies, and thus arrived with certain values. "Moral economy relationships in Kenyan floriculture are embedded in complex forms of relational dependency that extend beyond this industry, this place, and this temporal moment" (Styles 2019, 17–18). I argue that it is precisely this embeddedness that makes for a unique moral economy.

Within this moral economy, expectations workers have of their jobs based on their ethnic and class backgrounds and gendered positions all seem to work primarily towards consent, not resistance. Other scholars focused on the moral economies of peasants or wage labourers to shed light on moments of resistance. The Naivasha case shows that the moral-economy concept can also be used to interrogate relatively stable and harmonious labour relations. Both ethnic and gendered moral debates guarantee employees' consent to work on the farms and their preparedness to cope with high levels of discipline. Actors in the Naivasha flower industry have highly diverse backgrounds but they all come from strong and particular moral economies. For all, work is a "moral positive" (Fouksman 2020), if for different reasons.

Acknowledgements

My research in Naivasha was conducted in the context of the DFG-funded Research Unit *Resilience, Collapse and Reorganization in Social Ecological Systems of African Savannahs.* A first draft of this chapter was developed for and presented during the workshop *The Moral Dimensions of Economic Life in Africa,* held on June 14 and 15 back in 2018, at the Nordic Africa Institute Uppsala. I thank the organisers Tijo Salverda, Cristiano Lanzano, and Jörg Wieratz as well as the other participants in the workshop for their constructive feedback that helped me to eventually develop this chapter. Finally, I thank the diverse research participants from the Naivasha flower industry for sharing their views on labour relations with me.

References

Becht, Robert, Eric O. Odada, and Sarah Higgins. 2005. "Lake Naivasha. Experiences and Lessons Learned Brief." International Waters Learning Exchange and Resource Network, January 15, 2021. Accessed May, 15, 2023. http://www.worldlakes.org/uploads/17_Lake_Naivasha_27February2006.pdf.

Berman, Bruce, and Stephen Larin. 2016. "Introduction". In *The Moral Economies of Ethnic and Nationalist Claims*, edited by Bruce Berman, André Laliberté, Stephen Larin, and Gabrielle Lynch, 3–22. Vancouver: UBC Press.

Berman, Bruce, and John Lonsdale. 1992. *Unhappy Valley: Conflict in Kenya and Africa.* Vol. 1. 2 vols. Oxford: James Currey.

Besky, Sarah. 2014. *The Darjeeling Distinction: Labor and Justice on Fair Trade Tea Plantations in India.* Berkeley: University of California Press.

Carrier, James G. 2018. "Moral Economy: What's in a Name." *Anthropological Theory* 18 (1): 18–35. https://doi.org/10.1177/1463499617735259.

Clayton, Anthony, and Donald C. Savage. 1974. *Government and Labour in Kenya 1895–1963.* London: Frank Cass.

Cohen, David W., and E. S. Atieno Odhiambo. 1989. *Siaya: The Historical Anthropology of an African Landscape.* Eastern African Studies. London: James Currey.

Daily Nation. 2014. "Karuturi Staff Demonstrates for Salaries." April 13, 2014.

Dolan, Catherine S. 2007. "Market Affections: Moral Encounters with Kenyan Fairtrade Flowers." *Ethnos* 72 (2): 239–61. https://doi.org/10.1080/00141840701396573.

Dolan, Catherine, Maggie Opondo, and Sally Smith. 2002. *Gender, Rights and Participation in the Kenya Cut Flower Industry.* Report No. 2768. Chatham: NRI.

Droz, Yvan. 2006. “Street Children and the Work Ethic: New Policy for an Old Moral, Nairobi (Kenya).” *Childhood* 13 (3): 349–63. https://doi.org/10.1177/0907568206066357.

Faludi, Andreas. 2005. “The Netherlands: A Culture with Soft Spot for Planning.” In *Comparative Planning Cultures*, edited by Sanyal Bishwapriya, 285–308. New York: Routledge.

Fisher, Josh. 2018. “In Search of Dignified Work: Gender and the Work Ethic in the Crucible of Fair Trade Production.” *American Ethnologist* 45 (1): 74–86. https://doi.org/10.1111/amet.12600.

Fouksman, Liz. 2020. “The Moral Economy of Work: Demanding Jobs and Deserving Money in South Africa.” *Economy and Society* 49 (2): 287–311. https://doi.org/10.1080/03085147.2019.1690276.

Friedemann-Sánchez, Greta. 2009. *Assembling Flowers and Cultivating Homes: Labor and Gender in Colombia*. Lanham, MD: Lexington Books.

Gemählich, Andreas. 2022. *The Kenyan Cut Flower Industry and Global Market Dynamics*. Woodbridge: James Currey.

Gibbon, Peter, and Lone Riisgaard. 2014. “A New System of Labour Management in African Large-Scale Agriculture?” *Journal of Agrarian Change* 14 (1): 94–128. https://doi.org/10.1111/joac.12043.

Giger, Markus, Emily Mutea, Boniface Kiteme, Sandra Eckert, Ward Anseeuw, and Julie G. Zaehringer. 2020. “Large Agricultural Investments in Kenya’s Nanyuki Area: Inventory and Analysis of Business Models.” *Land Use Policy* 99 (December): 104833. https://doi.org/10.1016/j.landusepol.2020.104833.

Kabiru, Joseph G., Paul N. Mbatia, and Edward K. Mburugu. 2018. “Emerging Conditions of Labour in the Cut Flower Industry in Kenya.” *International Journal of Education and Research* 6 (5): 1–12.

Kanyinga, Karuti. 2009. “The Legacy of the White Highlands: Land Rights, Ethnicity and the Post-2007 Election Violence in Kenya.” *Journal of Contemporary African Studies* 27 (3): 325–44. https://doi.org/10.1080/02589000903154834.

Kofti, Dimitra. 2016. “Moral Economy of Flexible Production: Fabricating Precarity between the Conveyor Belt and the Household.” *Anthropological Theory* 16 (4): 433–53. https://doi.org/10.1177/1463499616679538.

Kuiper, Gerda. 2019. *Agro-Industrial Labour in Kenya: Cut Flower Farms and Migrant Workers’ Settlements*. Cham: Palgrave Macmillan.

Kuiper, Gerda. 2020. “‘They Just Move in with Relatives’: Translocal Labour Migrants and Transient Spaces in Naivasha, Kenya.” *Journal of Eastern African Studies* 14 (2): 227–49. https://doi.org/10.1080/17531055.2020.1730562.

Kuiper, Gerda. “The Living Wage Debate in the Kenyan Cut-Flower Industry.” *Die Erde*, forthcoming.

Kuiper, Gerda, and Andreas Gemählich. 2017. "Sustainability and Depoliticisation: Certifications in the Cut-Flower Industry at Lake Naivasha, Kenya." *Africa Spectrum* 52 (3): 31–53. https://doi.org/10.1177/000203971705200302.

Kuiper, Gerda, and Clemens Greiner. 2021. "Export Horticulture and Labour Migration in Kenya: Translocality and Transiency in a Secondary Town." *Geoforum* 122 (June): 1–10. https://doi.org/10.1016/j.geoforum.2021.03.013.

Lonsdale, John. 2008. "Soil, Work, Civilisation, and Citizenship in Kenya." *Journal of Eastern African Studies* 2 (2): 305–14. https://doi.org/10.1080/17531050802058450.

Lynch, Gabrielle. 2011. *I Say to You: Ethnic Politics and the Kalenjin in Kenya*. Chicago, London: The University of Chicago Press.

McEwan, Cheryl, Alex Hughes, and David Bek. 2017. "Fairtrade, Place and Moral Economy: Between Abstract Ethical Discourse and the Moral Experience of Northern Cape Farmers." *Environment and Planning A: Economy and Space* 49 (3): 572–91. https://doi.org/10.1177/0308518X16674722.

McIntosh, Janet. 2016. *Unsettled: Denial and Belonging among White Kenyans*. Ethnographic Studies in Subjectivity 10. Oakland: University of California Press.

Monteith, William, and Lena Giesbert. 2017. "'When the Stomach Is Full We Look for Respect': Perceptions of 'Good Work' in the Urban Informal Sectors of Three Developing Countries." *Work, Employment and Society* 31 (5): 816–33. https://doi.org/10.1177/0950017016650606.

Mwangi, Macharia. 2007. "Naivasha Town: Where Poverty and Affluence Live Side-by-Side." *Daily Nation*, August 18, 2007.

Mwathi, Mary, and Michael Njuguna. 1996. "Company Sacks Striking Workers." *Daily Nation*, April 7, 1996.

Neumark, Tom. 2019. "Inequality." In *A Research Agenda for Economic Anthropology*, edited by James G Carrier, 40–55. Cheltenham: Edward Elgar Publishing.

Obiria, Moraa. 2020. "It's Triple Work for Flower Farm Workers". *Daily Nation*, May 20, 2020. https://nation.africa/kenya/gender/it-s-triple-work-for-flower-farm-workers-309436.

Omondi, George. 2007. "Burying the Hatchet after a History of Feuds'." *Business Daily*, November 7.

Palomera, Jaime, and Theodora Vetta. 2016. "Moral Economy: Rethinking a Radical Concept." *Anthropological Theory* 16 (4): 413–32. https://doi.org/10.1177/1463499616678097.

Petrou, Kirstie, and John Connell. 2017. "Food, Morality and Identity: Mobility, Remittances and the Translocal Community in Paama, Vanuatu." *Australian Geographer* 48 (2): 219–34. https://doi.org/10.1080/00049182.2016.1204671.

Riisgaard, Lone, and Peter Gibbon. 2014. "Labour Management on Contemporary Kenyan Cut Flower Farms: Foundations of an Industrial-Civic Compromise." *Journal of Agrarian Change* 14 (2): 260–85. https://doi.org/10.1111/joac.12064.

Riungu, Catherine. 2004. “Code of Conduct on the Cards.” *Daily Nation*, August 29, 2004.

Salverda, Tijo. 2020. “Multiscalar Moral Economy.” *Focaal*: 1–14. https://doi.org/10.3167/fcl.2020.033101.

Sayer, Andrew. 2010. “Class and Morality.” In *Handbook of the Sociology of Morality*, edited by Steven Hitlin and Stephen Vaisey, 163–78. New York: Springer New York. https://doi.org/10.1007/978-1-4419-6896-8_9.

Scott, James C. 1976. *The Moral Economy of the Peasant Rebellion and Subsistence in Southeast Asia*. New Haven: Yale University Press.

Smith, Sally, and Stephanie Barrientos. 2005. “Fair Trade and Ethical Trade: Are There Moves towards Convergence?” *Sustainable Development* 13 (3): 190–98. https://doi.org/10.1002/sd.277.

Styles, Megan A. 2019. *Roses from Kenya: Labor, Environment, and the Global Trade in Cut Flowers*. Seattle: University of Washington Press.

Thompson, E.P. 1971. “The Moral Economy of the English Crowd in the Eighteenth Century.” *Past & Present* 71 (50): 76–136.

Tyce, Matthew. 2020. “A ‘Private-Sector Success Story’? Uncovering the Role of Politics and the State in Kenya’s Horticultural Export Sector.” *The Journal of Development Studies* 56 (10), 1877–1893. https://doi.org/10.1080/00220388.2020.1715944.

Ulrich, Anne. 2014. “Export-Oriented Horticultural Production in Laikipia, Kenya: Assessing the Implications for Rural Livelihoods.” *Sustainability* 6 (1): 336–47. https://doi.org/10.3390/su6010336.

Weiler, Anelyse M., Gerardo Otero, and Hannah Wittman. 2016. “Rock Stars and Bad Apples: Moral Economies of Alternative Food Networks and Precarious Farm Work Regimes.” *Antipode* 48 (4): 1140–62. https://doi.org/10.1111/anti.12221.

CHAPTER 9

Flourishing Flowers, Withering Livelihoods: Social Networks for Food Security in Naivasha, Kenya

Kariuki Kirigia

1 Naivasha: a Nerve Centre[1]

Arriving in Naivasha Town, one encounters a place full of activity. As clouds of grey dust rise with every howling gale, it is the bustle of people, bodabodas, and motor vehicles that remains ubiquitous throughout. Naivasha is a town on the move. Key roads connecting cities, towns, Kenya's countryside, and even neighbouring countries like Uganda converge in Naivasha. The numbers of hawkers, restaurants, food and clothing vendors, banks, and hotels add further hue to the picture of a town in motion. I arrived in Naivasha in January 2014 by matatu from Nairobi, at a time of the year when people are getting back to work after the Christmas holidays and the sun is usually scorching hot. I boarded another matatu heading towards Kwa Muhia, a small settlement area along the Moi South Lake Road. We set off only after the matatu had overly exceeded the allowed passenger capacity for public vehicles, despite other matatus waiting in line for their turn to ferry passengers. Squeezed into a seat in the middle of the vehicle, I noticed how the hurly-burly dissipated as we drove away from Naivasha Town. Soon after, multiple flower farms lined up along the Moi South Lake Road with greenhouses protruding above the fencing around the farms as if peeping over the veil meant to shield floriculture activities away from the public gaze. Groups of people sat down, as if in a meeting, at the entrances of some of the farms in search for employment. This scene pointed to the contingent nature of wage labour in the floriculture sector in Kenya, and East Africa in general (Kirigia et al. 2016). All the while, a certain underlying tranquillity embraced the Moi South Lake Road compared to Naivasha, except in small towns such as Karagita and DCK, but this serenity only masked the ongoing unremitting flower production in the background. This mask was pulled back in the evening as wage labourers left the flower farms for home either by bus,

1 First submission date: February 18, 2021. Acceptance date: March 13, 2023.

 | DOI:10.1163/9789004695429_010

public transport, bicycle, or on foot. As flowers are a commodity produced chiefly for the export market, the floriculture trade connects the local and the global economy principally through what Appadurai (1996) calls the financescape, further underlining the nature of Naivasha as a place on the move. These pull and push forces, whereby Naivasha attracts job seekers from around the country (Kuiper 2020; Styles 2019) whilst at the same time producing goods for markets further away, render Naivasha a nerve centre where forces converge and diverge simultaneously.

In this chapter I examine the dynamics of this motion as it occurs within the floriculture sector and the resultant impacts of wage labour on local livelihoods. I focus primarily on the food-security status of the farm labourers' households and advance that food constitutes a basic livelihood need that must be met as a bare minimum to ensure human survival. Due to its primacy for life, its complexity because of other competing household needs, and multiple aspects that influence and define food security, this study provides a productive window for gaining insights into the livelihoods of the wage labourers in the floriculture sector. Further, the strategies employed by wage labourers to secure the food security of their households illuminate the nature of work within the floriculture sector and agribusiness investments writ large, especially when the labourers are migrants from other parts of the country, as is the case for many workers in the flower farms in Naivasha (Kuiper 2020). It is in the realm of the strategies utilised to address food needs that social networks, expanded further in this chapter, are identified as a critical yet complex strategy for food security among wage labourers in Naivasha. I explore how social ties are built in an area where a wage labourer is not only foreign, but is also often faced with limited financial capital and time constraints to the extent that those ties become the avenue through which much-needed material benefits are obtained within an otherwise precarious economy.

2 Food Security and Sustainable Livelihoods

I explore and discuss food security by building on the following definition by the FAO:

> Food security exists when all people, at all times, have physical and economic access to sufficient, safe and nutritious food that meets their dietary needs and food preferences for an active and healthy life.
>
> FAO 1996

This definition outlines the multifaceted nature of food security by incorporating the key aspects of availability, accessibility, safety and nutritional value, and people's preferences. Within the Naivasha context, these aspects of food security are complexified by two key dynamics. First, the wage labourers in the floriculture sector, majority of whom are migrants, do not possess land on which to grow food. Second, flowers constitute an inedible, chiefly for-export product. In addition to rendering food availability and accessibility complex, the Naivasha context does not easily lend itself to the main ideas behind the food-sovereignty discourse (Edelman 2014; McMichael 2014; Patel 2009) where among the major expectations is that of securing individual food production. I attend to the safety and nutrition aspects that are requisite for a healthy and active life by examining the perceived demands of the floriculture industry on the wage labourers' bodies, the extent to which these demands are intertwined with food preferences, and the ways the preferred foods are accessed.

This chapter draws from the sustainable livelihoods perspective, defined as follows: "A livelihood comprises the capabilities, assets (including both material and social resources) and activities for a means of living. A livelihood is sustainable when it can cope with and recover from stresses and shocks, maintain or enhance its capabilities and assets, while not undermining the natural resource base" (Scoones 1998, 5). This definition is an adaptation by Scoones (1998) of the earlier explication advanced by Chambers and Conway (1992), who are credited with giving the sustainable livelihoods approach significant impetus in the past circa three decades. The use of livelihoods analyses to understand the status of food security among wage labourers in Naivasha is productive on several grounds. To start with, livelihoods perspectives have been around for a long time and, while Chambers and Conway (1992) have been rightfully credited with advancing this framework, Scoones (2009, 173) notes that these perspectives have "a rich and important history that goes back another 50 or more years where a cross-disciplinary livelihoods perspective has profoundly influenced rural development thinking and practice." Further, by "drawing on diverse disciplinary perspectives and cutting across sectoral boundaries, livelihoods perspectives provide an essential counter to the monovalent approaches that have dominated development enquiry and practice" (Scoones 2009, 183). For this reason, Scoones (2009, 183) observes that the concept of livelihoods has entered a rich discursive space where it has functioned as a "boundary terminology" that does both the work of breaking down and bridging divides. In this chapter, I use sustainable livelihoods perspectives to penetrate the discursive spaces that bridge the local, trans-local, and global spheres, by examining the multiple facets of food security and by reflecting on flowers as a global commodity that, by traversing vast geographies on its way from production to consumption, links the local and the global.

At the same time, livelihoods perspectives have come under criticism for, among other things, leaving out essential elements of the debate, often due to the assumptions contained within the normative framings used in livelihoods discussions, the tendency to contrast between what are considered ideal scenarios with their alternatives, and the lack of clarity on what essentially informs what is deemed as progress (Scoones 2009, 184). In addition, the strength of the livelihoods perspectives to cross disciplinary divides can become its Achilles heel in that, as a "malleable concept which opens up such rich diversity in empirical description, it can equally be squashed down into the narrow instrumentalism of log-frames and planning formats, or get deployed by particular political commitments, dominated in recent years by neo-liberal reform" (Scoones 2009, 185). I attempt to overcome these challenges by bringing forth the voices of the wage labourers, to contextualise the meaning of words as communicated in Kiswahili language, and, through thick description, to vividly convey the stories and perspectives of the interlocutors. Further, I remain critical of the modernist ideas of progress which, within the realm of agribusiness and the corporate world, have adopted the culture of measurement instruments as markers of accomplishing corporate social responsibility to the extent that these instruments become the telos of progress. I employ the household as the primary unit of analysis, as it mediates the agency/structure divide and enhances feasibility in collection of empirical data (De Haan and Zoomers 2005; Schmink 1984).

3 Research Methods

I conducted field research principally in Naivasha Sub-County, the hub of Kenya's floriculture sector, in the areas adjacent to Lake Naivasha, which serves as the primary source of water for farming in the area. Floriculture farms surround the lake to secure access to water like birds of prey readying for a fresh carcass. Wage labourers populate the small towns along the Moi South Lake Road as well as within the farms where worker settlements have been put up as to remain nearby their places of work. Shops and small kiosks selling foodstuffs and other household items, often packaged in small quantities and in virtuosic display, dominate business activities in these towns. Already at this early stage of my research, it was palpable how the floriculture sector through its wage labourers constituted the core of the economy of the small towns along the Moi South Lake Road. However, this is gradually changing over time as other economic activities, such as geothermal power production, have expanded over the past years, a dynamic that is discussed in more depth by Nweke-Eze and Adongo (this volume) and Kioko and Wanjala (this volume), and one that additionally underpins the argument for Naivasha as a nerve centre. I carried out further research in Nairobi as it is home to key actors in the floriculture sector including the Kenya Flower Council (KFC) and the Dutch Embassy. The short distance between Nairobi and Naivasha made it feasible to conduct research in these two sites, and it also underlined the geographical significance of the floriculture companies being able to access the Jomo Kenyatta International Airport relatively fast for an export industry dealing in a perishable commodity.

I carried out fieldwork in two phases: the first phase between January and April 2014 in Naivasha and Nairobi, and the second phase for four weeks between December 2014 and January 2015, in Naivasha, interspersed with key informant interviews and observations in the Netherlands, the primary first destination for flowers from Naivasha. During the first phase, I lived in a small rental compound in Kwa Muhia, one of the small settlement towns along the Moi South Lake Road, shown on Map 1.1 of the Naivasha area, among households of wage labourers in the floriculture industry, and business operators. This proximity to wage labourers and business operators facilitated close observation of the everyday lives of these households. I identified respondents working as wage labourers in the floriculture farms through purposive sampling and snowballing techniques, which proved to be an effective method in a setting where many of those working in the flower farms know each other. A second category of interlocutors was comprised of the owners and managers of floriculture farms, the Dutch embassy, the Kenya Floriculture Council, and other organisations, to whom I sent emails and made phone calls to request and arrange meetings and interviews. In addition to the wage labourers' households (35) and floriculture companies (5), other respondents included horticulture companies (2), NGOs and institutions (11), government officials (3), business operators (7), smallholder farmers (5), the Dutch embassy and a primary school teacher, and key informants (2) with detailed knowledge of the history of the Naivasha area. I collected data chiefly through semi-structured interviews, participant observation of everyday life around Kwa Muhia, DCK, Karagita, and Kamere – all small towns along the Moi South Lake Road – and informal discussions with residents in the area. I conducted interviews with the wage labourers both at their workplaces and at their homes, which provided insights into the dynamics of both settings wherein most of the labourers' livelihoods are situated.

4 Land and Water: A History of Politics of Access and Use

Floriculture in Naivasha began in the 1970s and underwent marked growth in the 1990s (Kuiper 2019; Styles 2019). At the time of carrying out this study, there were around 68 floriculture companies in the Naivasha area and, although flower farming has expanded to other areas such as in the Mount Kenya region in recent years, Naivasha remains the hub of the floriculture industry in Kenya (Kirigia et al. 2016). Floriculture companies possess on average 30 to 40 hectares of land, positioning them as medium-scale farms (Mwangi 2019). When considered altogether, however, the land area under floriculture increases significantly due to the high number of floriculture farms. Most of the land on which floriculture farms sit has either been bought from previous landowners or obtained through leasehold arrangements, with farms sometimes having both arrangements of tenure in place. According to Maurice Bolo (2005), the majority of the land around Naivasha was under white settlers and government officials, who later leased it to large-scale agricultural investors. For example, one farm in possession of 45 hectares had purchased 33 hectares from previous landowners and leased 12 hectares from a landowning group. These dynamics of landownership in Naivasha partly reflect the complex history of land politics in the area, which date back to the pre-colonial period when the Naivasha area served as an important pasture area for the Maasai pastoral peoples (Sankan 1965; see also Waller, this volume).

Huge swathes of land were expropriated during the colonial period, on which white settlers went on to establish large farms and ranches. Native Kenyans worked as labourers on these farms and ranches, sometimes settling there with their families (Kuiper 2019). Following Kenya's independence in 1963, some of the land belonging to white settlers was sold to local government officials and politicians (see Kuiper and Gravesen, this volume). This dynamic of land transfer through the market reflected the terms of independence agreed upon by the British colonial officials and the government officials of the newly independent Kenya (Kiamba 1989; Okoth-Ogendo 1991), which Wasserman (1973) has termed "independence bargain" for the way it favoured the British settlers, especially when it came to land that had been forcefully acquired from native communities. I followed up on the land question with two key interlocutors based at Kwa Muhia, both of whom had information about the history and the present state of land ownership and politics in the area. They informed me that land around Lake Naivasha was still in the hands of a few individuals, mainly white settlers, who had acquired it during the colonial period. The area has continued to witness the creation of enclosures and, for the Maasai, one of the more recent experiences of losing land through enclosures came with the establishment of the Hell's Gate NP. The eviction of the Maasai from the park was justified on the grounds of the need to conserve biodiversity in the area, yet in the past few years, investments in geothermal power production have taken precedence (Nweke-Eze and Kioko 2021; Rogei 2021; Nweke-Eze and Adongo, this volume). This historical unfolding in what is today's Hell's Gate NP demonstrates the prioritisation of capital investments in land above the said biodiversity conservation and, ultimately, land access by the Maasai pastoral community.

While Naivasha is a semi-arid area that receives low rainfall, the availability of water from Lake Naivasha renders land in Naivasha a highly productive resource, echoing the observation by Woodhouse (2012) that it is water that critically renders land arable. The politics of water use and access in Naivasha, therefore, are central to discussions about land-based activities, especially floriculture farming. The two key tributaries of Lake Naivasha are the rivers Malewa and Gilgil. As the rivers flow down from the highlands traversing plots of land and boundaries, they bring forth other actors into the resource-use debate in the form of upstream communities, whose activities influence water inflows into the lake and, subsequently, the volume of the lake, and water access (Everard and Harper 2002; Nyongesa et al. 2016). The interdependence between upstream and downstream communities, coupled with competition for water among various actors, especially the flower farms, have rendered management of the lake waters even more critical (see Harper et al., this volume, and Pacini, Giberti, and Harper, this volume). The Water Resources Management Authority (WRMA) is the government entity tasked with water management and, while such work is carried out in collaboration with other partners such as the Water Resource Users Associations (WRUAs), it is WRMA that retains overall authority over the management of water resources countrywide, including provision of certificates to water users such as the floriculture companies.

While WRMA monitors water uptake by floriculture companies, it has become imperative to incorporate upstream communities in the efforts to enhance the health of Lake Naivasha. Accordingly, WWF and CARE International put in place a PES-scheme in 2008. The scheme encourages upstream communities to adopt sustainable agricultural practices to prevent soil erosion and subsequent siltation of water inflows into Lake Naivasha, the main factors blamed for the dwindling lake water levels. These interventions in turn not only enhance the health of the lake, but also ameliorate livelihoods among upstream farmers and propel sustainable economic development (Sang et al. 2017). Following the severe drought of 2009 that led to extremely low water levels in Lake Naivasha, multiple stakeholders, especially from the floriculture industry, supported the PES-scheme. By 2012, the PES-scheme had 785 service providers, who however comprised only four per cent of the farmer households in the upstream communities (Sonneveld et al. 2018). WWF averred that the scheme spurred economic growth among upstream communities, whose agricultural production not only increased, but also generated surplus produce to sell in the market. The floriculture industry contributes financially to the scheme, but farmers obtain only US$17 per year. The vouchers for agricultural inputs are tightly regulated such that there are specific shops where farmers can obtain farm inputs. The tight regulation and compensation in kind are done to ensure the tokens are used only for agricultural purposes in the quest to attain water security for downstream communities (Boonstra 2010; Rouillé-Kielo 2019). The demand and competition for water has thus bolstered collective efforts to conserve the Lake Naivasha basin and, according to WWF, created a win-win situation for both upstream and downstream communities. As Naivasha grows as a nerve centre, such as with the entry of the Kenya Power Lighting Company, which is overseeing geothermal power generation in the Hell's Gate NP, demand for water continues to rise. Water governance in Naivasha will therefore only increase in complexity, making it interesting to see how responsibility is shared among different water users (Rouillé-Kielo 2019).

5 Precarity in Transitory Labourscapes

This study set out to investigate the status of food security within the households of wage labourers in the floriculture sector in the Naivasha area. A key feature of the floriculture industry is the extent to which it has spurred labour mobility by drawing workforce from near and far parts of Kenya (Kuiper 2019; Styles 2019), an illustration of the sector's popularity as a major source of employment, and of the continued growth of Naivasha as a nerve centre. Even then, Naivasha constitutes, as Kuiper (2020 and Styles (2019) observe, a transient place of work, as many wage labourers perceive working in the floriculture industry as a steppingstone to modes of livelihoods higher up the economic ladder. To further investigate this aspect of temporality, I inquired about the wage labourers' duration of stay in Naivasha. While this ranged from one year to 42 years, only two respondents out of 32 interviewees had lived in Naivasha for a relatively longer period: 30 and 42 years, with the majority having resided in the area for less than a decade. The short duration of the labourers' stays in Naivasha underpins the view that, for many labourers,

Naivasha is a transient workplace as they prepare to move to occupations higher up the socioeconomic ladder.

A youthful workforce predominates in the floriculture sector due to their ability to quickly accomplish labour-intensive work (Dolan, Opondo, and Smith 2002; Odhong and Omolo 2014). The respondents in this study were aged between 23 years and 51 years, which buttresses the observations by Dolan, Opondo, and Smith (2002) about the youthful nature of the workforce. Often characterised by tough physical work coupled with exposure to farm chemicals, the industry exerts exacting physical demands on the bodies of the labourers. As one respondent strongly asserted, "The work is really hard and sometimes you have to skip lunch because harvesting (of the flowers) has to be done in time … the conditions at work shorten one's lifespan, which means it is very difficult to reach 70 years of age".

While this assertion informs us about the physical demands of work in the floriculture sector, it also points to work conditions that at times require labourers to forego meals as a way of carving out time to keep the industry operational. The long-term effects of these seemingly temporary exertions on the bodies of the labourers foreclose any imagination of living a long life, a reality that stands in stark contrast to what is usually anticipated from employment: the means to improve one's quality of life. However, if rather than begetting consternation such a reality is understood as characteristic of the precarity of working within the floriculture industry, then it implores us to engage with floriculture production as a site of what Lauren Berlant calls "slow death" in reference to "the physical wearing out of a population and the deterioration of people in that population that is very nearly a defining condition of their experience and historical existence" (Berlant 2011, 95). Such a rendering of employment in turn calls for a study of work not as an end, or simply as a means of earning an income, but rather an engagement within which lie a complex of relations between the worker and the work done, and where the worker, by virtue of being replaceable by a more youthful labourer, is rendered disposable. All the while, the floriculture industry continues to grow and even expand to other spaces that promise higher profits or lower costs, such as in Mount Kenya and Ethiopia, respectively (Kirigia et al. 2016). It is by analysing these relations between the labourer and the labour provided, or the worker and the work conditions, then, that we can comprehend the short- and long-term impacts of varied forms of wage labour. Arguably, the coercion to work in an industry of which one is aware of the associated debilitating outcomes, where the imaginaries of dystopian futures prevail, reflects the dearth of alternative options to meet livelihood needs in a Kenyan context steeped in high poverty levels and unemployment rates.

One of the key pull forces of the floriculture sector is its demand for low-skilled labour. While this aspect of floriculture labour has opened the door to those with lower-level educational backgrounds in the past (Bolo 2006; Styles 2019), these patterns are changing, with labourers having higher educational levels mainly due to the high unemployment rate in Kenya especially among the youth (Kabiru, Mbatia, and Mburugu 2018). I gathered information about educational backgrounds from 28 respondents; of these, 5 had only attended primary school and 17 had a high school education, while only 6 had gone to college. None of the respondents had university- or higher-level education. Within the Kenyan education system, high school or secondary education follows a general curriculum that culminates in a final national exam that determines the educational path one can take, with those with higher scores being able to attend university, while many others join colleges and technical and vocational education training centres. The level of education correlated with monthly income above KSh 21,000,[2] with all those earning more than that amount having either high school or college education. While less emphasis on employees' educational backgrounds opens a wide door to employment in the floriculture sector, one employee noted that the abundance of cheap labour increases job insecurity among the employed. As I observed during fieldwork, there were always people waiting outside the gates of the floriculture farms in the hope of securing employment. My interview with the Workers Union representative in Kwa Muhia added important perspective to the insecurity of floriculture wage labourers by noting that attempts by the labourers to enhance their job security by joining the Workers Union were often thwarted by their employers. The wage labourers registered with the union were often targeted for unfair blame and accusation at their place of work in an attempt to establish grounds to terminate their contracts. I found this form of precarity among waged labourers in Uganda as well when I carried out an extended study of the Eastern African floriculture sector (Kirigia et al. 2016), an occurrence which suggests that one of the risks of seeking institutional security or negotiation power, whilst in a precarious job, is, ironically, added precarity for the labourer. It can be surmised that the employment insecurity among the wage labourers in the floriculture sector in Naivasha coerces the labourers to withstand difficult work conditions in the view that it is only temporary if working in the sector is indeed, as Styles (2019) and Kuiper (2020) assert, seen as a transitory occupation.

---

2 At January 1, 2014, KSh 1 was equivalent to $0,0114 (source for the exchange rate: https://www.oanda.com/currency-converter/en/). Hence, KSh 21,000 was equivalent to roughly $239 at the time of research.

The transitory nature of work in the floriculture sector is underscored further by the temporary mode of settlement among many wage labourers. Of the 34 interlocutors that participated in this study, 31 lived in rental houses outside of the flower farm premises, while only three lived in housing provided by the employer, and none lived in individual-owned housing. To understand the supply side of housing in the Naivasha area, I interviewed five flower companies, none of which provided housing for the wage labourers. A monthly house allowance, in addition to the monthly wage, was given to those residing outside a company's premises. This allowance ranged from KSh 1,500 to KSh 10,000,[3] and differed depending on an employee's position at work. For example, two of the participants in this study working as a production manager and laboratory technician received KSh 3,000 and KSh 3,600 per month[4] respectively. The higher house allowance, however, did not translate to more expensive housing, as the laboratory technician, for example, spent only KSh 1,200 per month on rent. Out of 31 respondents, only nine indicated that they spent more on monthly house rent than the house allowance provided by their employer. One of the reasons for this housing-cost dynamic could be that the housing market in the areas around the flower farms is largely for low-income households and as such, there are few options for what could be considered high-income households. Looking ahead, it will be interesting to see how housing arrangements are likely to change as Naivasha grows as a nerve centre, which could escalate housing costs as demand for housing rises.

6 Food for Work

The physical and time demands of work in the floriculture sector constitute two key factors that influence food consumption among the wage labourers. After long days at work, the wage labourers were left with limited time to carry out other activities, including preparing food. Households thus consumed food that could be easily and quicky prepared, and these comprised three main foods: *ugali*,[5] rice, and *githeri*.[6] Of these, *ugali* was seen as the best meal for it was both filling and a great source of bodily energy required to work for long hours. The exposure to agrochemicals at the workplace, especially among those working in the spraying department, further influenced the type of food

3 Approximately between $17 and $114, based on the exchange rate mentioned in footnote 2.
4 Approximately $34 and $41 respectively, based on the exchange rate mentioned in footnote 2.
5 *Ugali* is a local dish made from maize flour.
6 *Githeri* is a local dish that is prepared from maize and beans.

consumed. To withstand such a chemical-laden environment, the labourers consumed what they termed "highly nutritious food". A young man in his twenties working in the spraying department said, "I have to make sure I eat well. For instance, I consume eggs and milk because the chemicals used at the flower farm are quite strong (devastating) and, if I have to protect my health, then it is important that I eat well". Another interlocutor, a woman who worked as a picker in one of the farms, complained that while they should only enter the greenhouses after a specific number of hours following spraying, these guidelines were not followed when there was higher demand for flowers, such as on Valentine's Day. On those days, the pickers had to enter the greenhouses much sooner after spraying had been completed. As such, while the sprayers may be the group that appears to be exposed to chemicals and, therefore, the ones in need of "highly nutritious food", the experiences of others such as the pickers demonstrate that exposure to agrochemicals is high not only among the sprayers, but also among many others working in the flower farms. Thus conceived, the impact of working in a floriculture farm on one's body and health is a concern shared by floriculture labourers across the board. And, if these stories reveal food as a strategy for persevering in the face of ruinous work conditions, it is the extent to which labourers put their lives on the line even when doing so means that they compromise their ability not only to work in the future, but also to continue to live, that paints the picture of the coercive nature of capitalist exploitation at the base of the global flower value chain.

According to Kenya's Ministry of Health (RoK 2006), one needs a balanced diet containing carbohydrates, proteins and vitamins, to ensure a healthy life. I gathered information from 32 households about the food they consumed on a daily basis and, after classifying the foods into the categories spelled out by Kenya's Ministry of Health, only five households consumed a balanced diet daily. A third of the households consumed a balanced diet several times in a month, three households indicated that they rarely consumed a balanced diet, and one household did not consume a balanced diet at all. The inability to consume a balanced diet on a daily basis was largely attributed to financial constraints. As one interlocutor said, "It is expensive to consume a balanced diet every day, and my level of income does not enable me to do so". While some respondents stated that fruits were expensive and bordered on luxury, it was key sources of proteins such as meat and milk that were deemed more costly.

Without the means to produce food, households obtained food from the local markets in the small towns along the Moi South Lake Road. Financial capital, therefore, became the primary means through which the wage labourers accessed food. Food supplies to the local markets came from Maiella, Moi-Ndabi, and Ndabibi, which are the key food-producing areas adjacent to the floriculture areas. Food is usually available in the local markets throughout the year, though with varying supply depending on weather conditions in the food-producing areas where agriculture is predominantly rain-fed. As one of the employees said, "During the dry seasons food prices rise because it is more expensive to produce agricultural products". In addition to food-production challenges, a market operator noted that during the dry seasons they are forced to get food from areas further away, sometimes as far as Tanzania, which increases the price of food significantly. Further, food prices in the local markets are affected by weather changes such that during the rainy seasons, the dirt roads to the food-producing areas become impassable, and consequently the transport costs rise and are reflected in increased food prices. Thus, the weather affects food security both in terms of food availability and accessibility, which, for the wage labourers, are both reflected in the local food prices.

Despite wages being the main source of income for the households of the wage labourers, the respondents felt that the wages were too low to support a decent livelihood, a challenge also identified by Dolan, Opondo, and Smith (2002) and Styles (2019). Many households depend only on wages to address household needs. Reported monthly household income ranged from KSh 5,500 to KSh 40,000, with a single case of a farm manager reporting a much higher monthly income of KSh 130,000. The average monthly income among the respondents in this study was KSh 24,211[7] and, upon excluding the outlier, KSh 130,000, the average household monthly income was 17,988.[8] This amount falls below the living wage amount, KSh18,542 as estimated by Anker and Anker (2014). To gain insight into the relationship between the amount of money households spent on food and the ideal amounts that they would have required to ensure their households were food secure, I carried out a comparative analysis of the figures, as displayed in Figure 9.1.

Figure 9.1 shows that the majority of the households spend less on food than they would in an ideal scenario, implying that household food needs are not adequately met on a daily basis. The ideal daily expenditure on food follows current expenditure, although in a less linear way, to suggest that households spending relatively less amounts on food would require spending a relatively higher amount to adequately meet their food needs. The spiky shape of the ideal food expenditure graph indicates that some of the households currently spending much less on food would in an ideal scenario spend much more than those already spending higher amounts. The spiky nature of the ideal food expenditure graph suggests that some households spending much less on food would ideally spend much more than those already spending higher

7 Approximately $276, based on the exchange rate mentioned in footnote 2.

8 Approximately $205, based on the exchange rate mentioned in footnote 2.

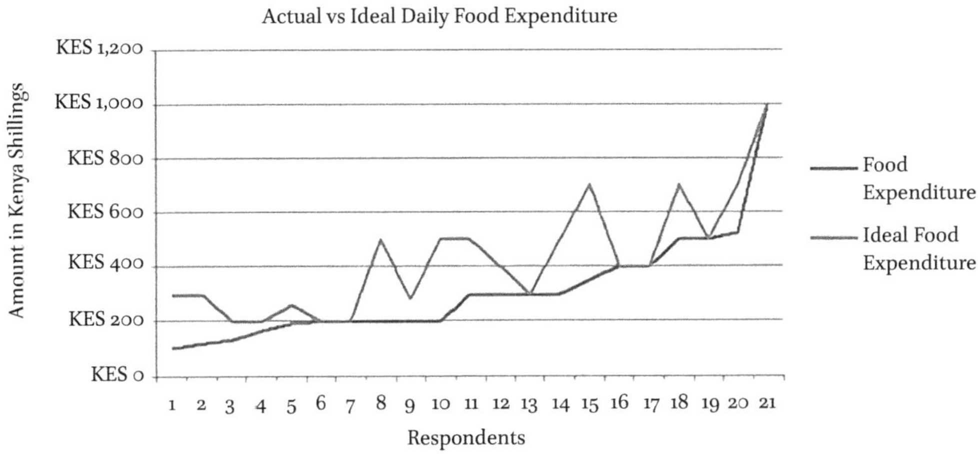

FIGURE 9.1 A comparison of actual and ideal expenditures on food
SOURCE: FIELD DATA

amounts. In-depth interviews revealed the varied ways households would use additional income to address the food-security gap. While some respondents stated that they would buy higher quantities of food, others noted that they would increase the varieties of food consumed to improve their nutrition, and a few others would buy food that was currently unaffordable on a regular basis. These different ways to address food needs show that income is only a means to accessing food, beyond which varied routes can be taken towards food consumption.

In the face of limited resources and unmet livelihood needs, employees devised innovative ways of addressing their food needs. Investing in one's social capital emerged as one of the key strategies for overcoming barriers experienced due to low wages. For migrant workers whose employment is relatively temporary and often insecure, building strong social networks outside of the workplace can be a daunting task, yet this is what is vital for one to be able to navigate the contours of the life of a low-wage labourer in an unfamiliar workplace distant from home. The workplace is the primary space where such social networks bud and grow from spending majority of the time with colleagues. To obtain financial capital to meet household needs, employees borrow money from colleagues with the assurance that they will reimburse the money once wages are paid. Secondly, employees establish their own financial savings associations, often in the form of SACCOs[9] or ROSCAs,[10] at the workplace, where agreed-upon monthly financial contributions are made,

9 Savings and Credit Cooperative Society.
10 Rotating Savings and Credit Association.

and from which they can obtain loans in times of need at a certain rate of interest. Aware of their financial limitations, some employees follow market dynamics to buy food in bulk when prices are lower. A wage labourer living within the floriculture premises told me that, "I buy non-perishable foodstuff when the supply is high and store it for future consumption. When I have some disposable income, I do shopping that can last for some time because I know there will come a time when I will not be able to do shopping due to other financial demands".

Outside the workplace, employees establish strong social relationships with food vendors and business operators in their settlement areas from whom, on the basis of trust, they can obtain foodstuffs and other household items on credit. As one of the food vendors at Kwa Muhia told me, most of her customers who were also employees in the floriculture sector often obtained foodstuffs from her stall on credit and paid either midway through the month when they received advance wages, or at the end of the month when wages were paid. In her view, the labourers often live off credit due to the low wages paid in the floriculture industry and, even then, they go for the cheapest foodstuffs which are of lower quality, and in small portions. As a close observer of the life lived by many low-wage labourers in the floriculture sector, the vendor felt that there were limited possibilities for such employees to climb up the economic ladder due to their low wages. Having had long-term interactions with floriculture employees, the vendor said she always sympathises with them and could not turn down their requests to obtain foodstuffs on credit. This assertion by the vendor shows that the relationships established with the wage labourers are not purely commercial but also empathic in nature. The assertion that the wages in the floriculture sector are too low to climb up the economic ladder echo those of other residents in Kwa Muhia and other settlement areas working outside the floriculture sector, and arguably allude to the reluctance to seek jobs in the flower farms by those from around the Naivasha area.

While owning a plot of land in Naivasha is uncommon among wage labourers, some have been granted small gardening plots within the quarters of the floriculture farms, and others own plots of land outside Naivasha from which they obtain food. The acreage owned by interlocutors in this study ranged between $7m^2$ to seven acres, with slightly more than half of the respondents possessing an acre or less, while only one respondent had seven acres. Having a plot of land far away from the workplace influences the type of crops that one cultivates, with one respondent stating that, "I chose to cultivate maize and beans because I have to do farming from afar and these crops do not require much attention". Maize and beans are the staple food in Kenya, and they are often harvested all at once, meaning one can afford to visit their distant farm only a few times over the course of a crop cycle. For those whose spouses reside in the countryside where they hail from, they bring food back to Naivasha, usually non-perishable foodstuffs, after their occasional visits back home. The alternative possibility of accessing food from one's plot of land functions as a diversification strategy that eases financial overdependence on wages.

One of the NGOs working on the conservation of Lake Naivasha felt that the floriculture industry had diminished land for other potential activities, including possible food production, as the farms are nested around the lake for the ease of access of water for irrigation. The notion of diminished land for food production was however strongly challenged. As I learned from a research officer of the Lake Naivasha Riparian Association (LNRA), the soils around Lake Naivasha are not particularly fertile and have an extremely low water-holding capacity, qualities that make the land unsuitable for crop cultivation. A manager in one of the flower companies affirmed that soil quality in the Naivasha area is too poor to cultivate flowers, and that therefore pumice – a volcanic rock substance obtained from the areas around Mount Longonot, Kenya – is used in flower cultivation. As such, it is access to water from Lake Naivasha rather than access to rich soils that underpins the establishment of the floriculture industry in the area, in addition to Naivasha's geographical nearness to Nairobi. The Naivasha area receives low rainfall that is inadequate for practising rainfed agriculture throughout the year, a climatic fact that renders agricultural production an extremely expensive venture requiring irrigation equipment and application of significant external inputs.

An important aspect of food security, and one which poses an enormous challenge for Naivasha despite efforts by floriculture companies to provide clean water at the workplace, is access to clean drinking water. Most of the households living outside the floriculture companies depend on tap water which, as I observed from living in a shared compound with some households of the wage labourers and others pursuing different economic activities such as shopkeeping, flows for just a few days per week and, sometimes, only once in a week,[11] and is unsafe for drinking. The intended benefits from the provision of clean drinking water at the workplace are therefore reversed as the labourers consume untreated water at their homes. The scarcity of water at home results in poor sanitation as there is rarely water to flush toilets, on top of which the sewage systems are often dysfunctional. The extremely poor state of sanitation, such as in the compound I lived in, poses significant health risks for everyone, especially children. These experiences in the dwelling places

11 I observed these dynamics while living in one of the rental housing compounds occupied by wage labourers from the floriculture industry.

of wage labourers demonstrate that corporate interventions to improve food security and livelihood conditions only in the workplace, while positive, stand to be reversed by the poor quality of life at home.

7 Corporate Interventions

I have titled this section *corporate interventions* to initiate a shift of thought away from the generally accepted and utilised concept of corporate social responsibility (CSR). Such a shift inspires rethinking corporate activities targeted at employees and, at times, adjacent local communities, not simply as appendages, but rather as critical interventions that constitute vital means to address basic livelihood needs in settings like Naivasha. In many countries in the Global South, corporate interventions have become more critical due to a dearth of provision of public services by the state, in part viewed as constitutive of the rolling-back of the neoliberal state and, at other times, viewed as the results of private indirect governance where public resources are corruptly appropriated by individuals at the expense of the public, essentially breaking the social contract between the citizenry and the state (Ferguson 2006; Mbembe 2001). Further, corporate interventions drive us to reflect on the extent to which they amount to a known telos that is to be attained through preconceived, instrumentalised measures and formats. As such, our attention is drawn not only to the tangible deliverables that can be measured by the certification bodies, for example, but also the ideas behind such an imagined world where the corporate is tasked with procuring public goods and services.

The floriculture industry in Naivasha has enacted corporate interventions mainly under the tenets of CSR, wherein interventions are understood as the pursuit of responsible business that contributes to the welfare of the employees and local communities (Heemskerk 2012). Among the CSR initiatives implemented by floriculture companies in Naivasha, which I now term corporate interventions, are provision of meals at the workplace, educational support through infrastructural development, bursaries, school meals, employment of teachers, provision of clean drinking water, and healthcare services at the workplace. The manager of one of the companies that provided clean water to the public explained the significance of the initiative in the area:

> We provide clean water to the community around here. There is a tap outside the flower farm that is accessible to everyone. This has been operational since 2005. It was set up following discussions with the community leaders, who identified access to clean drinking water as a major

> problem for the area. Water provision reduces risks and burdens among the community members since they do not have to encounter hippos as they go to fetch water. The employees get out of work late and those times can lead to human encounters with the hippopotamuses.

The impact of this water project arguably underlines the importance of community participation in the decision-making processes about the kinds of interventions desired by local communities. At the same time, following earlier observations about continuing water problems in Naivasha, this experience shows that there is need for co-ordination and advocacy for such corporate development projects to reach broader populations across Naivasha, such as through enhanced public private partnerships that push the state to be more accountable to the citizenry.

In addition to corporate interventions, it has been suggested that foreign agribusiness investments in the Global South facilitate the transfer of agricultural and technical knowledge to smallholder farmers (Attah 2021). This proposition is often underpinned by the view that such investments possess more advanced agricultural technologies and knowhow vis à vis the smallholder farmers in the Global South. However, interviews with local farmers in Maiella, one of the agricultural areas adjacent the South Lake area, revealed that knowledge transfer from the floriculture industry to the farmers was non-existent. The farmers instead acquire agricultural knowledge primarily through individual farming experience and farmer-to-farmer knowledge exchange. The floriculture farms, maintain the farmers, operate as enclaves and the farmers have little knowledge about what happens behind the closed gates of the farms. What the farmers deemed a positive contribution was not directly attributed to the floriculture industry as the proponents of technological and knowledge transfer would have it, but, rather, indirectly through the expansion of the market for their farm produce from the large population of wage labourers. As one of the farmers noted, "When they [floriculture companies] employ people, there is money circulating in the local economy which is good for business [increasing demand for our produce]. I think this is the only way that the flower farms have contributed to agriculture around here". This observation illustrates that wage labourers in the floriculture sector are thus integral in making Naivasha a nerve centre by creating a market for agricultural produce from the surrounding areas and, overall, contributing to the larger economy of Naivasha.

While the imagined trajectory of knowledge transfer to local farmers is confounded, I found out that the wage labourers gain agricultural knowledge from the workplace which they in turn apply on their own plots, usually at their places of origin. As one of the respondents said, "I did not have any agricultural knowledge when I started working at the flower farm. But, right now, I have gained so much knowledge that I view farming as one of the options should I stop working at the flower farm". This futuristic hope of pursuing farming on an individual plot constitutes what Styles (2019) terms "rosy aspirations" that are built on the view of employment in the floriculture sector as the means to better life pursuits.

The limited impact of the floriculture industry on local agriculture and communities can be traced back to the industry's operation as an agribusiness enclave, which inhibits not only physical access but also the transfer of knowledge and other potentially positive externalities. This apparent disconnect, particularly with the farmers in adjacent localities, challenges the proposition that the presence of foreign agribusiness investments results in the transfer of agricultural technology and knowhow. When I posed the question about the contributions of the floriculture sector to the Naivasha area to the local chief, he asserted that the sector had not contributed much more than offering employment opportunities in Naivasha. This view not only echoed that of the farmers, but also underscored the perspective that any benefits from the floriculture industry principally accrued to the wage labourers and not to the local communities in general. The extent to which potential impacts have been hamstrung by the enclave nature of the floriculture industry underlines the need to establish corridors for facilitating the flow of direct and indirect benefits to local communities. Ferguson's (2006, 38) observation that capital hops onto localities of high potential whilst "neatly skipping over most of what lies in between" underlines the need for such deliberate efforts to address the enclave nature of the floriculture industry.

According to Maurice Bolo (2008), flower production on a single hectare requires approximately US$50,000. This cost does not include the additional expenses of off-farm activities such as fortifying marketing networks. These hefty investments in cultivating flowers are, however, offset by the high returns from the sale of the flowers. My interviews with two floriculture companies revealed that the industry is highly lucrative since flower exports accumulate revenue returns from having most of the processes in the commodity's value chain undertaken locally at a significantly lower cost. The highest costs in flower production are incurred during the purchase of external inputs, which suggests that it is imperative to have significant financial capital for one to engage in floriculture production, a requirement that largely locks out smallholder farmers. Further, as one investor in Naivasha told me, it takes around eight years for a floriculture company to break even and recover its costs before profits can be realised. While floriculture is a lucrative undertaking, therefore, a business needs to be in operation for a long time before profitable returns can be realised, pointing to further lofty financial demands for engaging in medium- to large-scale flower farming. An evaluation on the differential contributions by the floriculture companies to corporate development ought therefore to factor in the period for which a floriculture farm has been in existence, and/or whether corporate interventions are already deemed costs of operation.

8 Conclusion

This study set out to uncover the linkages between the floriculture industry and the status of food security among wage labourers' households in the Naivasha area. Employment within the floriculture industry, which acts as a pull force for labour from different parts of Kenya, functions as the primary link between wage labour and food security. The low wages earned, however, mean that the labourers' households are unable to adequately meet their food and livelihood needs. An ingenious strategy to overcome these financial limitations is the investment in social networks that are empathetically woven together by trust, which enable employee households to access food from the market on credit. These social networks constitute vital social capital in securing food for the wage labourers' households. However, difficulties in accessing food still persist despite their utilising social networks, thereby revealing the limits of social capital in meeting food and livelihood needs. Further, as households still have to pay for the goods accessed on credit, the need for financial capital remains paramount. In this regard, social networks function as implements for "buying time" among wage labourers.

Previous debates on foreign agribusiness investments have focussed almost entirely on land acquisitions, and in the process shrouded the role of water. The case of Naivasha underscores the central role of water and uncovers fundamental interlinkages between agribusiness investments and local livelihoods and food security. Natural resources, such as water in this case, bring together different players, including foreign agribusinesses dependent on water for farming. As a necessary resource, water has coerced the floriculture industry to implement a PES-scheme targeting upstream farmers, whose activities influence the health of Lake Naivasha, the primary source of water for farming. As such, it is in the entanglements of interests that agribusinesses appear to engage with local communities, which suggests that building such spaces of engagement between agribusinesses and local communities has the potential to generate more interventions that address local challenges.

The implementation of corporate interventions, through what is often termed CSR, has shown its potential to impact the livelihoods of wage labourers. However, these interventions are found to be limited when it comes to impacting local communities beyond those working in the floriculture industry, a disconnect that has rendered the floriculture industry an enclave wherein those outside know little if anything about what transpires within the farms. As a result, the impact of the floriculture industry on local farmers is transmitted through the indirect link mediated by the wage labourers who, through their wages, have created a market for local farmers' produce. Further, as a highly capital-intensive industry, the operationalisation of the floriculture sector differs from the practices of smallholder farmers. Rather than assuming agricultural knowledge is automatically transferred from foreign agribusiness investments, therefore, deliberate linkages have to be created to facilitate that kind of knowledge transfer. In addition, rather than treating such knowledge transfer as a one-sided occurrence, the agency of local farmers ought to be recognised so that consultations are conducted to learn about the agricultural practices and needs of local farmers, as opposed to positioning them simply as recipients of "modern" technological knowledge presumed to be the preserve of foreign investments. A livelihoods approach can facilitate such engagement that brings together emic and etic perspectives, as in the case of the water project implemented by one of the floriculture companies. At the same time, there is need for attention to be paid to the limits of corporate interventions and the potential for the state to roll back from the provision of public services requisite for ensuring basic livelihoods. In order to create a platform for a sustainable livelihoods approach, it is critical that private and public responsibility are viewed not as substitutes for or alternatives to one another, but rather as complementary. The nature of food security, weaving together multiple livelihood dimensions that include food, water, sanitation, and health, presents an opportunity to examine the efficiency and effectiveness of wage labour and corporate interventions in addressing the precarity of life in Naivasha.

Acknowledgements

This research was supported with funding from Utrecht University and the Sasakawa Young Leaders Fellowship Fund (SYLFF). I would like to acknowledge critical feedback obtained from colleagues at the Concordia Ethnography Lab at Concordia University in Montreal, Canada, and the interlocutors in Naivasha and Nairobi that participated in this research study.

References

Anker, Richard and Martha Anker. 2014. *Living Wage for Kenya with Focus on Fresh Flower Farm Area near Lake Naivasha.* Report prepared for Fairtrade International, Sustainable Agriculture Network/Rainforest Alliance and UTZ Certified.

Appadurai, Arjun. 1996. *Modernity at Large: Cultural Dimensions of Globalization.* Minneapolis: University of Minnesota Press.

Attah, Noah E. 2021. "Behind Accumulation and Dispossession: State and Large-Scale Agricultural Land Investments in Nigeria." In *The Transnational Land Rush in Africa,* edited by Logan Cochrane and Nathan Andrews, 113–136. New York: Springer.

Berlant, Lauren G. 2011. *Cruel Optimism.* Durham: Duke University Press.

Bolo, Maurice O. 2005. "Agricultural Systems of Science, Technology and Innovation: The Case of Kenya's Floriculture Industry." Technical Center for Agriculture and Rural Cooperation (CTA), The Netherlands.

Bolo, Maurice O. 2006. *Knowledge, Technology and Growth: The Case Study of Lake Naivasha Cut Flower Cluster in Kenya.* World Bank Institute.

Bolo, Maurice O. 2008. "The Lake Naivasha Cut Flower Cluster in Kenya." In *Knowledge, Technology, and Cluster-Based Growth in Africa,* edited by Douglas Zhihua Zeng, 37–52. Washington: World Bank.

Boonstra, Marijke. 2010. *Sustaining and Enlarging the Payment for Watershed Services (PWS) Program in the Naivasha Catchment.* Amsterdam: Institute for Environmental Studies (IVM).

Chambers, R. and G. Conway. 1992. *Sustainable Rural Livelihoods: Practical Concepts for the 21st Century.* Brighton: Institute of Development Studies (UK).

De Haan, Leo, and Annelies Zoomers. 2005. "Exploring the Frontier of Livelihoods research." *Development and Change* 36 (1): 27–47. https://doi.org/10.1111/j.0012-155X.2005.00401.x.

Dolan, Catherine, Maggie Opondo, and Sally Smith. 2002. *Gender, Rights and Participation in the Kenya Cut Flower Industry.* Report No. 2768. Chatham: NRI.

Edelman, Marc. 2014. "Food sovereignty: Forgotten Genealogies and Future Regulatory Challenges." *The Journal of Peasant Studies* 41 (6): 959–978. https://doi.org/10.1080/03066150.2013.876998.

Everard, Mark, and David M. Harper. 2002. "Towards the Sustainability of the Lake Naivasha Ramsar Site and its Catchment." *Hydrobiologia* 488: 191–203. https://doi.org/10.1023/A:1023390430571.

FAO. 1996. "Rome Declaration on World Food Security and World Food Summit Plan of Action". World Food Summit, 13–17 November 1996, Rome, Italy.

Ferguson, James. 2006. *Global Shadows: Africa in the Neoliberal World Order.* Durham: Duke University Press.

Heemskerk, Lisanne. 2012. "How Responsible is Responsible Business? An Analysis of the Drivers and Effects of the Responsible Business Practices of Dutch Enterprises Operational in Kenyan Agribusiness." Master's thesis, University Utrecht. https://studenttheses.uu.nl/handle/20.500.12932/16411.

Kabiru, Joseph G., Paul N. Mbatia, and Edward K. Mburugu. 2018. "Emerging Conditions of Labour in the Cut Flower Industry in Kenya." *International Journal of Education and Research* 6 (5): 1–12.

Kiamba, Makau. 1989. "The Introduction and Evolution of Private Landed Property in Kenya." *Development and Change* 20 (1): 121–147.

Kirigia, Evans, Gemma Betsema, Guus van Westen, and Annelies Zoomers. 2016. "Flowers for Food? Scoping Study on Dutch Flower Farms, Land Governance and Local Food Security in Eastern Africa." Utrecht: Food & Business Knowledge Platform, LANDac, and IDS, Utrecht University.

Kuiper, Gerda. 2019. *Agro-industrial Labour in Kenya: Cut Flower Farms and Migrant Workers' Settlements.* Cham: Palgrave Macmillan.

Kuiper, Gerda. 2020. "'They Just Move in with Relatives': Translocal Labour Migrants and Transient Spaces in Naivasha, Kenya." *Journal of Eastern African Studies* 14: 227–249. https://doi.org/10.1080/17531055.2020.1730562.

Mbembe, Achille. 2001. *On the Postcolony.* Berkeley: University of California Press.

McMichael, Philip. 2014. "Historicizing Food Sovereignty." *The Journal of Peasant Studies* 41 (6): 933–57.

Mwangi, "Nungari." 2019. *The Power to Flourish: Unearthing the Roots of Kenyan Flower Producers' Market Access Strategies.* PhD-Diss., University of Cambridge.

Nweke-Eze, Chigozie, and Eric M. Kioko. 2021. "'But We Cannot Do It All': Investors' Sustainability Tensions and Strategic Selectivity in the Development of Geothermal Energy in Kenya." In *Sustainable Development in Africa. Fostering Sustainability in One of the World's Most Promising Continents*, edited by Walter L. Filho, Rudi Pretorius and Luiza O. de Sousa, 385–404. https://doi.org/10.1007/978-3-030-74693-3_21.

Nyongesa, Josephat M., Hillary K. Bett, Job K. Lagat, and Oscar I. Ayuya. 2016. "Estimating Farmers' Stated Willingness to Accept Pay for Ecosystem Services: Case of Lake Naivasha Watershed Payment for Ecosystem Services Scheme-Kenya." *Ecological Processes* 5: 1–15. https://doi.org/10.1186/s13717-016-0059-z.

Odhong, Emily A., and Jacob Omolo. 2014. "An Analysis of the Factors Affecting Employee Relations in the Flower Industry in Kenya: A Case of Waridi ltd, Athi River." *International Journal of Business and Social Science* 5 (11): 147–160.

Okoth-Ogendo, Hastings W.O. 1991. *Tenants of the Crown: Evolution of Agrarian Law and Institutions in Kenya.* Nairobi: ACTS Press.

Patel, Raj. 2009. "Food Sovereignty." *The Journal of Peasant Studies* 36 (3):663–706. https://doi.org/10.1080/03066150903143079.

Rogei, Daniel S. 2021. "Mediating *Maendeleo*: Examining the Nexus Between Geothermal Extraction, Wildlife Conservation and Community Well-being in Olkaria-Suswa, Southern Kenya." PhD-diss., Carleton University.

RoK. 2006. Kenya National Guidelines on Nutrition and HIV/AIDS. Ministry of Health, Republic of Kenya.

Rouillé-Kielo, Gaële. 2019. "Distributing Responsibilities in an Agricultural Ecosystem: Insights from the Lake Naivasha Water Basin in Kenya." *Nature and Culture* 14 (3): 251–270. https://doi.org/10.3167/nc.2019.140303.

Sang, Joseph, Iddi Mwanyoka, Josephat Nyongesa, Dosteus Lopa, and John Mwangi. 2017. "Case Studies of Water-related PES schemes in East Africa." In *Co-investment in Ecosystem Services: Global Lessons from Payment and Incentive Schemes*, edited by Sara Namirembe, Beria Leimona, Meine van Noordwijk, and Peter Minang, 105–116. Nairobi: World Agroforestry Centre (ICRAF).

Sankan, S.S. Ole. 1965. *The Maasai*. Nairobi: East African Literature Bureau.

Schmink, Marianne. 1984. "Household Economic Strategies: Review and Research Agenda." *Latin American Research Review* 19 (3): 87–101. https://doi.org/10.1017/S002387910002149X.

Scoones, Ian. 1998. *Sustainable Rural Livelihoods: A Framework for Analysis.* IDS Working Paper 72. Brighton: IDS.

Scoones, Ian. 2009. "Livelihoods Perspectives and Rural Development." *The Journal of Peasant Studies* 36 (1): 171–196. https://doi.org/10.1080/03066150902820503.

Sonneveld, Ben G.J.S., Max D. Merbis, Amani Alfarra, Olcay Ünver, and Maria A. Arnal. 2018. *Nature-Based Solutions for Agricultural Water Management and Food Security.* FAO Land and Water Discussion Paper. Rome: FAO.

Styles, Megan A. 2019. *Roses from Kenya: Labor, Environment, and the Global Trade in Cut Flowers.* Seattle: University of Washington Press.

Wasserman, Gary. 1973. "The Independence Bargain: Kenya Europeans and the Land Issue 1960–1962." *Journal of Commonwealth & Comparative Politics* 11 (2): 99–120. https://doi.org/10.1080/14662047308447181.

Woodhouse, Philip. 2012. "New Investment, Old Challenges. Land Deals and the Water Constraint in African Agriculture." *The Journal of Peasant Studies* 39 (3–4): 777–794. https://doi.org/10.1080/03066150.2012.660481.

CHAPTER 10

Export-Crop Production in Kinangop (Kenya): Exploring Small-Scale Farmers' Tactics and Strategies under Contract Farming

Gaële Rouillé-Kielo, Bernard Calas and Sylvain Racaud

1 Introduction[1]

Contract farming describes a situation in which processing companies contract out the production of agricultural goods to farmers. The development of contract farming has gained support among development agencies, who view it as a potential lever for increasing the productivity and income of small farmers in developing countries. However, the scientific literature on the subject offers many counterpoints to this idealised vision of contract farming in the Global South. Studies have pointed to (i) its detrimental effects on the concentration of wealth by improving the incomes of the better-off, while marginal and poor farmers remain mostly excluded from contract schemes (Bellemare and Bloem 2018; Ray, Clarke, and Waley 2021; Ton et al. 2017); and (ii) the lack of empowerment and the various pressures experienced by contract farmers, including those who own their land, on the management of their production (Pérez Niño 2016; Ray, Clarke, and Waley 2021). The example of Kinangop, explored in this chapter, contributes to questioning the opportunities and pitfalls of contract farming for small farmers. Indeed, since the early 2000s, the production of export crops (vegetables, summer flowers) on small-scale rainfed farms has rapidly expanded on the Kinangop Plateau (Nyandarua County, Kenya), located in the direct vicinity of Naivasha – Kenya's main horticultural hub. Most of this production is done under contract, structuring the production and marketing practices of small-scale farmers. Despite the strict standards imposed by external actors, this paper shows that farmers seek to take advantage of an externally oriented market. We argue that the insertion of Kinangop Plateau agriculture into global trade networks results in farmers developing individual and collective tactics to make the most of contract farming or to emancipate themselves from it.

1 First submission date: February 22, 2021. Acceptance date: March 13, 2023.

 | DOI:10.1163/9789004695429_011

The capacity for initiative on the part of small-scale producers depends on many external factors, including market opportunities and constraints (remunerative prices, stringent standards, etc.). In this chapter, we are also interested in the capacity of small-scale farmers to act and react in the face of this new and constantly evolving system. Following Michel de Certeau (1990), we analyse the "art de faire" of small-scale farmers to increase their room for manoeuvre in the export-crop value chain, where powerful actors are able to impose their vision of the market (e.g. by imposing standards). This is expressed opportunistically through "tactics" – which can be seen as means of resistance to institutionalise power by using the flaws in the system – and "strategies" that aim to fulfil long-term ambitions (de Certeau 1990).

This chapter is mostly based on interviews with small-scale farmers and on farm visits between 2015 and 2016, and on complementary data (farm visits and interviews) in 2020. Reference will also be made to documents consulted at the Kenyan National Archive in Nairobi. It is structured as follows: in the next section, we highlight the main physical and demographic characteristics that make Kinangop a predominantly rural and agricultural area. The third section traces back the major historical and political shifts that occurred over the colonial and post-colonial periods in this area, which resulted in successive transformations of the productive orientations and structures. The fourth section will explore the introduction of the export system in Kinangop. The last section discusses the tactics and strategies used by small-scale farmers under contract farming to overcome the constraints of this system where they are in a disadvantaged bargaining position.

2 Case-Study Description: Kinangop, a High-Altitude Rural and Agricultural Area

While previous research has given insights into the formation of the Naivasha horticultural hub (see for instance Calas 2013; Kuiper and Gemählich 2017; Harper and Styles, Müller-Mahn and Gemählich, and Kuiper, all this volume) and consequently its role as a "nerve centre", little attention has been paid so far to the development of export-crop production and to the specific economic dynamics in Kinangop.[2] More broadly (and critically), very few studies in the social sciences have been carried out in this area, at least since the well-known

2 With the exception of the recent doctoral work by Nungari Mwangi (2018), which was partly conducted in Nyandarua County.

research work by Christopher Leo about land-allocation processes in this area (Leo 1978, 1984; Leo and Lonsdale 1988).

Although located in the close vicinity of Lake Naivasha, the Kinangop Plateau has very distinct physical characteristics, which are reflected in the differences in agronomic features. The latter are mainly explained by altitude contrasts: the lake is located in the Rift Valley, at about 1,880 metres above sea level; the Kinangop Plateau (see Figure 10.1), separated from the Rift by an escarpment, rises between 2,400 metres westwards and 2,700 metres on its eastern edge. These differences are mirrored by the very different climatic conditions in these two areas, aridity being much more prevalent in the Naivasha region while Kinangop is much more humid and colder. In Kinangop (unlike along the shores of Lake Naivasha), the rainfall regime makes it possible to cultivate without resorting to irrigation. On the other hand, cold and waterlogging can be limiting factors for some crops. In the *Farm Management Handbook of Kenya*, the influence of the proximity of the Nyandarua Mountains is highlighted: at night, heavy cold air can reach the plateau and therefore

FIGURE 10.1 East-west view across the Kinangop Plateau from a settlement village farm in the foothills of the Nyandarua Mountains. Tulaga, North Kinangop. In the background, Mount Longonot, south of Lake Naivasha
PHOTO: GAËLE ROUILLÉ-KIELO 2016

"decrease the mean minimum temperature for (*sic*) almost 3° C" (Jaetzold et al. 2010, 43). These climatic conditions can be detrimental to certain crops such as maize, while others, such as potatoes, are resistant to low temperatures (*idem*). Rainfall conditions differ across the plateau, with a longer rainy season in the south-eastern part, and a potentially more intense aridity in the western part of the plateau during the dry season. This influences the ability of soils to remain moist throughout the year, and thus the possibility of planting certain annual crops (the report gives the example of pyrethrum).

While Lake Naivasha and its riparian zone are characterised by a great diversity of land uses and economic activities (agro-industry, geothermal electricity production, commercial activities and services, tourism, and conservation), the economy of Kinangop is still largely dominated by small-scale agriculture. A reported 65% of the population of Nyandarua County is engaged in animal husbandry, and 75% in land cultivation (Nyandarua County Government 2018). In fact, most people are involved in mixed farming. The main crops in terms of land use and market value in the county are potatoes,[3] maize, peas, cabbage, and carrots (see Figure 10.2). These crops are mainly intended for the national market. Only a few crops (i.e., cut flowers and snow peas) are for export to the European market. Nonetheless, they represent a relatively high market value compared to the small area occupied.

Kinangop is part of Nyandarua County, which presents a relatively low urbanisation rate compared to neighbouring counties, such as Nakuru County where Naivasha is located. According to the most recent national census, almost 90% of the population of Nyandarua County live in rural areas (KNBS 2019). The rural population density is quite high, with 245 inhabitants per square kilometre in Kinangop Sub-County and 209 inhabitants/km^2 in Kipipiri Sub-County, for example.[4] On the plateau, very few places exceed the population threshold of 2,000 that would allow them to be classified as "urban centres". However, these few urban centres have experienced significant growth in recent years. For instance, the population of the small urban centre of Engineer has more than doubled over the last decade (KNBS 2010; Nyandarua County Government 2018). This demographic growth has had an accelerating effect on a trend towards increasing scarcity of land resources that started in the 1960s.

3 Nyandarua County is ranked first for potato production in the country.

4 This is much lower than Naivasha Town's density of 1350 inhabitants per square kilometre (KNBS 2019).

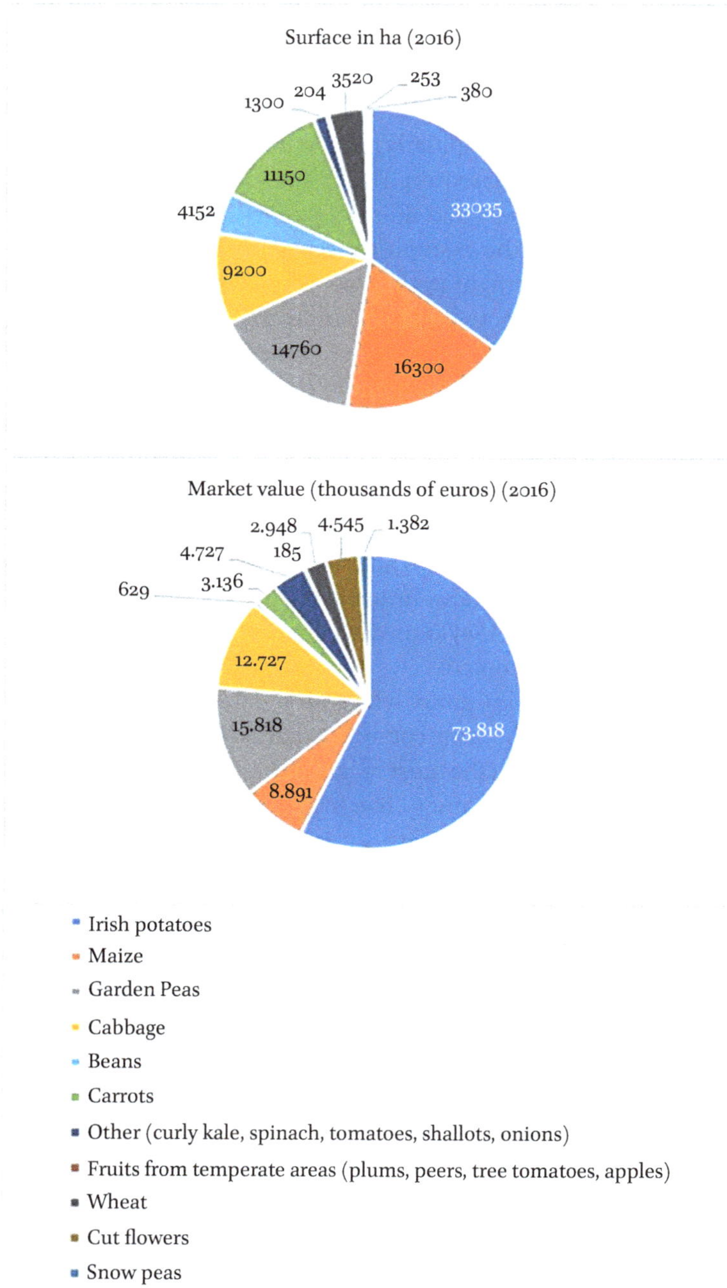

FIGURE 10.2 Main crops in Nyandarua County, by area and value, 2016
DATA SOURCE: NYANDARUA COUNTY GOVERNMENT (2018, 26)

3 A Historical Glance at Kinangop: Major Shifts in Land Use Throughout the Twentieth Century

This section examines the main historical shifts that helped to shape the contemporary agricultural system on the Kinangop Plateau during the twentieth century. As in most places in Sub-Saharan Africa, the agricultural history of Kinangop has beeen as much shaped by political events and decisions as by agronomic conditions. This history is made up of changes in land ownership that brought about major transformations in land occupation and use. Three major shifts can be identified: i) the colonial grab between 1900 and 1934; ii) the settlement schemes during the post-independence period; and iii) the political and economic liberalisations of the 1990s. These shifts largely determine the current agrarian system[5] present on the Kinangop Plateau.

Before the European colonisation of Central Kenya, the grassy Kinangop Plateau was part of the Maasai influence area, partly shaped by the browsing of cattle herds and bush fires (see Waller, this volume). They occupy this area seasonally, in complementarity with the Naivasha area, as explained by Claire Médard:

> The Keekonyokie would occupy the Kinangop Plateau during the dry season (January to March), and the Naivasha plains during the rainy season (April to June). Then between July and September, during the cold and dry season, they returned to Kinangop or to the Ewuaso Ewuaso region and finally in November-December, during the rainy season, they went back to the Rift Valley plains again.
>
> MÉDARD 1999, 48 – TRANSLATED FROM FRENCH

The Maasai were expelled from the plateau in the aftermath of the 1904 agreement between their representative – Laibon Lenana – and the representative of the British settlers (see Waller, this volume). This paved the way for the appropriation of land by the white settlers, henceforth barring the Maasai from accessing it. The latter only left behind a few toponyms (such as Njabini) as testimonies of their past (and seasonal) presence on the Kinangop Plateau.

5 An agrarian system ["système agraire" in French] is "a mode of exploitation of the milieu, historically constituted and enduring, adapted to the bioclimatic conditions of a given area, and responding to the social conditions and needs of the moment" (Mazoyer 1987, 11 – translated from French). Such a system can cover an area ranging from the size of a village to that of a region (Cochet 2011).

3.1 *Kinangop in the Colonial Period*

European farmers settled on the Kinangop Plateau very gradually during the first half of the twentieth century. It all started with the decision taken in 1902 to open the land to the settlement of European settlers in the highlands of Kenya. Initially, few Europeans settled there (their numbers reached about 1,000 on the eve of the First World War in 1914) and they concentrated around the nucleus formed by Nairobi. In 1915, a new Crown Lands Ordinance clarified the terms of settlement by defining the boundaries of the estates more precisely and extending their allocation to 999 years for the pastoral estates. This led to the development of (particularly) large estates around Lake Naivasha. On the Kinangop Plateau, the arrival of Europeans accelerated with the allocation of land to former soldiers as a reward for their services during the First World War.

During the 1920s, the colonial estates began to specialise and the 125 farms of the Naivasha district mostly engaged in ranching activities. On the Kinangop Plateau, the natural temperate-type grassland was progressively turned into extensive fields of cereals and pyrethrum[6] (Morgan 1963, 150). The relative importance of the Kinangop production area increased after the Second World War, particularly in the wake of the agricultural development Swynnerton Plan (1954), which made it one of the pivotal areas for the intensification of the colony's production. As pointed out by Odingo (1971, 55):

> It is on the Kinangop Plateau of the Rift valley that the second most important arable farming area in the Highlands is to be found. On this plateau in 1960 the arable percentage ranged from 12 to 38 percent and in North Kinangop it was as high as 51 percent.

Mixed farming was common on the plateau: alongside the production of pyrethrum and cereals (wheat, barley, rye), many farms also engaged in cattle and sheep rearing.[7] In the early 1960s, the authorities encouraged the diversification of agricultural production, which led to the development of dairy production on the plateau (*ibid*).

The share of pyrethrum produced in the total volume of exports from the Kenyan colony increased in the 1940s. By 1960, pyrethrum covered more than 20% of the cultivated areas (Odingo 1971). On some properties, such as that of Ndiara (810 hectares), pyrethrum represented up to 60% of the agricultural cover (*see above*). This crop was increasingly combined with wheat production or with dairy farming and fodder plants. A temperate plant, in Kenya

6 Pyrethrum is a perennial plant commercially used to produce natural insecticide.
7 Sheep farming was more prominent in the warmer lands to the west of the plateau.

FIGURE 10.3 Abandoned house of a former settler (Martin Lay, locally nicknamed "Kinyaweh" which gave the area its vernacular name), Gathara, North Kinangop
PHOTO: GAËLE ROUILLÉ-KIELO 2016

pyrethrum requires high altitudes (above 2,000 meter), temperatures below 25 °C and rainfall between 1,000 and 1,250 millimetres. Flowering lasts from nine to ten months per year, during which it is possible to collect the flowers every two or three weeks. This long flowering period made the Kenyan highlands one of the most productive areas for this crop (more than 226 kg of flowers per acre on average on Kinangop). Harvesting pyrethrum was very labour-intensive, and many workers (most of them women and children) were needed (see Figure 10.4). This fits in well with the "squatter" system already well established on the plateau, with the arrival of Kikuyu workers mostly from the 1950s onwards. In 1961, Kenya was the world's largest producer of pyrethrum. Its production was supervised by a Pyrethrum Marketing Board.

3.2 *Settlement Schemes and a New Productive System*

After the disruptive agricultural development at the beginning of the twentieth century on the Kinangop Plateau, the second major historical shift occurred at independence with two decisive decisions taken at the time. First, it was decided that the Kinangop Plateau would be part of the Central Province,

FIGURE 10.4 Pyrethrum picking, Rift Valley Province, Kenya, 1945
WITH PERMISSION FROM THE UK NATIONAL ARCHIVES, REF CO1069/137 (55)

leading to the administrative separation of the Lake Naivasha basin (Lake Naivasha being part of the Rift Valley Province).[8] The second decisive decision was to relegate the Kinangop Plateau to the One-Million-Acre Scheme that was to bring about the subdivision of the former European estates bought by the government.

It was notably in response to the land claims expressed during the Mau Mau uprising[9] that, from the 1960s to the 1970s, land-allocation programmes for previous European-occupied land were set up in Kenya. They involved 500,000 African Kenyans out of a total population of 11.2 million at the time (Boone 2014). Two categories of allocation programmes, the so-called "settlement schemes", were implemented: "low-density schemes" and "high-density schemes" (Boone 2014; Leo 1984; Oucho 2002):

> Land was allocated in average parcel sizes of 20 to 40 acres (8 to 16 hectares) in low-density schemes that were intended to be run as businesses

8 According to Hornsby, the 1962 decision to allocate the Nakuru district (where Naivasha is located), which was demographically dominated by the Kikuyu ethnic group, to the Rift Valley rather than to the ethnically homogeneous Central Province (Kikuyu) could be explained by the desire to provide more security for European farmers and to ensure the maintenance of landholdings in this area (Hornsby 2012, 74).

9 The Kinangop Plateau and the nearby Aberdare (Nyandarua) Mountains were key spots of the Mau Mau uprising. For more documentation, see Kanogo (1987) and Charton (2011).

> and generate commercial returns. In high-density schemes where the creation of a stable peasantry was the government's goal, the average landholding size was 10 to 15 acres (4 to 6 hectares) in areas with good farming potential, and 27 to 40 acres (11 to 16 hectares) in less favourable areas.
>
> BOONE 2014: 144

The One-Million-Acre Scheme implemented in the Nyandarua District was funded by loans and grants from the UK and German governments. It involved 1.5 million acres and 35,000 former squatter families (Hornsby 2012; Leo 1984). It was a "high-density scheme", aimed specifically at Africans perceived to have less agricultural experience, including former squatters (Leo 1984; Leo and Lonsdale 1988; Lukalo and Odari 2016).

Among the new settlers of the Million-Acre Scheme, a relatively large proportion had already been residing in the district for several years at the time of land allocation, where they were working as squatters for European farmers. A 1963 Ministry of Agriculture document estimated that 2,130 families identified as future settlers were already living on three of the farms planned to be divided and distributed under the Settlement Scheme in North Kinangop.[10]

The Kinangop Plateau was divided into different schemes (see Map 10.1). Among the first established, in 1962, those on the northern part of the plateau extended over 400 to 500 hectares (410 for Wanjohi and 444 for Kipipiri), while those on the southern part of the plateau were more than ten times larger (5,283 hectares for Njabini, for example) (Lukalo and Odari 2016).

According to local inhabitants who took part in the process, the allocation of plots was done randomly (by drawing lots during a "ballot") and led to the moving of many beneficiaries, sometimes from one end of the plateau to the other, depending on the location of the allotted plot (Rouillé-Kielo 2020). The former squatters who did not get land during this first settlement scheme, stayed in the villages in Kinangop. Subsequently, to satisfy their appetite for land, new land-allocation schemes were put in place under the Jomo Kenyatta government. The "Haraka Schemes" (1967–1969) on forest reserves (degazetted for this purpose),[11] extended over a large area from the southern edge of Mount Kipipiri

10 Document "*Hand over notes – Nyandarua District*", April 16, 1963, Ministry of Agriculture, Kenya National Archives (KNA).

11 A total of 6,100 hectares of forest reserves were degazetted by the Kenyatta government (Lambrechts et al. 2003). In Mikaro, according to the archival documents found, 1,874 plots of land have been allocated (mostly to men) in this area (File "*Geta scheme*", Special Commissioner – Squatters, Ministry of Lands and Settlement, KNA). This would represent about 3,000 hectares.

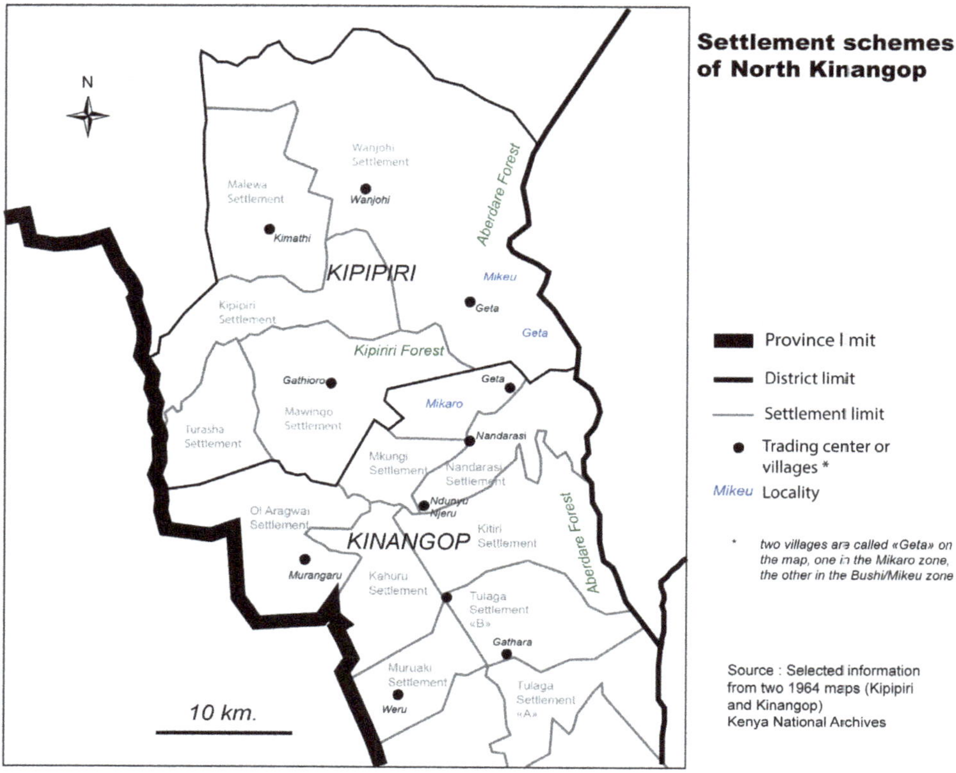

MAP 10.1 Location of the high-density settlement schemes, Nyandarua
DATA SOURCE: "SETTLEMENT SOUTH KINANGOP. KIPIPIRI/WANJOHI/MALEWA", MARCH 1963–JUNE 1964, PROVINCIAL AGRICULTURAL DEPARTMENT, GOVERNMENT OF KENYA, KENYA NATIONAL ARCHIVES

to the "corridor" between this massif and the Aberdare (Nyandarua) Range, on sometimes very steep land. The new holders were granted 4 acres each.

From the 1980s to the early 2000s, in the context of the introduction of multi-party politics, the allocation of land in the public-land domain became a powerful lever of political patronage. However, after the subdivision of former settlers' holdings, and then of post-colonial estates, the Moi regime had only the State domain at its disposal as a tool for land redistribution serving clientelist and electoral strategies (Klopp 2012). This State domain comprised of experimental farms, forest reserves, prison land, or unregistered community land, most often located in pastoral areas. In Kinangop, this strategic allocation of land by authorities occurred in the so-called Gathara village. After a long administrative struggle, on the eve of Kenya's first general multi-party elections

TABLE 10.1 Demographic growth in Nyandarua District, 1962–2019

Date	1962	1979	1999	2019
Population	117,000	233,000	479,000	638,000
Density of population (inhabitants/km²)	36	66	145	194

DATA SOURCES: MINISTRY OF ECONOMIC PLANNING AND DEVELOPMENT (1965); KNBS (1981, 2001, 2019)

in December 1992, the villagers were granted individual title deeds over about 100 acres of public land (Rouillé-Kielo 2020). Each new owner then received three quarters of an acre on a markedly steep land on the western fringe of the Nyandarua Mountains. Between the 1960s and the 1990s, the distribution of land on the plateau as well as the rapid subdivision of land (Kikuyu inheritance norms dictate that each son receives a parcel of land) led to a transformation of the land-tenure structure with a multiplication of small properties.

The redistribution of land in the 1960s led to strong population growth in Nyandarua District: as shown in Table 10.1, the population doubled within less than twenty years between 1962 and 1979 following the first waves of land allocation. The population has grown steadily since then,[12] albeit at a slower pace over the last two decades.

3.3 *From a Highly Regulated to a Liberalised System*

Farmers in North Kinangop were encouraged to plant pyrethrum and to produce milk. In a document produced by the Nandarasi cooperative society, it is noted that "the economy is based on the cultivation of pyrethrum, potatoes, butter and wool".[13] This productive orientation was combined with the creation of cooperatives – one per scheme – whose main function was to collect the farmers' produce and to deliver it to the Kenya Creameries Cooperative[14] for milk and the Pyrethrum Board of Kenya for pyrethrum. Between 1963 and 1966 nearly 200 cooperatives were registered every year. They aimed to "fulfil"

12 This is in line with the country-wide population growth, which has increased from 8.1 million in 1963 (at the time of independence) to 47.5 million in 2019 (the date of the last census), an increase of about 6 times.

13 "*Nandarasi Farmers Cooperative Society*", 1964–1968, KNA, Nairobi.

14 The cooperative was established in 1925 by colonialists. After independence, it maintained its monopoly on milk collection, processing and marketing (Leksmono et al. 2006; Wangu, Magnus, and van Westen 2021).

the government's "development promises" and "almost all small-farm coffee, pyrethrum and dairy production went through the cooperatives" (Hornsby 2012, 136). In 1966 these cooperatives were then nationalised under the Cooperative Act. They then became essential components of the technical and political supervision of the countryside.

In Kinangop, part of the proceeds from the sale of agricultural products were used to repay the 30-year loan that each farmer subscribed to when receiving his or her plot of land. This was intended to repay the subsidy granted by the British government for the payment of financial compensation to white settlers who left their properties behind.[15]

After 1976, because of the boom in synthetic substitutes, world pyrethrum prices declined, and the 1975 production record was no longer surpassed. Furthermore, the Pyrethrum Council was unable to pay producers due to management problems and collapsed in 1983. Although market liberalisation led to a slight recovery in the mid-1980s, the decline was irreversible and Kenya, which supplied 75% of the world's pyrethrum in 1998, produced only 2% by 2015; the nu3mber of producers fell from 300,000 to less than 30,000.

In the 1980s–1990s, the liberal tide of the "Washington Consensus" that swept through Kenya led to a series of bankruptcies of public and parastatal enterprises. These severely affected the market opportunities for small farmers in Kinangop. In the context of donor-imposed multi-party competition and popular protest and its Kenyan translation (the Gado Democracy), President Moi, anxious to maintain his power despite the elections, opportunistically followed the dictates of international donors and attacked cooperatives and marketing boards. The coffee and milk sectors, public breweries and parastatal supermarket chains were particularly targeted and gradually dismantled (Grignon 1996). This highly political liberalisation left farmers in Central Province, including those in Kinangop, vulnerable to market forces.

In the early 1990s, Kenya quickly became "a genuine capitalist open market economy (…) one of the most liberalised in Africa" (Hornsby 2012, 567). The liberalisation of the economy ("Soko Huru" in Swahili) particularly affected the agricultural sector. The Sessional Paper No. 6 of 1997 on "Cooperatives in a Liberalized Economic Environment" aimed "to make cooperatives autonomous, self-reliant, self-controlled and commercially viable institutions" (Wanyama 2009, 4). The government no longer subsidised some services (such as insemination services to dairy farmers) and their support to cooperatives declined

15 The purchase price that was planned for the area, according to 1961 documents cited by Christopher Leo, amounts to £24 per acre (Leo 1984, 101).

(Nyoro and Ngugi 2007). Furthermore, cooperatives lost their monopoly in the agricultural sector and were to compete with private actors (Wanyama 2009).

Liberalisation led to the collapse of several structuring institutions in rural areas such as the Kenya Creameries Cooperative, the Pyrethrum Board and the Kenya Farmers Association. New actors then came to fill the "vacuum" that was left. In Kinangop, they were "Kenyan entrepreneurs" or intermediaries working for resellers (brokers), who could directly negotiate with farmers without price controls. On the one hand, farmers gained access to better selling prices. On the other hand, they lost their collective bargaining capacity, which they were previously able to mobilise through the boards and the home cooperative. Farmers were now free to join the cooperative of their choice (those of neighbouring schemes or newly created private cooperatives, such as "Muki", founded in the mid-1990s in Kinangop). The opening up to competition rapidly eroded the number of cooperatives operating in the area. In North Kinangop, there were sixteen cooperatives, created at the time the settlement scheme was set up, before the liberalisation of the sector. By the mid-2010s, only two of them (Tulaga and Kitiri) had survived. The disappearance of the very strong supervision exercised by the cooperatives (in terms of production orientation and prices) also means the opening up of opportunities for new markets and new players.

4 Contract Farming in Kinangop

For a long time (especially after the fall of the pyrethrum market), crops grown in Nyandarua were of relatively low added value (unlike in other counties also dominated by small-scale farming such as Nyeri or Kiambu, where export crops such as coffee or tea were grown). Over the last two decades, higher-value crops have started being produced in Kinangop, such as off-season vegetables for the European market, or cut flowers. For instance, in the mid-2010s, Nyandarua ranked second nationwide (after Meru) for the production of snow peas. On the farms visited, different pea cultivars were grown, namely garden peas and two types of edible peas with pods: (sugar) snap peas and snow peas. The plots were either dedicated to the growth of a particular crop or integrated into a crop-rotation or crop-association system.[16] On smallholdings where flowers were grown, farmers produced "outdoor flowers" or "summer flowers"

16 A common rotation was observed: that of potatoes/peas, with peas allowing good nitrogen fixation which favours the growth of potatoes. In most cases, an intercropping system was adopted, with maize and peas being the dominant combination.

(for instance agapanthus, craspedia or ammi). In this section, the reasons for the emergence and development of export crop production in Kinangop will be scrutinised.

4.1 *The Emergence of Export Crop Production in Kinangop: A Regionally Declined National Initiative?*

The production of agricultural products for European markets in Kenya can be traced back to the development of tea and coffee crops on the fertile land of central Kenya during the colonial era. More recently, Kenya has become one of the new production centres located in the tropics for fresh (sometimes off-season) horticultural produce for Europe. The rapid growth of the cut-flower sector from the 1980s onwards was a major milestone in the rise and structuration of the Kenyan horticultural industry.

In Kinangop, as in other rural areas of Kenya, the production of cut flowers by small-scale farmers is confined to summer flowers, which can be grown in the open field (Mwangi 2019). The production achieved by small-scale farmers in the flower sector is marginal, and represents 5 to 10% of the total volume of flowers produced in Kenya. This phenomenon is not specific to Kenya and has also been observed in other tropical countries (e.g. Uganda, Tanzania, Ecuador) (Bolo 2010). Kenyan small-scale farmers, including in Kinangop, have also been engaged in other "non-traditional agro-exports" (i.e. agricultural products that were not previously exported from this country) in recent years. These include vegetables such as snow peas, garden peas and French beans (Mannon 2005; Okello 2011). The production of vegetables for export has particularly developed from the end of the 1980s onwards. Throughout the 1990s, small-scale farmers produced a significant share of the total volume of vegetables exported from Kenya (40–70%). Thereafter, their contribution fell sharply (around 20%) (Dolan and Humphrey 2000) due to two interrelated causes: the growing importance of supermarkets – with higher standards – as buying points for fruit and vegetables in Europe; and the increasing involvement of Kenyan exporters backwards in the farm production to gain greater control and reduce rejects (Nielsen and Gibbon 2011).

In the late 1990s, the main explanations given by external observers (NGOs, donors, some academics) for the difficulties encountered by small-scale farmers was their inability to comply with importing countries' regulations and expectations regarding chemical use or quality consistency (Nielsen and Gibbon 2011). Against this background, the Kenyan authorities, supported by foreign actors, developed an action plan in the early 2000s to improve the inclusion of small farmers in the global value chain. This plan was based on two major elements: EurepGAP – GAP standing for "good agricultural practices" – and contract farming. EurepGAP aims to help actors involved in the value chain (farmers, exporters, retailers) to meet the ever-higher expectations regarding food safety of European consumers. It sets a number of rules to reduce the presence of pesticide residues and other chemical inputs. All the regulations are laid out in a manual to guide the practices. Quality standards in flower production are less widespread than for fruit and vegetables (see Mwangi 2018, 207) but some certifications require the producers to comply with GlobalGAP and to meet environmental standards (Mwangi 2018; Riisgaard 2009).

According to an official of the Horticultural Board in Nairobi (met in February 2016), one of the reasons why Kinangop has long been excluded from the production of high-value and export horticultural products was the poor quality of the roads that had long prevailed in the region. It could have adversely affected transport time and therefore the cost and quality (freshness) of the products. This situation has greatly improved with the creation of a tarmac road across the plateau to Nairobi.

Conversely, a major comparative advantage for the development of the production of certain export crop vegetables was the climatic conditions on the Kinangop Plateau. For instance, snow peas need temperate temperature to develop well. Another advantage of the county is the comparative availability of land in Nyandarua compared to other counties (such as Nyeri or Murang'a). As a consequence, it was one of the first counties targeted for the development of snow peas after Nyeri in the 2000s. The French beans came later when the demand for this off-season product skyrocketed. The introduction of these export crops and the development of this type or production overall has "boosted the economy of the county" (Interview Feb. 2016, Horticultural Board, Nairobi).

To support the development of export crops in the county, a local office of the Horticultural Crops Directorate (HCD)[17] was created in Miharati (Northern part of the Kinangop Plateau) in 2007. According to its director, who one of the authors met several times in 2016, the aim was to better cover Nyandarua, considered at the time as a "really potential area" for the development of horticulture, more precisely for export-crop production. The creation of a local office would have helped provide assistance to farmers newly engaged in this activity.

Farmers engaged in the production of export crops in Kinangop remember the first public meetings (*barazas*) that took place around 2004 in the area, with the involvement of local extension officers and authorities, in churches or on

17 Formerly known as the Horticultural Crops Development Authority.

public grounds. During these meetings, representatives of export companies explained that they were looking for "entrepreneurs", as one former producer recalls:

> There were officials of exporters and officials of a company which supplies the chemicals, and about 20 people from various places. These companies introduced themselves. They explained that they've found a high potential for growing snow peas, sugar snaps and garden peas in the area. We did not know these crops before they told us about them. Then, they told people to form a group so as to be trained and to be provided the seeds. They explained to us about the chemicals to be used, the grading and collection place.

In this account, the very official nature of the introduction of these new crops to the farming community (as evidenced by the presence of the local authorities) can be perceived. The main part of the farmer's account concerns the particularities of the technical itinerary to be followed to produce crops that were previously completely unknown to them. This technical itinerary is very precise and seemed already designed by the companies (from the seeds to the type of chemical fertilisers, including the buildings where certain tasks will be carried out – grading, collection). The respect for this process forms the main challenge – and pitfalls – of this new production for small-scale farmers.

4.2 *Small-Scale Production: An Adjustment Variable to Fluctuating Demand?*

Involvement in horticultural export activities requires farmers to align with the standards of the countries of destination, in this case those of the European Union defined in the EurepGap by the end of the 1990s, later on defined as the GlobalGap. These standards require "the implementation of detailed and formalised control systems and record keeping at farm level" (Nielsen and Gibbon 2011, 81) and a high initial cost to achieve compliance (Muriithi, Mburu, and Ngigi, 2011). The expected outcome of this very controlled process is a low rate of production rejection. But with limited access to, and therefore little visibility in the marketing channel, contract farmers are notified by retailers if their production has been rejected or accepted when the control takes place at the airport. Common reasons for the production to be rejected are the acknowledged excessive use of chemicals or the presence of pests on the products.

According to the representative of Horticultural Board in Nairobi, meeting the requirements would not be a challenge for small-scale farmers:

> It's like when you have only one son instead of 3 children, your attention is focused on this one. It is not diverted. The same happens for small-scale farmers. They have only ¼ acre or 50 square metres. So it's not difficult for them to meet the requirements.
>
> FEB. 2016

However, several local informers disagree with this statement. Many farmers in Kinangop would have stopped producing export crops due to the numerous rejections of their production on the grounds of non-compliance with standards.[18] For instance, a self-help group launched in 2004 in order to gather together small-scale farmers engaged in snow-pea production stopped its activity in 2007 in the aftermath of several rejections.

4.3 *Contract Farming vs. Selling to Brokers: Cons and Pros*

According to the local branch of the HCD, two main types of farms in Nyandarua can be distinguished. "Large-scale farmers" can be defined as producers with greenhouses. "Small-scale farmers" are those who must pool their production (by forming registered groups of about ten farms in general) in order to reach a volume of production of one tonne minimum. The latter are bound via a system of contract farming to export companies – they are the ones we mostly met and interviewed. The HCD ensures that the export companies respect their contractual commitments to small-scale farmers, specifically the provision of technical advice, seeds, harvesting equipment and the transport of the production to Nairobi with refrigerated trucks. In 2016 there were reportedly about 20 exporting companies that were active in Nyandarua County, the most important ones being Wamu Investments (a Kenyan company active in different parts of the country) which worked with about seven farmer groups in Nyandarua County at that time, Kenya Fresh (six groups), AAA Growers (two groups) and Finlays.

With regard to vegetable production, contract farming is not the only value chain. Producers also sell (part of) their produce to brokers. The latter are usually locals who approach farmers to buy their crops. They usually operate on foot or by motorbike taxi (*boda boda*). According to the farmers, the advantage of selling to brokers is that they offer a better purchase price in times of peak demand. Also, they pay "cash" at the time of purchase rather than every

18 We have only been able to discuss cases of rejections with vegetable farmers, but such difficulties also affect small-scale flower growers, as shown by the mention of rejections of cut flowers produced by outgrowers in 2007 in Kiambu County by Nungari Mwangi (2018).

fortnight as is the case for companies. Contract farming, on the other hand, allows farmers to guarantee (in theory) an outlet for their production. It is therefore a choice between profit maximisation (with brokers) and security (with contract farming).

The marketing system for flowers differs from that of vegetables. The production is not bought at a fixed price but according to the demand. This results in a great fluctuation in prices throughout the year (with peaks in demand and price at Valentine's or Mother's Day). Small-scale farmers did not mention brokers being involved in the marketisation of flower production, but they highlighted the possibility of selling to "large-scale farmers", industrial farms that have greenhouses and the mandatory export licence. In this instance, small-scale farmers act as outgrowers. Apart from these "standard" configurations, small-scale farmers deploy different tactics to get the most out of the recent integration of Kinangop into global trade networks.

Contracts between producer groups and export companies commit both parties to a medium-term agreement – that of a crop/year/season (to be verified). It is not uncommon for contracts to be broken by the exporter when the terms are not respected, for example in terms of quantity or quality. Furthermore, growers do not systematically accept all the terms of the contract, such as the obligation proposed by Finlays to its suppliers to plant only snow peas for export. The growers want to continue to plant garden peas for the local market. This choice illustrates a strategy of limiting the risk – for example of rejection by the European market – by opting for a national market where prices and standards are better controlled. They try to play with the complementarities of crops and markets. However, the loss of market share during rejections must be qualified since brokers buy, albeit at a lower price, the produce rejected by the companies. Brokers also make it possible to sell surpluses during periods of overproduction. The agro-export model integrates formal and parallel circuits.

The different market opportunities lead to different tactics among producers. While some prefer to engage with a single export company, others diversify their customers and crops. Some produce snow peas without contract farming. The initiative of producers allows them to create networks parallel to the companies, for example with brokers. In an incentive-based political and economic environment (role of public policies and a liberalised market), the lower-level actors regain power over the upper-level economic actors through opportunistic bypass tactics. This agro-exporting model makes it possible to read the interactions between individual and collective economic behaviours and the rationality of external actors.

The range of possible choices is linked to professional trajectories, with those who are better off in terms of training often doing better. The economic opportunities linked to the development of export crops can be levers for social ascension: economic, social and economic capital are complementary. Individuals manage to position themselves at the interface between producers and companies. These positions offer them a positional rent. This issue of benefiting from better positions echoes what Auffray (2018) has shown about peasantry in Mount Elgon: groups of farmers rely on individuals who have specific resources and use this bias to convert and grow them. Nevertheless, these actors at the interface, integrated into their locality, must deal with the strategies of other members of the group for the implementation of projects driven by external actors. In Kinangop, group or cooperative leaders capitalise on their favourable socio-economic position, but they must come to an agreement with the producers they need to, for example, respect the quantities defined in the contract with the company. It is therefore a question of constant negotiation with the members of the group. Does this mean that the model formalises longer-term clientelistic dynamics? In any case, the rent captured by these local-interface actors is reinvested locally.

4.4 *Diversification and Cash Generation in a Context of Land Scarcity*

The involvement of small-scale farmers in Kinangop in export-crop production can be explained by the pressing need to generate more income from farming land, while the latter is becoming more and more scarce and expensive. This is one of the notable strategies implemented in this area to gain higher incomes while cultivating very small plots, such as zero grazing or planting fruit trees.

The size of the plots obtained during the first settlement schemes (Million-Acre and Haraka) ranges from 4 (for those located on former forest reserves) to 40 acres (for flat areas of the One-Million-Acre Scheme). In the Kikuyu inheritance system, each son is supposed to receive a plot from his father, and since the adoption of the 2010 Constitution, daughters are also deemed to inherit. Subdivisions of the plots obtained in the 1960s were carried out at a relatively rapid pace. In the flat area of the plateau, the critical size below which subdivision is no longer possible is more rarely reached than in the sloping areas where the size of the holdings was smaller (especially in the case of late allocation programmes or in the Settlement Village previously mentioned). Most farmers we met owned their land,[19] but there is reportedly also land rented out (especially in steep areas) or used for roaming animals in the absence of the owners and of cultivation (both in steep and flat areas).

19 This is due to the context of the research that was carried out by Gaële Rouillé-Kielo, which focused on the farmers involved in a program of "payments for environmental services" who were all owners (a prerequisite for participation).

TABLE 10.2 Comparison of potatoes and snow peas production performances on one acre of land in Mikaro

	Potatoes	**Snow peas**
Costs of production	KSh 60,000	KSh 70,000
Yields	40 to 80 bags (120-130 kg per bag)	2.5 tons
Sale price	KSh 16 / kg	KSh 100–250 / kg

DATA SOURCE: INTERVIEWS WITH FARMERS

In this context, it is also worth noting the significant increase in land prices in recent years reported by the farmers interviewed. In the study area, one acre was worth on average between KSh 700,000 and KSh 1 million – a theoretical average of KSh 850,000.[20] A very influential factor in setting the price is the proximity to traffic routes, particularly tarmac roads. This very high price makes it difficult for existing farmers or those who cannot inherit to buy new land.

This context of increasing land scarcity and rising land prices is influencing farmers' engagement in the production of export crops. In the Mikaro Scheme (one of the "Haraka Schemes") where the size of the plots is often below one acre, 40% of farmers produced export vegetables (mainly sugar peas) according to a local extension officer working for the Ministry of Agriculture interviewed in 2016. While the area occupied by these export crops has increased rapidly, farmers continue to grow vegetables for the local or national market (potatoes in particular). This serves a strategy of diversifying sources of income and of maximising the productivity of the land.

5 Discussion

The export-farming model exposes producers to numerous production and marketing constraints, but it is flexible enough for producers to find room for manoeuvre at the local level. The initiative capacities of producers in Kinangop echo the behaviours of farmers who turned to commercial food crops in the 1980s, as a complement to "classic" export crops (Chaléard 1996). However, the model of former export agriculture (coffee, cocoa, etc.) was vertical and not very flexible, contrary to what is observed in Kinangop. Export production

20 At January 1, 2016, KSh 1 was equivalent to $0,0096 (source for the exchange rate: https://www.oanda.com/currency-converter/en/). Hence, KSh 700,000 was equivalent to roughly $6,720, KSh 1 million to $9,600 and KSh 850,000 to $8,160 at the time of research.

is undeniably based on decisions taken by actors coming from outside the plateau, but the actors on the plateau have the capacity to take initiatives and to circumvent the system.

The export-crop model, while not obliterating asymmetrical power relations, also relies on the goodwill of small producers. Through opportunistic tactics, they exploit the relative horizontality of the model by infiltrating its flaws. This "art of doing" (de Certeau 1990) is a product of producers' rationality that integrates the advantages and constraints of the agro-export model. Their tactics are based on complementarities. In Kinangop, as in other African rural areas, agricultural societies take advantage of complementarities, they increasingly exploit the diversity of locations at several scales (Charlery de la Masselière 2013). Smallholders are trying to use the complementarity of crops, terroirs and marketing channels to reduce risks and to benefit from market opportunities. Small farmers are no longer cut off from the cities and involved only in cultivating the land. On the contrary, especially at the household level, there is a very significant diversification of income sources (whether in relation to the sector of activity from which the income is derived, or to the area, rural or urban, where it is generated). As noted by Nungari Mwangi about small-scale farmers involved in flower production in Nyandarua County and in other parts of Kenya: "Diversification is the common factor in producers' strategies for securing new lucrative export opportunities in a bifurcated market" (Mwangi 2018, 244).

The agro-export model of the Kinangop Plateau illustrates how actors of the Global South, including those apparently positioned at the bottom of the production system, are playing an increasingly important role in defining the new contours of globalisation (Horner and Nadvi 2018). This example supports Horner and Nadvi's (2018) argument that producers provide an increasing diversity of markets at multiple scales. Indeed, small-scale producers in Kinangop, via crop rotations, supply the national market (e.g. potatoes, carrots, and garden peas) and the global market (snow peas, sugar snaps). Small-scale producers, interface actors, involved in several value chains, in several markets with respective requirements, contribute to positioning a Kenyan countryside region within global trade networks and thus to reconfiguring the more polycentric world trade.

6 Conclusion

The Naivasha plains and the Kinangop Plateau were part of the same pastoral system, that of the Maasai, before the major disruptions brought about by the alienation of the land by European settlers. During the colonial period, particularly from the 1950s onwards, agricultural development driven by the colonial authorities transformed the Kinangop Plateau into a key mixed-farming area of the colony. Many (Kikuyu) squatters from other parts of the White Highlands, including the large ranching farms near Naivasha, came to work on the Kinangop Plateau. They were, later on, the first beneficiaries of the redistribution of land from European settlers from 1962 onwards. To satisfy the land appetite of the population that could not benefit from the first wave of settlement schemes, further land-allocation programmes took place, encroaching on forest reserves and offering very small parcels of land to new owners. The Kinangop Plateau became a new Kikuyu land horizon, while the land tenure system and agricultural orientations of the land around Lake Naivasha showed little evolution. The divide between these two agricultural areas, characterised by very different climatic and agronomic conditions, was reinforced by a growing demographic differentiation.

The development of the horticultural industry, which represented a major disruption in Naivasha, was initially spatially concentrated around the shores of the lake, with the installation of numerous greenhouses and new housing areas for workers. The demand for labour from the horticultural industry in Naivasha did not pull in the neighbouring rural population of the Kinangop Plateau. The development of horticulture took a different route to reach this area, and complementarities with Naivasha developed in a different way. The climatic conditions and the agrarian system of the plateau (characterised by the presence of many small-scale landowners, accustomed to crop diversity) have made it possible to develop new productions (summer flowers, off-season vegetables), not well suited to the agronomic centre of Naivasha.

Acknowledgements

Gaële Rouillé-Kielo's thesis was supported by Région Ile-de-France. The empirical research of all contributors was partially supported by IFRA-Nairobi. The authors would like to thank Léa Benoît for her valuable contribution to this chapter.

References

Auffray, Manoel. 2018. «Utilité et privilèges de l'interface: deux groupes locaux d'agriculteurs sur le Mont Elgon (Ouganda).» *Les Cahiers d'Outre-Mer* 278: 291–315. https://doi.org/10.4000/com.9201.

Bellemare, Marc F., and Jeffrey R. Bloem. 2018. "Does Contract Farming Improve Welfare? A Review." *World Development* 112: 259–71. https://doi.org/10.1016/j.worlddev.2018.08.018.

Bolo, Maurice O. 2010. "Learning to Export: Building Farmers' Capabilities through Partnerships in Kenya's Flower Industry." Montpellier: ISDA 2010.

Boone, Catherine. 2014. *Property and Political Order in Africa: Land Rights and the Structure of Politics*. Cambridge: Cambridge University Press. https://doi.org/10.1017/CBO9781139629256.

Calas, Bernard. 2013. "Mondialisation, clusterisation et recyclage colonial. Naivasha: Laboratoire d'une Afrique émergente." *EchoGéo* 26. https://doi.org/10.4000/echogeo.13586.

Chaléard, Jean L. 1996. *Temps des villes, temps des vivres. L'essor du vivrier marchand en Côte d'Ivoire*. Paris: Karthala.

Charlery de la Masselière, Bernard. 2013. «Systèmes spatiaux, systèmes de ressources et identités: Redéfinir les contextes des dynamiques territoriales.» In *Dynamiques rurales dans les pays du Sud: L'enjeu territorial*, edited by Bernard Charlery de la Masselière, Bénédicte Thibaud, and Virgine Duvat, 9–18. Toulouse: Presses Universitaires du Mirail.

Charton, Helene. 2011. "Acteurs, victimes et témoins de la violence dans l'histoire. L'exemple Mau Mau (Kenya)." *Cahiers d'études africaines* 51 (201): 169–92.

Cochet, Hubert. 2011. "Origine et actualité du « Système Agraire »: Retour sur un concept." *Revue Tiers Monde* 207: 97–114. https://doi.org/10.3917/rtm.207.0097.

de Certeau, Michel. 1990. *L'invention du quotidien* (Vol. 1, Arts de Faire). Paris: Gallimard.

Dolan, Catherine, and John Humphrey. 2000. "Governance and Trade in Fresh Vegetables: The Impact of UK Supermarkets on the African Horticulture Industry." *The Journal of Development Studies* 37 (2): 147–76. https://doi.org/10.1080/713600072.

Grignon, Francois, Gérard Prunier. 1996. *Le Kenya contemporain*. Paris: Karthala.

Horner, Rory, and Khalid Nadvi. 2018. "Global Value Chains and the Rise of the Global South: Unpacking Twenty-first Century Polycentric Trade." *Global Networks 18* (2): 207–37. https://doi.org/10.1111/glob.12180.

Hornsby, Charles. 2012. *Kenya: A History since Independence*. London: Bloomsbury.

Jaetzold, Ralph, Helmut Schmidt, Berthold Hornetz, and Chris Shisanya. 2010. *Farm Management Handbook of Kenya*. Vol. I-II. 2nd ed. Nairobi: Ministry of Agriculture in cooperation with the German Agency for Technical Cooperation (GTZ).

Kanogo, Tabitha. 1987. *Squatters and the Roots of Mau Mau, 1905–1963*. London: James Currey.

Klopp, Jacqueline M. 2012. "Deforestation and Democratization: Patronage, Politics and Forests in Kenya.". *Journal of Eastern African Studies* 6 (2): 351–70. https://doi.org/10.1080/17531055.2012.669577.

KNBS. 1981. *Kenya Population Census 1979*. Nairobi: KNBS.

KNBS. 2001. *Kenya Population Census 1999*. Nairobi: KNBS.

KNBS. 2010. *Kenya Population Census 2009*. Nairobi: KNBS.

KNBS. 2019. *Kenya Population Census 2019*. Nairobi: KNBS.

Kuiper, Gerda, and Andreas Gemählich. 2017. "Sustainability and Depoliticisation: Certifications in the Cut-Flower Industry at Lake Naivasha, Kenya." *Africa Spectrum* 52 (3): 31–54. https://doi.org/10.1177/000203971705200302.

Leksmono, Cokro, John Young, Nick Hooton, H. G. Muriuki, and Dannie Romney. 2006. "Informal Traders Lock Horns with the Formal Milk Industry: The Role of Research in Pro-poor Dairy Policy Shift in Kenya." Working Paper 266. London and Nairobi: Overseas Development Institute and International Livestock Research Institute, Nairobi. Accessed September 12, 2023. https://hdl.handle.net/10568/1692.

Leo, Christopher. 1978. "The Failure of the 'Progressive Farmer' in Kenya's Million-Acre Settlement Scheme." *The Journal of Modern African Studies 16* (4): 619–38. https://doi.org/10.1017/S0022278X00021960.

Leo, Christopher. 1984. *Land and Class in Kenya*. Toronto: University of Toronto Press.

Leo, Christopher, and John Lonsdale. 1988. "Correspondence [On Lonsdale's Review of Leo's Book 'Land and Class in Kenya']." *African Affairs* 87 (349): 613–16.

Lukalo, Fibian, and Samuel Odari. 2016. "Exploring the Status of Settlement Schemes in Kenya." NLC Working Papers Series 3. National Land Commission.

Mannon, Susan E. 2005. "Risk Takers, Risk Makers: Small Farmers and Non-Traditional Agro-Exports in Kenya and Costa Rica." *Human Organization* 64 (1): 16–27. https://doi.org/10.17730/humo.64.1.m4020qqetn4pte8n.

Mazoyer, Marcel. 1987. *Dynamique des systèmes agraires. Rapport de synthèse présenté au Comité des Systèmes Agraires*. Paris: Ministère de la Recherche et de la Technologie.

Médard, Claire. 1999. "Territoires de l'ethnicité: Encadrement, revendications et conflits territoriaux au Kenya." PhD diss., University of Paris I.

Ministry of Economic Planning and Development. 1965. *Kenya Population Census 1962*. Nairobi: MEPD.

Morgan, William T.W. 1963. "The 'White Highlands' of Kenya." *The Geographical Journal* 129 (2): 140–55. https://doi.org/10.2307/1792632.

Muriithi, Beatrice Wambui, John Mburu, and Margaret Ngigi. 2011. "Constraints and Determinants of Compliance with EurepGap Standards: A Case of Smallholder French Bean Exporters in Kirinyaga District, Kenya." *Agribusiness* 27 (2): 193–204. https://doi.org/10.1002/agr.20261.

Mwangi, Nungari. 2019. *The Power to Flourish: Unearthing the Roots of Kenyan Flower Producers' Market Access Strategies*. PhD-Diss., University of Cambridge.

Nielsen, Hendrik, and Peter Gibbon. 2011. *The Kenya Horticultural Development Programme (KHDP). DIIS Report No. 15*. Danish Institute for International Studies.

Nyandarua County Government. 2018. *The Socio-Economic Transformative Agenda for Nyandarua County. Nyandarua County Integrated Development Plan 2018–2022.*

County Government of Nyandarua. Accessed October 20, 2023. https://repository.kippra.or.ke/bitstream/handle/123456789/663/NYANDARUA%20COUNTY%20CIDP%202018-2022.pdf?sequence=1&isAllowed=y.

Nyoro, James K., and Isaac K. Ngugi. 2007. "A Qualitative Analysis of Success and Failure Factors of Agricultural Cooperatives in Central Kenya." In *Decentralization and the Social Economics of Development: Lessons from Kenya,* edited by Christopher B. Barrett, Andrew G. Mude, and John M. Omiti, 13–32. Cambridge, MA: CABI. https://doi.org/10.1079/9781845932695.0012.

Odingo, Richard. 1971. *The Kenya Highlands: Land Use and Agricultural Development.* Nairobi: East African Publishing House.

Okello, Julius J. 2011. "Exit, Voice and Loyalty in Kenya's French Bean Industry: What Lessons Can We Learn from Smallholder Farmers' Past Response to International Food Safety Standards?" *African Journal of Food, Agriculture, Nutrition and Development* 11 (1): 4460–75. https://doi.org/10.4314/ajfand.v11i1.65874.

Oucho, John. 2002. *Undercurrents of Ethnic Conflict in Kenya.* Leiden: Brill.

Pérez Niño, Helena. 2016. "Class Dynamics in Contract Farming: The Case of Tobacco Production in Mozambique." *Third World Quarterly* 37 (10): 1787–808. https://doi.org/10.1080/01436597.2016.1180956.

Ray, Nabati, Graham Clarke, and Paul Waley. 2021. "The Impact of Contract Farming on the Welfare and Livelihoods of Farmers: A Village Case Study from West Bengal." *Journal of Rural Studies* 86: 127–35. https://doi.org/10.1016/j.jrurstud.2021.06.003.

Riisgaard, Lone. 2009. "Global Value Chains, Labor Organization and Private Social Standards: Lessons from East African Cut Flower Industries." *World Development* 37 (2): 326–40. https://doi.org/10.1016/j.worlddev.2008.03.003.

Rouillé-Kielo, Gaële. 2020. "Traduction du concept de Paiements pour Services Hydriques, politiques de l'eau et processus de territorialisation au Kenya." PhD diss., Université Paris-Nanterre.

Ton, Giel, Sam Desiere, Wytse Vellema, Sophia Weituschat, and Marijke D'Haese. 2017. "The Effectiveness of Contract Farming for Raising Income of Smallholder Farmers in Low- and Middle-income Countries: A Systematic Review." *Campbell Systematic Reviews* 13 (1): 1–131. https://doi.org/10.4073/csr.2017.13.

Wangu, James, Ellen Mangnus, and August C. M. van Westen (Guus). 2021. "Recognizing Determinants to Smallholders' Market Orientation and Marketing Arrangements: Building on a Case of Dairy Farming in Rural Kenya." *Land* 10 (6): 572. https://doi.org/10.3390/land10060572.

Wanyama, Fredrick O. 2009. "Surviving Liberalization: The Cooperative Movement in Kenya." IDEAS Working Paper Series from RePEc, Federal Reserve Bank, St. Louis. Accessed September 12, 2023. http://search.proquest.com/docview/1698577844/?pq-origsite=primo

PART 4

Resource Conflicts and Co-Existence in and around Naivasha

CHAPTER 11

Frontier Dynamics: Cross-Cutting Ties, Conflict and Contestation on Agricultural and Conservation Hinterlands of Lake Naivasha

Marie Müller-Koné and Eric M. Kioko

1 Introduction[1]

The agro-industrial nerve centre of Lake Naivasha in Kenya's Rift Valley has profoundly influenced some of its hinterlands over time (see Harper and Styles, this volume). This chapter focuses on the transformation of the hinterlands of the lake basin bordering Narok County, namely the Enoosupukia and Maiella settlements, which are hidden from the view of Lake Naivasha, hidden from the horticulture, geothermal and tourist developments along the shore of the lake (see Figure 11.1). These hinterlands have been transformed mainly through investment in market-oriented agriculture, starting with pyrethrum production by European settlers in Maiella in the early twentieth century. Agricultural intensification, particularly market-oriented food production, is believed to have picked up rather profoundly from the mid-twentieth century in these settlements supplying food produce to the fast-growing population of flower farm workers and the general population in and around the nerve centre.

We refer to Enoosupukia and Maiella settlements as frontier zones, which are perceived as "empty spaces", although used mainly as hunting grounds and dry-season grazing areas by Maa-speaking groups from the eighteenth century and yet later transformed through agricultural entrepreneurship by individuals and groups from outside the area. The process of transformation of these rural frontiers of Enoosupukia and Maiella is largely linked to demographic and economic changes at the nerve centre. The process is also characterised by struggles over control and use of land between foraging, cattle husbandry, and farming involving "il Torobo" (Dorobo), Maasai and Kikuyu groups respectively, but also the by the nurturing of mutually beneficial ties and networks across social groups. Scientific investigation of the process of transformation of the frontier has mainly focused on the characteristic social and political conflict in

1 First submission date: February 24, 2021. Acceptance date: March 13, 2023.

 | DOI:10.1163/9789004695429_012

FIGURE 11.1 View of greenhouses and geothermal plants in Naivasha from Maiella
PHOTO: MARIE MÜLLER-KONÉ 2018

the 1990s (Klopp 2001; Matter, 2004; Hornsby 2012). Patterns of intercommunal cooperation and cross-cutting ties as well as new dynamics of land claims have received less scientific attention. This chapter pays closer attention to these developments of the twenty-first century and their role in shaping the Naivasha nerve centre.

Specifically, we examine the ways in which frontier dynamics – the development of an agricultural frontier and a recent history of violence and eviction – play out in present inter- and intracommunal relations on the outskirts of the nerve centre. We understand frontiers as "unruly" places that are structurally marginalised, in need of development in the eyes of developers, yet endowed with an abundance of resources – in the present case, fertile land – to be exploited (Geiger 2008; Schmitt 1997, 62; Tsing 2003). We hypothesise that such frontier situations emerge at the fringes of such nerve centres as the hyper-globalised Naivasha floriculture complex.

Foraging livelihoods, pastoralism and, later, farming, have historically characterised Enoosupukia and Maiella. Apart from issues of livelihoods, resource access and spatial competition and negotiation, the frontier also becomes an ethnic and minority battleground where specific groups negotiate and protect their interests using autochthony and marginalisation discourses (e.g. Scott 2004; Waller, this volume). Contestations, negotiations, disputes, and conflicts form part of the social order. As with most frontier zones, the influence of people from outside the area is substantial, and is a contributing factor in conflict.

The settlements of Enoosupukia and Maiella are situated roughly 12 km from Kongoni in the south lake area (see Map 11.1 and Figure 11.1). The hilly landscape of Enoosupukia, situated on the higher escarpments of the south-eastern fringes of the Mau Forest Complex, features a majority of pastoralist Maa speakers (different sections of the Maasai group – Purko, Keekonyokie,

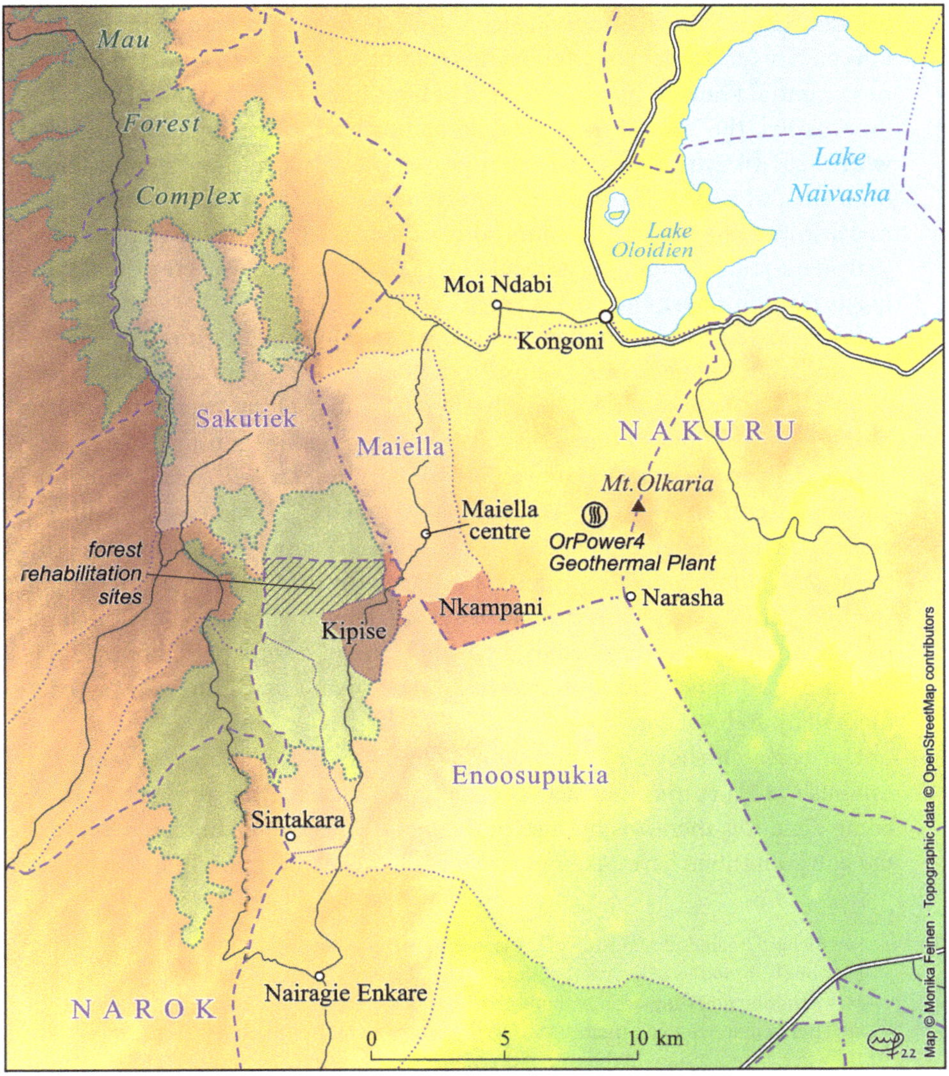

MAP 11.1 The Maiella/Enoosopukia area
CARTOGRAPHER: M. FEINEN

and Damat[2]) and Dorobo,[3] former hunter-gatherers who progressively adopted the Maasai culture and language (see Cronk 2004). Enoosupukia used to be a forest and hunting ground for the Enoosupukia Dorobo, who claim autochthony in that place (Blackburn 1982; Matter 2010; see the landscape in Enoosupukia in Figure 11.2). By contrast, Maiella, situated closer to Lake Naivasha on the way up to Enoosupukia, is predominantly a Kikuyu-speaking settlement and a former colonial settler estate. It was directly integrated into the settler frontier that developed around Lake Naivasha around the turn of the twentieth century (cf. Kuiper 2017; Waller, this volume). The settlers attracted Kikuyu from Central Kenya, who were in need of land, as squatters on their farms. Yet even earlier, the Maasai from Laikipia had arrived in the Enoosupukia area with their Kikuyu friends and relatives, who henceforth drew more Kikuyu from Central Kenya. The Maasai, who controlled the Rift Valley around Naivasha in pre-colonial times, claim to have ancestral ties to the Maiella area (Hughes 2006; Lawren 1968; Muriuki 1974; Tignor 1976; Berntsen 1976; Galaty 1993b, 187–190; Spear and Waller 1993).

While the history of conflict and cooperation at the Enoosupukia-Maiella frontier stretches back into the nineteenth century, we provide an overview of how the agricultural frontier evolved in the course of the twentieth century to then analyse how its impact is felt in the twenty-first century, specifically during the period from 1990 onwards, looking at interethnic social, economic and spatial contestations and cooperation. We will do so by drawing on ethnographic data collected in Maiella and Enoosupukia between 2013 and 2014 (by the second author), and between 2018 and 2020 (by the first author). Data from the first period of research combined survey data, qualitative interviews and participant observation, whereas the data from the second period of research encompassed qualitative interviews and observations. Both were supplemented by archival data.

Our research shows two competing findings: (1) that groups continue to establish cross-cutting ties across ethnic divides, and that the will to overcome divides is there, visible also in the fact that no large-scale intercommunal confrontations have taken place since the 1990s. The peaceful conditions

2 See Richard Waller's description of these groups in this volume.

3 Dorobo (Il Torobo) is a derogatory term used by Maasai to refer to their foraging neighbours. Many Dorobo nowadays refer to themselves as Ogiek, which is their own denomination. Usually, the people who call themselves Ogiek speak Kalenjin as their first language and live close to Kalenjin groups in the area of the Mau forest complex (Blackburn 1982). The Dorobo in Narok (like those in northern Kenya), who have lived in proximity to Maasai groups, however, speak Maa rather than Kalenjin, and often still refer to themselves as Dorobo – in our case, the Enoosupukia Dorobo – which is why we use the term Dorobo throughout the text.

that are formed around cross-cutting ties and non-violent conflict resolution are necessary for the continued supply of food produce to the Naivasha nerve centre and related market exchanges between there and the frontier zones. (2) Elements of the past violence are still present in ongoing efforts by displaced Kikuyu and Maa-speaking groups to recuperate lost land in Enoosupukia they still feel entitled to, and can be seen to potentially impact future imaginations of residents of Maiella and Enoosupukia. Specifically, the role of some state individuals and institutions in the land claims creates legal uncertainties and confusion by seemingly appearing to support some claims against others depending on the ethnicities of the local claimants. We argue that these contestations and power struggles are shaped by the growing demand for farmland in the frontier zones due to the increasing demand for food produce at the agro-industrial nerve centre.

The chapter is structured as follows: first, we briefly discuss frontier theory and then describe the pre-colonial, colonial and post-colonial history of territorial occupation of and relationships in Maiella and Enoosupukia. Thereafter we describe how this history of conflict and cooperation manifests in present-day life and how it shapes future relationships between local groups (Maa-speakers and Kikuyu), state actors (lands office in Narok and key elite), and the nerve centre (Lake Naivasha area). The conclusion revisits the theory with a discussion on how contemporary precarious relationships between and within frontier communities are affected by and affect the future of the wider Naivasha basin.

2 Violence and Cross-Cutting Ties at Resource Frontiers

Frontiers are spaces of opportunity – for land appropriation, for resource exploitation, and for new economic ventures (Imamura 2015, 69). The concept of a resource frontier entails the notion that some kind of "free land" exists to be occupied, and this land is typically located on commons or state-claimed lands. This "unused nature" is both an ideological and a physical condition (Hall 2013; Korf, Hagmann, and Doevenspeck 2013; Kröger and Nygren 2020, 376; Nelson 2011). The notion of empty space is also associated with an "ideological project of civilizing the not yet civilised or reputedly 'barbarian' populations inhabiting such contested regions" (Korf, Hagmann, and Doevenspeck 2013, 10), as the original, highly problematic use of the notion of frontiers by Frederick Jackson Turner (1893) indicates. The term "frontier" enshrines the teleological belief of a linear forward progression from a known, ordered, territorialised space into an unknown, empty space, which has to be incorporated into the former (Schetter and Müller-Koné 2020, 2021). The frontier thus denotes a situation of rigorous change: "… a zone of not yet – not yet mapped, 'not yet' regulated" (Tsing 2003, 5100). Consequently, frontiers are often spaces of social dispossession and of violence (Geiger 2008).

The area under study here lends itself to the frontier perspective. The British colonial office saw the Lake Naivasha area and its surroundings as "empty" of "natural African tenants" (Kuiper 2017, 56) despite its occupation by pastoralists and hunter-gatherer groups (see Section 11.3). The British territorialised this frontier space through the railway (see Waller, this volume), settler farms and ranches that developed over the first decades of the twentieth century (cf. Kuiper 2017, 56–59). The lake basin and its closer surroundings were integrated into the "White Highlands", developing into a territorialised space and eventually a hyper-globalised floricultural complex – a nerve centre that continues to draw surrounding frontier spaces into its reach. The area of research at the borderlands between Lake Naivasha and Narok County, the former southern reserve, epitomises the progressive incorporation of frontier spaces into imagined "ordered" territory as mentioned above above (see view from frontier space onto nerve centre Naivasha in Figure 11.1).

The physical changes in the landscape initiated by turning nature into a resource are forest degradation, forest fragmentation, transformations in land use and changes in land cover. The "discovery" of resources can take many forms: the clearing of forests and cultivation, ranching (Baretta and Markoff 1978), logging (Tsing 2003), or mineral extraction (Werthmann and Grätz 2012); even scenery can become a resource through strategies of conservation (Rasmussen and Lund 2018, 388; Guyot 2011; Harper and Styles, this volume). Unlike the cases in Latin America, Asia and some parts of Africa where the state initiated or encouraged agricultural settlement in minority regions in the second half of the twentieth century (Geiger 2008), the frontier that developed in Enoosupukia from the mid-1900s emerged through occupation by agricultural entrepreneurs who hired local labour (known popularly as squatters) in forested areas that are far removed from control by the state apparatus, as discussed in the next section.

Frontiers like Enoosupukia are "contact zones" where different social orders interact. Following changing land use and socioecological transformation, they tend to be marked by violent contestations (Geiger 2008; Hoefle 2006; Korf, Hagmann, and Doevenspeck 2013; Rasmussen and Lund 2018; Reid 2011). Endeavours to turn nature into exploitable economic resources generate conflicts, not only about the question of direct access to the valued resources, but also over the definition of what constitutes a resource in the first place, and over the way in which resources are to be exploited (Hall 2013; Schmink and Wood 1992). Where land access and use rights are contested, notions of autochthony may arise in the effort to protect claims (Larsen 2015; Little 2001; Radcliffe 2019), although some sections among local inhabitants also become part of the frontier dynamics, and sometimes even initiate frontier processes of capitalist transformation (see Li 2014, on indigenous frontiers). It is therefore important to underline that frontiers and ethnic territories are "created by diverse subjects engaged in situated struggles over categories, recognition and boundaries" (Hoffmann and Anthias 2020, 2).

Being spaces of contact, frontiers may also encourage the formation of ties and networks across social groups. For the studied areas, such cross-cutting ties have the potential to develop into conflicting loyalties (Kioko 2016). The theory of cross-cutting-ties/conflicting loyalties argues that cross-cutting ties between communities lead to conflicting loyalties with a number of actors; these actors in turn will attempt to prevent violence between communities (e.g. Colson 1953; Radcliffe-Brown and Forde 1950).[4] Intermarriages and negotiations over access to land are thus part of a web of social-economic relations that holds residents of an area together through practices, even if the discursive positionings may be more confrontational.

The nerve centre at Naivasha has historically influenced the frontier spaces on the margins, which, in turn, develop and transform through processes of contact, conflict and cooperation, all of which feed back to the nerve centre.

3 Shifting Inter- and Intra-Group Relationships: Conflict, Coexistence, and the Shaping of an Agricultural Frontier

This section provides a brief historical overview of intergroup relations in the precolonial, colonial and postcolonial eras. It discusses the development of the concomitant evolution of the agricultural frontier in Maiella and Enoosupukia and the violent confrontations that ensued, including the eviction of residents on both sides of the conflict. We will highlight the governmental interventions on the frontier, namely the transformation of customary into private, statutory land tenure (through land-adjudication areas) and the declaration of protected areas.

4 However, in some instances, cross-cutting ties may even escalate conflicts (Schlee 1997), because ties may easily become fragile, collapse and/or break down, thus explaining conflict (e.g. Fukui 1994). Still, actors have the capacity to reorganise, strengthen and/or form new alliances or try to sustain existing ones (Kioko 2016, 39).

3.1 *History of Inter-Group Relations in the Narok-Naivasha Borderlands*

Interdependent relationships between Maa-speaking groups (including Dorobo hunter-gatherers) and their neighbours, particularly Kikuyu farmers, have existed prior to their contact with Europeans. The relationships can be described in terms of three aspects: (1) social-cultural: intermarriage, adoption, and ceremonies such as initiation; (2) economic ties: trade, patron-client services, and cooperative resource use; and (3) political networks: military alliances against common enemies, and anti-colonial campaigns (see Lawren 1968; Muriuki 1974; Tignor 1976; Berntsen 1976; Galaty 1993b, 187–90; Spear and Waller 1993; Blackburn 1996).

Dundas (1908, 136–37) notes the possibility that the Dorobo, Kikuyu, and Maasai share a common ancestral tribe called the Endigiri, and that their ancestors came from beyond Mount Kenya. The Kikuyu claimed descent from the Shagishu and Ngembe groups, which are believed to have originally been Dorobo, and Karurui' (Dundas 1908, 136–37). The Kikuyu expanded southwards from Central Kenya in precolonial times, and when they crossed the Chania River, probably around 1700, they entered territory that was occupied by the Dorobo hunter-gatherers. Through mutual adoption ceremonies, they bought land individually from the Dorobo and integrated those tracts of land into their clan land over time. The Kikuyu had reached the edges of Maasai territory near the Ngong Hills when the first Europeans entered the area in 1887 (Leakey [1977] 2007).

The Maasai initially saw neither the Dorobo nor the Kikuyu as a threat. The Maasai adopted Kikuyu into their families (Waller 1993) and allowed them to clear some forests for gardens (Blackburn 1982, 295). The Dorobo had a more symbiotic relationship with the Maasai, which was based on mutual interdependence. The Maasai depended on the honey that Dorobo collected for honey beer, and relied on them as circumcisers for initiation ceremonies (Blackburn 1982; Kenny 1981). The list of items the Dorobo traded with Maasai is very long, and in excess of their respective needs (Blackburn 1982, 299; Kenny 1981, 481). The Dorobo adopted the Maasai language, initiation ceremonies, ornaments and dress, and gave their wives to Maasai husbands, so that in appearance, they are nowadays difficult to distinguish from Maasai (Kioko 2016; Kratz 1980).

Cooperation between these groups not only revolved around social and economic ties, but was also a response to natural disasters like droughts and diseases. The rinderpest epidemic of the 1890s (Waller, this volume) forced erstwhile herders to work (mostly temporarily) on Kikuyu farms, while some were absorbed into the Kikuyu households, but such adaptive ties did not stop the occurrences and continuation of inter-group raids (Berman and Lonsdale 1992, 27; cf. Waller, this volume). Inter-group ties between Kikuyu, Maasai and

Dorobo may also be seen during and after the "great migration" – following the removal of northern Maasai from Laikipia between 1904 and 1911 (Hughes 2006). The Maasai moved with "their Kikuyu" and arrived at the southern reserve which at the time was already occupied by Purko and Loita Maasai in large numbers, as well as by groups of Kaputiei, Sigirari, Matapato, Lo' Dogolani and Kak-o-Nyuki (Keekonyokie) (Spear and Waller 1993).

Dorobo welcomed northern Maasai and Kikuyu who arrived in Enoosupukia and its environs. These groups continued to admit new Kikuyu arrivals from central Kenya, some of whom were fleeing from white settlers following the Mau Mau uprising of the 1950s, while some were in search of settlement and farmland following British occupation of their regions (Anderson 2005, 21). Kikuyu often received land in return for labour or through intermarriages between their daughters and Maasai and Dorobo men (Blackburn 1996; Klopp 2001; Matter 2010; Waller 1993).

The post-colonial period has seen increasing numbers of land-seeking individuals and groups in Enoosupukia, some of whom come from the nerve centre after failing to get jobs in the flower farms or for purposes of diversifying their livelihoods owing to low wages and unpredictable working conditions at the agro-industrial hub (Kioko 2016). The influx of farmers and consequent agricultural intensification on former pastoral and hunting grounds led to disputes between Maasai, Dorobo, and Kikuyu regarding access to and control over resources, particularly in the postcolonial period (Kioko 2016, 80–84).

Nowadays, in the hills and plains adjoining the Naivasha basin, agro-pastoralism has progressively replaced specialised pastoralism and hunting, which were the main economic activities in the early twentieth century. Today, Maasai and Dorobo also cultivate, just like their Kikuyu neighbours (see Figure 11.2). Many, however, rent out their land to Kikuyu and other migrant tenants. According to Matter (2010, 253) and a number of elders we interviewed, Dorobo took up small-scale farming and usually hold a small number of cattle as well, as they were barred from hunting since the colonial period. Due to changes in land-use patterns, hunting and gathering practices are no longer tenable; only a handful of Dorobo families have one or two bee hives today (Kioko 2016). All communities also practice a mixture of off-farm activities, such as petit commerce, butchering, or transport.

The history of relations between Maa speakers and Kikuyu is summarised succinctly by one of the author's research participants:

> The Maasai [referring to Maa-speakers] and Kikuyu are one large 'family' and the conflicts between them can be taken as any normal conflicts in a family setting, except that politics knocks the heads of two brothers

> against each other, as it was in 1993. The survival of one group is largely dependent on the other.[5]

3.2 *The Agricultural Frontier in Maiella and Enoosupukia*

Enoosupukia is characterised by good agricultural soils and abundant rainfall, which allows for up to three harvests per year for some staples including vegetables: Irish potatoes (locally known as *shangi*), onions, cabbages, carrots, various types of leafy vegetables, and legumes such as beans and peas (*Pisum sativum*) (see Figure 11.2). According to the 2019 census, Enoosupukia Location hosts 20,961 residents with a density of 96 people per km^2, whereas Maiella Location hosts 14,646 inhabitants, with a density of 112 people per km^2, and Maiella trading centre with a density of 138 per km^2 (KNBS 2019, 164, 169).

As the study area is in close proximity to Lake Naivasha's agro-industrial hub, this hub provides a ready market for food produce from Enoosupukia and Maiella. This booming market for food produce has progressively drawn large numbers of youth (between 20 and 40 years of age) into commercial cultivation through leasehold arrangements in Enoosupukia and its environs. Apart from Naivasha, other markets for food produce have progressively opened up, including Nairobi, Machakos, and Mombasa, among other towns in Kenya. Some produce also crosses into Kampala, Uganda (Kioko 2016, 46).

FIGURE 11.2 A small-scale farm in Enoosupukia hills
PHOTO: MARIE MÜLLER-KONÉ 2018

5 Interview with Shushu/elderly woman, Maiella trading centre, August 21, 2013.

This influence of developments in the Naivasha basin in the hinterlands at the Nakuru/Narok border had set in early in the twentieth century, when Naivasha's settler economy reached up to Maiella. The area known as Maiella today, bordering Enoosupukia, was until 1964 a settler farm (Maiella Estate) constituting some 16,338 acres of land. According to residents, an Italian commercial farmer who acquired the estate from a British settler gave it its name (Kioko 2016, 43).[6] He had developed a pyrethrum plantation in Maiella, which was a lucrative business in the Naivasha area in the colonial period (Kuiper 2017, 56–64), employing Kikuyu squatters on his farm. Since then, the development of agricultural activities has spread further, into Enoosupukia. The agricultural frontier was hence initiated by frontier entrepreneurs from the Naivasha area, which mobilised additional frontier entrepreneurs who ventured further into the perceived "empty" land. We thus observe a cascading influence of developments in the Naivasha basin on the agricultural frontier at the Nakuru/Narok boundary.

Enoosupukia is part of the Melili forest, which is part of the greater Mau Forest Complex. As late as 1962, Enoosupukia, with a few other forested areas in Narok District, were among the only forests in the entire country not yet officially reserved for protection (note by J.P.W. Logie and W.G. Dyson in their report "Forestry in Kenya: A Historical Account of the Development of Forest Management in the Colony" in the 1962 issue of *Colony and Protectorate of Kenya* [quoted in Matter 2010, 214]). Attempts by the colonial administration to put Enoosupukia under the authority of the Forest Department failed due to resistance by the Maasai Native Council, who wanted to keep the forests under their control (Matter 2010, 217).

As Maiella was bordering the fertile Enoosupukia area, Maiella Estate became an ideal entry point connecting land-seeking clients from the Lake Naivasha area and persons from other parts of Kenya, to land opportunities in Enoosupukia and other parts of Narok. This process had started already in colonial times through intergroup land transfers (Klopp 2001, 151), when Dorobo and Maasai transferred rights of access to land through sale and/or exchange in the form of gifts to "newcomers", mostly of Kikuyu descent. These early transfers of rights to land were informal and involved word of mouth or, in only a few instances, hand-written agreements as evidence of transfer (e.g. Kioko 2016, 246). The agreements increasingly grew into monetary transactions over the decades.[7]

Land transfers in Enoosupukia increased after the many Kikuyu labourers on Maiella Estate managed to buy the land from the Italian settler around

6 Rhodes House, Oxford/ Handing Over Report/Mr. A.D. Galton-Fenzi to Mr. R.A. Jeary/Narok District/1st March 1957.

7 Interview with elderly Kikuyu woman, Maiella, 4 June 2019.

the time of independence in 1963, by forming a cooperative called the Ng'ati Farmers Cooperative Society Limited and by mobilising their relatives from the Kikuyu heartland in Central Kenya to join the cooperative and offer financial support. They obtained a title deed for the farm in July 1974 (Kioko 2016, 97). The Maasai, who claimed this as their ancestral land, were not successful in regaining possession of Maiella Estate at that moment (Matter 2010, 133), which led to disputed claims between the two groups as discussed later on.

The state of Enoosupukia as a forest, a hunting ground, pastureland, or as agricultural land remains contested not only between residents but also between these communities and the state. While acknowledging the water-catchment function of forests in Narok in the execution of the 1972 Narok District Development Plan, E.C. Trump highlighted the agricultural potential of the area in a report for the FAO entitled "Vegetation and Land Use Survey of Narok District": "The Narok forested and ex-forested areas contain the only undeveloped arable lands remaining in Kenya (apart from areas in other districts where irrigation will be required)" (quoted in Matter 2010, 219). Consequently, the Development Plan designated the larger proportion of the Melili forest (including Enoosupukia) for mixed farming, officially opening up an agricultural frontier characterised by uncertainty over titling and conflicts over use of, access to and control of land.

In 1977, an adjudication section was declared in one part of the Enoosupukia area called Kipise Section. The whole of Enoosupukia, however, was unofficially divided into individual land holdings. In the 1970s and 1980s the process of land acquisition by Kikuyu and other migrant farmers accelerated. The terms and legitimacy of such agreements are largely disputed today. Kikuyu and Maa-speakers disagree on whether these transactions constituted permanent or temporary transfers of rights, and whether the transferred rights indicated use rights only, or actually meant a full transfer of ownership rights (Matter 2010, 138–221; Kioko 2016, 95–102). In addition, the registration process of the land adjudication was only completed officially in October 2007, leaving those in possession of allotment letters in a legal limbo for decades. During the 60-day objection period, more than 1,000 complaints were filed, compared with 1,850 parcels listed in the register. Many complaints related to contested transactions between Maa-speakers and Kikuyu (Kioko 2016; Matter 2010, 146). As titles can only be issued when all disputes on a parcel are settled, the first titles for Kipise Section were only issued from 2013 onwards.[8]

8 Interviews with displaced Kikuyu in Maiella, September 2018, June 2019. In interviews with Dorobo elders in Sintakara, March 2019, it was suspected that this was due to the death in 2016 of Maasai leader William Ole Ntimama, who had tried to prevent Kikuyu from getting land titles.

Consequently, in the last four decades, rapid agricultural intensification in Enoosupukia through land transfers and later leasehold arrangements between Maa-speaking landowners and land-seeking clients has changed and shaped the landscape enormously. The result is the massive conversion of the Enoosupukia forest through market-oriented cultivation. While Enoosupukia was still largely forested at the time of independence, the forest cover nowadays is so much reduced that the green meadow hills of Enoosupukia do not even fulfil the ecological definition of a forest by the FAO of an area with canopy cover of over 10 percent (Matter 2010, 233).

In the following section, we zoom in on the contested land transfers mainly to Kikuyu migrants and ensuing violent conflict in the early 1990s which drastically disrupted food supply and movement between the nerve centre and the frontier.

3.3 *Violent Conflict in Enoosupukia Since the Early 1990s*

Beginning in early 1993, some political figures and administrators in Narok made moves to evict highland residents by reclassifying Enoosupukia as a forest, although parts of the Maasai elite and officials in Narok District had participated in the extra-legal land market in Enoosupukia (Kioko 2016; Matter 2010, 221). Central to this endeavour of gazetting Enoosupukia as a forest, and particularly getting rid of the Kikuyu population in the area, was the Maasai area Member of Parliament, William Ole Ntimama. Ntimama used the deforestation narrative, blaming forest destruction on the Kikuyu during public meetings and arguing that their farming tendencies had contributed to reduced water and interfered with dry-season grazing areas. In addition, the potential legal ownership of large parts of Enoosupukia by Kikuyu reduced the opportunities for the Maasai political elite to let non-local Maasai political cronies exploit the resources of the area. Hence, Ntimama and the Minister for Environment and Natural Resources, John Sambu, called upon the Narok District Council to declare Enoosupukia (outside the adjudication area Kipise Section) a District Forest, which the Council did in September 1993. The political background to that move was that multi-party elections in 1992 challenged the decade-long rule of the governing KANU party.

The conversion of Enoosupukia into a protected forest had been prepared in meetings with KANU followers in the course of 1991. As a severe drought hit the lowland areas at the same time as the declaration of Enoosupukia as a forest, by mid-October 1993, Maasai pastoralists from the lowlands coming to graze and Kikuyu residents in Enoosupukia had started to clash (Matter 2010, 223–32). In October 1993, an organised group of Maasai warriors descended upon and attacked Kikuyu farmers and their Maa-speaking affiliates (Dorobo and members of the Maasai group sympathetic to the

Kikuyu) in Enoosupukia. Thousands of Kikuyu and their affiliates were forcefully moved from Enoosupukia, while close to 50 people (both Maasai and Kikuyu) lost their lives (Hornsby 2012, 548; Kioko 2016, 105; Klopp 2001, 163; Matter 2010).

Although Enoosupukia officially became a County Council forest, the exact size and boundaries of the forest are still not clearly determined. The forest status continues to be contested by Enoosupukia residents who had unofficially demarcated their individual plots in the forest in the 1970s. After the 1993 clashes, many Maasai and Dorobo residents thus remained in Enoosupukia, both in the Kipise adjudication area and in the designated forest area. In 2005, the Narok District government once again decided to dispel people from the area, this time the Dorobo and Maasai people inhabiting the area that was now officially a forest. With just a few days' notice, people were brutally evicted from the forest area by several security agencies. On the evening of February 28, 2005, up to 400 soldiers – a combined force made up of Narok County Council forest and game rangers, Administration Police, and members of the paramilitary GSU – drove residents out, systematically setting fire to fields of maize and potatoes and demolishing houses. The media reported 1,200 displaced residents and 210 houses destroyed (Matter 2010, 236–37). In interviews of the first author, some former evictees remembered brute violence by some security forces, including rape.[9] The violence cut transport links between the nerve centre in Naivasha and the frontier, making movement and supply of food produce impossible.

Instrumental in the government's decision on the evictions were Kenyan state and non-state environmental groups with international support – including branches of the IUCN, the FAO, DFID, UNDP, and UNEP. The Kenya Forests Working Group (KFWG) had promoted sustainable forest management across the county since 1995 and lobbied the government intensively in the months preceding the eviction (Matter 2010, 239–40). Eviction and resettlement have characterised the social order to date as disgruntled farmers and pastoralists continue to push for their rights to portions (and sometimes the whole) of Enoosupukia forest, which will be elaborated below.

The agricultural frontier in Enoosupukia displays characteristics that are foregrounded in the literature but is also more complex. As highlighted by the frontier literature, the Kikuyu farmers who ventured into Enoosupukia with state support after independence, reinforced by the declaration of an adjudication section, were accused of environmental degradation and displaced by

9 Interviews with Enoosupukia residents, March–July 2019.

sections of the state apparatus in the 1990s. Some of these frontier entrepreneurs, however, had started making contacts and acquiring tracts of land from Dorobo and Maasai long before state institutions were established. As Dorobo and Maasai had started ceding land to Kikuyu outsiders before the state officially encouraged migrant farmers to cultivate the area, it is not straightforward to call this process of land transfers a process of dispossession. While the acquisition of land by European settlers in the Naivasha and Maiella area clearly dispossessed Maasai pastoralists and Dorobo hunters of grazing and hunting grounds, the story of Enoosupukia is more complicated. Here, the process more resembles an "indigenous frontier" (Li 2014), where local inhabitants join in the process of land-use transformation, as more and more Dorobo and Maasai turned to agricultural activities as well.

The land-tenure regime was only half-heartedly transformed by state institutions: a process of land titling was started but left dragging on for decades, so that extra-legal land transfers continued unabated, with parts of the political elite joining in. When the area was finally declared a state forest, the rehabilitation and protection of the area was also not effectively enforced by state institutions. Legal uncertainty thus remains a central characteristic of the frontier space that suits certain state actors, who are able to manipulate the issue of land access for their own political ends.

In what follows, we show how individuals and groups on the frontier navigate a post-conflict situation, conflicting claims, and uncertainties over land use and ownership.

4 The Post-Conflict Situation: Cross-Cutting Ties through Intermarriages and Leasehold Arrangements

Despite those discursive positionings that run along ethnic divides, an ethnographic study by Kioko (2016) in Maiella and Enoosupukia shows that in their everyday lives, cross-cutting ties are forged between the Kikuyu and Maa-speakers. Indeed, ties and networks which the two groups had formed and sustained before the violence in 1993 were instrumental in their return to peace and the post-conflict reconstruction including repossession of lands. Ties are nurtured through intermarriages and livestock (and other) economic transactions, as well as leasehold arrangements between Maa-speaking landowners and Kikuyu land-seeking clients.

Based on a 2014 survey (Kioko 2016), a sample of 140 marriages documented for Maiella and Enoosupukia recorded 48 (34%) intermarriages between Maasai and Kikuyu. Out of the 48 intermarriages recorded, about 13 occurred prior to the 1980s. Since the late 1980s intermarriages have increased, with Maasai men habitually marrying Kikuyu wives, following the rules of exogamy in which Maasai men often marry Kikuyu neighbours with whom they have constant interaction in the shared frontier. A considerable number of children result from these intermarriages. It is these children who embody the conflicting loyalties of their parents and kin. In situations of violence (such as was the case in October 1993), it is the actors who have ties with both conflicting groups that will first and foremost engage themselves in peaceful intercommunity relations.

Among the Maa-speakers and Kikuyu of Maiella and Enoosupukia, relationships which result from cross-cutting ties are often accompanied by significant material transfers and engagements, thereby fostering and giving meaning to concepts of sharing and interdependence (see Kioko and Bollig 2015), eventually developing into allegiances or conflicting loyalties. According to Kioko (2016) intermarriage unites in-laws and their larger social networks of both Kikuyu and Maasai kin and friends. In-law bonds facilitate peaceful relations between the families tied in kinship, but may also influence their friends and neighbours in a similar direction. Through in-law relations, the wider network of friends gains opportunities to negotiate for access rights to land. In-laws and their close core of friends receive land as gifts and/or through purchase, constituting a reciprocal social relationship where each party is indebted to the other – usually, an exchange of daughters for land. Unlike their Kikuyu neighbours, Maasai put forward a strong myth-based reason against marrying members groups such as the Luo, whom they accuse as not favouring circumcision of their sons, an important rite of passage among the pastoral Maasai. The myth is so strong that Maasai women married into the Luo community are considered outcasts by their own families.

Intermarriage between Maa-speakers and Kikuyu plays a crucial role in the context of access to land and settlement. Possible disputes over land thus become a family matter rather than a cause of disagreement between ethnic groups. This makes disputes easier to manage and averts instances of large-scale rivalry. As noted by Flap (1997, 209), persons that find themselves in situations of conflicting loyalties frequently and for long periods will gradually develop strong self-discipline and tolerance. Apart from access rights to land, intermarriage strengthens personal security, involves a wide range of socially meaningful material flows, and grows economically meaningful ties.

Apart from intermarriage, land rentals or leasehold arrangements between the two groups constitute a vital aspect of cooperation. Land rentals mainly from Maa-speaking landowners to Kikuyu land-seeking clients are accompanied by a multitude of interactions, negotiations and friendships, and they also build trust between those involved, and thus reinforce the economic value of the land. Landowners habitually transfer user rights to land-seeking clients and retain rights of disposal. Land rentals or leasehold arrangements are founded on trust between actors irrespective of their ethnic backgrounds. Strategies for maintaining trust and friendship involve presenting gifts to landowners (such as shoes, foodstuffs, blankets, and mobile phone airtime, etc.), inviting them to social events and ceremonies (e.g. circumcision parties, marriage and church meetings), chatting (physically or via WhatsApp), adhering to the agreed payment models and honouring debts. Some tenants help landowners with menial tasks in their farms or at their homesteads. Interestingly, some tenants also give or sell crop residue to landowners for their herds to eat, thereby cementing their arrangement. Notably, negotiations over access to land and the ensuing transfer of user rights from Maa-speaking landowners to mainly Kikuyu tenants make land a shared resource despite the existence of interpersonal land-related claims and disputes.

Despite the role of cross-cutting ties and conflicting loyalties in nurturing a cohesive social fabric and promoting non-violent behaviour, land claims still persist at both individual and group levels on the frontier. In the following, we describe the role of the state and elites in the present claims.

5 Present Land Claims: a Class and State-Community Struggle

While land has become a shared resource, it remains contested. Though crucial in redefining social relationships and access to land between diverse ethnic groups, the existence of ties and networks between families and friends does not put an end to historical land claims. The only difference is that these ties and networks facilitate a conducive environment for peaceful negotiations at individual and group levels while limiting the outbreak of violence.

Nevertheless, our analysis reveals that recent land claims often run along the lines of class and state-community opposition. This section looks at three cases of recent land claims in Maiella and Enoosupukia. First, in the Maiella Estate, intra-community wrangles about the distribution of land of the Ng'ati farm dominate, and especially take a new turn following expansion of geothermal exploration in the Naivasha area, on the lands adjacent to Olkaria including parts of the Maiella Estate (precisely, Narasha).[10] Second, in Enoosupukia, the Maa-speaking population of Enoosupukia forest that was evicted from the

10 See also Nweke-Eze and Adongo (this volume).

official forest area in 2005 contests the status of Enoosupukia as a protected forest and their own dispossession related to it in a state-community conflict. They have continuously challenged state institutions over the forced evictions and insist on compensation either monetarily or through allocation of other land, claims that are based on their loss of settlement and landholdings for a rehabilitation exercise. Thirdly, the attempts to retrieve land in Enoosupukia by some Kikuyu currently located in Maiella, who were displaced by the 1993 violence,[11] can equally be seen through the lens of state-community struggles: the Narok Lands Office irregularly allocated many of the plots in favour of Maa-speaking individuals. Both the dispossessed Kikuyu and displaced Maa-speakers from Enoosupukia forest are confronted with the shadows of past violence and injustice done to them, along with continuing injustices. At stake in all cases is the question of who has the authority to determine the status of land tenure, and on what grounds.

Class relations play an important part in present land claims on the Ng'ati estate in Maiella. A lot of disputes on Ng'ati land arose among Kikuyu themselves after conflicts with neighbouring Maasai were resolved in court by attributing 4,207 acres of the Ng'ati Estate to the Maasai community in 1996 (Republic of Kenya 2019, ix). The 581 Ng'ati members who bought the land from the settler in 1964 had initially received 2.5 acres each, and would subsequently get an allocation of 5.5 acres when the first part of the land was subdivided in 1994. This accounted to roughly 60% of the total Ng'ati Estate, and it meant that land parcels progressively became too small to be economically viable once parents subdivided and passed them on to their children and grandchildren or subdivided them for sale following increasing numbers of land-seeking members of Kikuyu immigrants.[12] Hence, many Kikuyu in Maiella have small landholdings which are not sufficient for market-based farming and hence some seek out rental land elsewhere, in Enoosupukia or in Nkampani, the Maasai area of Ng'ati estate.

The management of Ng'ati estate introduced social differences among the Maiella residents because only original members were able to receive plots in the first subdivision process. Those who arrived in Maiella after the foundation

11 See Section 11.3.2 on the filing of over a thousand complaints in Narok courts after the registration of land titles in Kipise in 2007. The post-conflict situation saw many Kikuyu reclaim land in Enoosupukia in front of Narok courts. The struggle to reclaim lands that by now have fallen on the hands of the younger Maasai generation, who sometimes do not respect old land-transfer agreements between their parents and Kikuyu friends or relatives, is still ongoing to date, especially in cases where the individuals involved in the early transactions are either dead or very old, to confirm these claims.

12 Interviews with Chief and Assistant Chief of Maiella, May and September 2018, and members of Ng'ati farmers association, May, September 2018, June 2019.

of the Ng'ati cooperative, among them the many displaced Kikuyu families from Enoosupukia, depended on Ng'ati members to lease or sell them land. Tensions rose during the second phase of subdivision that started in 2009 in other parts of the estate, when the Ng'ati management allocated the best land to themselves and to some government officials (Republic of Kenya 2019, x, 21–24), and irregularly increased the Ng'ati members from the initial 581 to 2,000. In addition, in 2014, the Ng'ati management sold some 3,093 acres of the 16,338 acres to Kenya's largest energy regulator, KenGen, for the extension of geothermal development close to Lake Naivasha at Narasha/Olkaria area (OrPower4), without consulting its members.[13] During this period, meetings of the assembly of the Ng'ati society were impeded by the administration on security grounds (Republic of Kenya 2019, 15–17). The sale sparked massive intra-communal conflict, which escalated to the burning of homes belonging to some managers of the estate, while a number of managers and their acquaintances were violently evicted from Maiella in 2014 (Kioko 2016, 261–67). A new management committee was formed, which is still in dispute in court with the former management. The government even called in a Commission in 2018 to resolve the dispute (Republic of Kenya 2019).[14]

While geothermal-related land struggles continue to characterise the Maiella Estate, Enoosupukia has witnessed conservation-related struggles over land at least since 2005. After the 2005 eviction from Enoosupukia forest, evictees sent delegations to Narok officials, including Ntimama, to reverse the eviction decision. Through years of lobbying, they finally succeeded in having a settlement scheme agreed for them called Sintakara settlement (Matter 2010). The settlement scheme was surveyed, parcel owners identified by a Land Committee set up in 2009 and title deeds issued for the residents in 2011. However, legal uncertainty remains for the Sintakara settlement, as the current Narok County government disputes the legality of the settlement and upholds that it is in the confines of the official county forest.[15] Some influential non-local Maasai managed to receive parcels of land in the scheme, and long-term Enoosupukia residents suffered substantial losses in terms of landholdings.[16]

While the county government had declared Enoosupukia a forest in 1993, it did not take the necessary measures to rehabilitate and protect the forest.

13 For details of the transaction, the money given to some members as returns from the sale and the aftermath of the conflict see Kioko (2016, 261–67).

14 Interviews with Chief and Assistant Chief of Maiella, May and September 2018, and members of Ng'ati farmers association, May, September 2018, June 2019.

15 Interviews with county government official, Nairobi, 26 April 2019, and former surveyor, telephone, 12 April 2019.

16 Interviews with Sintakara residents, March- July 2019, April- May 2020.

FIGURE 11.3 Remains of tree plantations by the Green Belt Movement in Enoosupukia
PHOTO: MARIE MÜLLER-KONÉ 2019

Since the 2000s, the prestigious Green Belt Movement, an NGO linked to Nobel Laurate Wangari Maathai, led efforts to rehabilitate the forest, although the area still presents a picture of a "forest with no trees"[17] as shown in Figure 11.3.[18] Moreover, Enoosupukia residents were never consulted on the rehabilitation exercise, but were merely hired as paid labour force, receiving KSh 500 (roughly $5) a day.[19] The state-led violence they endured during the 2005 eviction and the loss of their habitat and landholdings for a rehabilitation exercise that they are not involved in are clearly associated with a feeling of alienation, or even asphyxiation, as this quote from an interview with a Dorobo elder in Sintakara (March 2019) illustrates: "What we were given is just a rope around our neck".

Consequently, residents of Sintakara settlement adjacent to the protected areas still reclaim the forest. While some residents referred to Enoosupukia forest as a protected forest, many referred to it as community land, as the vignettes below show:

17 The state recently contracted Principles for Trees, an NGO, to rehabilitate the forest with indigenous Dombeya trees among others, which seems to fare better.

18 Interviews with community members and state authorities (county government, Water Towers Agency, National Environmental Management Agency (NEMA), Narok (Deputy and) County Commissioner, Kenya Forest Service, county rangers), March–July 2019.

19 Interviews with elders in Sintakara settlement, March–April 2019. At January 1, 2019, KSh 1 was equivalent to $0,0097 (source for the exchange rate: https://www.oanda.com/currency-converter/en/).

> Enoosupukia was our reserve; it was not a forest as designated by the state now:
>
> ELDERLY MAASAI RESIDENT, SINTAKARA SETTLEMENT, MARCH 2019

> Enoosupukia is not a state forest. We were alienated, because it was our land.
>
> ELDERLY MAASAI RESIDENT, SINTAKARA SETTLEMENT, MARCH 2019

To justify their claims, some Maa-speakers have settled close to the forest, even though the state allocated them land elsewhere in neighbouring Sakutiek and Moi Ndabi, which they consider less suitable for agriculture due to the relatively hot and dry periods experienced in the lowlands of Naivasha compared to Enoosupukia. Some Enoosupukia residents imagine claiming back the land of Enoosupukia forest as community land: "We're hopeful that God will bring change and throw out the government".[20]

When asked about the future of inter-group relations, residents of the Sintakara settlement express a certain unease with recently arrived communities, stating: "We lived harmoniously but now there is problem with new tribes", "I am not used to the new communities compared to the old ones", or "Now we have new neighbours unlike there before".[21] However, looking more closely, it becomes apparent that the Sintakara residents not only refer to migrant communities, like the Kikuyu, but also to other Maa-speaking people who have come to the area more recently. Statements like "I have no option since they are also allocated land just as I am", or "I just accept them, for they are also members"[22] must refer to non-local Maasai who have been allocated plots in the new Sintakara settlement scheme, due to their good political connections or corruption rather than by virtue of ancestry in the place (cf. Matter 2010). Thus, behind the expression of unease about newcomers may lie a hidden critique of state agents who have taken advantage of the new settlement scheme to irregularly settle some of their clients.

In addition to land struggles surrounding Maiella Estate and conservation in Enoosupukia, community-state relations are also at stake in land claims that find their way to the Narok Land Office. Encouraged by a nation-wide land-titling programme initiated by President Kenyatta in 2013, some Maiella

20 Elderly Dorobo woman, Sintakara settlement, March 2019.

21 Interviews in Sintakara settlement, April–May 2019.

22 Interviews in Sintakara settlement, April–May 2019.

residents began to pursue previously acquired lands in Enoosupukia and which they allegedly lost in the 1993 violence. They attempted both formal and informal avenues, going through the Chiefs, the Narok Lands Office, District Officers, Courts, and engaging local peace committees and dialoguing with the current occupants. Many of the Kikuyu interviewed were not successful in retrieving the land in Enoosupukia, but a few were. In the two successful cases, the former landowner, who had sold the land, had testified against a new occupant in court in one case, and in the other case, the occupant could not provide sufficient evidence to prove the land was his. The neighbours in Enoosupukia who know the Kikuyu claimants from before the 1993 violence are reported to have a welcoming attitude towards the Kikuyu claimants, some even supporting their claims to parcels of land they lost during the 1993 violence (Kioko 2016).[23]

The reasons why many Kikuyu had difficulties reclaiming land is that some descendants of previous landowners who had ceded the land to the Kikuyu do not recognise the land transaction, or that current, sometimes recent occupants refuse to leave the premises or have rented it to other tenants. When Kikuyu claimed such parcels at the Narok land office, they often found that the parcel was already registered in the name of somebody else through fraudulent transactions involving the Land Officers. When inquiring further, the Kikuyu claimants learnt that the minutes on their files stated that they had failed to appear thrice before the authorities – without them having been summoned at all. In addition to this recurrent vice, recent occupants tend to be well connected to influential Maasai figures as a Kikuyu testified: "These Maasais have paid him a lot of money and some have given him cows and goats or sheep so that he can testify lies against us".[24]

These different land claims and contestations all indicate an underlying tension arising from power differences, which leave many of the frontier's residents in a weak position in the face of well-connected individuals and society managers who use fraudulent means to allocate and/or sell individual or communally-claimed parcels of land for selfish gains.

The clamour for title deeds has intensified as both Maa-speakers and Kikuyu seek to secure their holdings. Evidently, confirmation of rights to land through individual title deeds gives one the confidence to invest in market-oriented food production and hopefully to escape future property-related disturbances to livelihoods and settlement. Such confidence is necessary to ensure continued supply of food produce to the fast-growing populations of the nerve centre

23 Interviews with Kikuyu claimants, Maiella, September 2018, June 2019, July 2020.

24 Interview with displaced Kikuyu claimant, Maiella, June 2019.

in Naivasha and beyond the local markets. We argue that negotiations and contestations surrounding the use and ownership of land in the agricultural frontier on the fringes of Lake Naivasha is strongly motivated by rising demand for farm produce at the nerve centre.

6 Discussion and Conclusion: Centre-Periphery Relationship and the Future of the Lake Naivasha Basin

What is the role of the nerve centre (Lake Naivasha) in all these dynamics? How does the nerve centre influence the growth of agricultural frontiers on the fringes of the land? How does the nerve centre and its emerging influence on mobility and commercial exploitation of the frontier zone affect intergroup relationships outside of the lake? How do different ethnic groups control their respective futures either violently or through non-violent cooperative strategies? The discussion here centres on the role of the nerve centre as a catalyst for change in the margins, as well as the interdependence of the nerve centre and the periphery (frontier relations, economy, and the linked role of state).

The nerve centre – represented by Lake Naivasha and its immediate surroundings including the town and the agro-industrial hub – provides a market for the agricultural produce coming from the frontier in Enoosupukia and Maiella, including meat and farm produce. In terms of other business-related transactions, the centre supplies goods in wholesale for the frontier consumers, e.g. sugar, cooking oil etc. In addition, as exposed in the literature on frontiers, the Lake Naivasha region provided the impetus for migrants to settle at the frontier, starting with entrepreneurs, who in addition to converting formerly hunting and grazing fields into large-scale plantations and livestock-production areas, began to bring squatters to work on those farms. The squatters would eventually purchase the settler estate, which heralded the arrival of more local farmers. To date, the nerve centre continues to be a sending area for people seeking to diversify into farming, those who do not get jobs in the floriculture sector, and those who lead precarious lifestyles due to low wages and seasonal employment at the flower farms.

Land claims and contestations form part of the social order on the frontiers. Respective groups negotiate access to land from different angles. Many Kikuyu argue they had bought land in Enoosupukia, whereas Dorobo and Maasai claim to have birth rights to the place. While the latter still accommodate Kikuyu as land leasers in Enoosupukia, it seems there is no going back to the situation pre-1993 clashes, where both lived together in Enoosupukia, under evolving terms and conditions. The 1993 violence that pitted Kikuyu and Maa-speaking people against each other is thus still permeating intergroup relations and circumscribing aspirations for a common future, as captured in this statement: "I can't live with people with whom we have ever quarrelled over land. Those people are not happy for me to get my land back. What if someone comes to kill me overnight?"[25]

Despite such attitudes from some individuals, ethnic relations seem to grow stronger, as evidenced by widespread social and economic cross-cutting ties including intermarriage, livestock trade, and leasehold arrangements. Moreover, cooperation and non-violent conflict resolution between groups is evidenced in the fact that there has not been a repeat of interethnic land-related violence since 1993, but only intracommunal attacks directed by Kikuyu against an elite perceived to be corrupt – the Ng'ati farm directors – rather than ethnic "Others". Further, behind the unease by Enoosupukia residents about newcomers lies a hidden critique of powerful Maasai outsiders – not Kikuyu.

Land claims remain one of the greatest challenges on the frontier. We, however, see a shift from violent regulation over claims towards a cooperative approach in pursuit of property rights to land. Using their marginalised position brought about by state actors and corrupt elite, a majority of residents in Maiella and Enoosupukia are united through collective action to push for formal title deeds with the hope of securing their futures and, as elsewhere, to avoid being expropriated by the state and its allies (Toulmin, Delville, and Traoré 2002, 4). Surprisingly, even title deeds are not secure in the end. State institutions intervene to territorialise the frontier at certain moments in time – through land adjudication and titling or the gazetting of forests – but they do so half-heartedly and thus leave the frontier in a state of legal uncertainty.

In conclusion, we observe the following: the current demands for food produce at the nerve centre coupled with agricultural intensification at the frontier suggests possibilities for deepened exchanges of people, goods and services between the lake and the hinterlands. Increasing populations at the nerve centre coupled with few and unpredictable job opportunities in the flower sector (as demonstrated by the closure of farms and termination of contracts for many flower workers during the COVID-19 pandemic) suggests a possibility for pursuit of alternative livelihoods elsewhere and the frontier thus becomes an immediate option. Ongoing land claims and the uncertainty over tenure paints a grim picture of the future of the frontier and the nerve centre. While cross-cutting ties and conflicting loyalties give hope for a peaceful future on

25 Maiella resident who had successfully recuperated land in Enoosupukia, September 2020.

the frontier, interpersonal and intergroup ties can easily become fragile and collapse or break down, thus explaining conflict or violence (e.g. Fukui 1994; Schlee 1997). However, we understand peace not simply as the absence of violence, but as the capacity for and practice of non-violent cooperation in the face of pertinent challenges (Kioko 2016, 38). Therefore, the future of the frontier and the nerve centre as well as their interdependence will largely depend on their capacity to deal with existing challenges.

Acknowledgements

This chapter is the result of research done in the context of the DFG-funded Collaborative Research Centre *Future Rural Africa: Future-Making and Social-Ecological Transformation*. We thank our research participants in Kenya as well as all the persons and institutions that facilitated our research for their time, efforts and support. In particular, we are grateful to the tremendous work done by our research assistant Sarah Njeri (Soila) and other assistants who prefer not to be mentioned by name. Without their invaluable contributions this work would not have been possible.

References

Anderson, David. 2005. *Histories of the Hanged: British Dirty War in Kenya and the End of the Empire*. London: Phoenix Orion Books Ltd.

Baretta, Silvio D., and John Markoff. 1978. "Civilization and Barbarism. Cattle Frontiers in Latin America." *Comparative Studies in Society and History* 20 (4): 587–620.

Berman, Bruce, and John Lonsdale. 1992. *Unhappy Valley: Conflict in Kenya and Africa*. Book one: State and Class. Athens: Eastern African Studies, Ohio: Ohio University Press, London: James Currey, Nairobi: Heinemann.

Berntsen, John L. 1976. "The Maasai and Their Neighbors: Variables of Interaction." *African Economic History* 2: 1–11. https://doi.org/10.2307/3601509.

Blackburn, Roderic H. 1982. "In the Land of Milk and Honey: Okiek Adaptations to their Forests and Neighbours." In *Politics and History in Band Societies*, edited by Eleanor Leacock and Richard Lee, 283–305. Cambridge: Cambridge University Press.

Blackburn, Roderic H. 1996. "Fission, Fusion, and Foragers in East Africa: Micro- and Macroprocesses of Diversity and Integration among Okiek Groups." In *Cultural Diversity among Twentieth-century Foragers: An African Perspective*, edited by Susan Kent, 188–212. Cambridge: Cambridge University Press.

Colson, Elizabeth. 1953. "Social Control and Vengeance in Plateau Tonga Society." *Africa* 23 (3): 199–212. https://doi.org/10.2307/1156280.

Cronk, Lee. 2004. *From Mukogodo to Maasai. Ethnicity and Cultural Change in Kenya.* Boulder: Westview Press.

Dundas, Kenneth R. 1908. "Notes on the Origin and History of the Kikuyu and Dorobo Tribes." *Man* 8: 136–39.

Flap, Henk D. 1997. "The Conflicting Loyalties Theory." *L'Année sociologique* 47 (1): 183–215.

Fukui, Katsuyoshi. 1994. "Conflict and Ethnic Interaction: The Mela and their Neighbours." In *Ethnicity and Conflict in the Horn of Africa*, edited by Katsuyoshi Fukui and John Markakis, 33–47. London: James Currey.

Galaty, John. 1993. "'The Eye That Wants a Person, Where Can it Not See?' Inclusion, Exclusion, and Boundary Shifters in Maasai Identity." In *Being Maasai: Ethnicity and Identity in East Africa*, edited by Thomas Spear and Richard Waller, 174–94. London: James Currey.

Geiger, Danilo. 2008. *Frontier Encounters: Indigenous Communities and Settlers in Asia and Latin America.* Bern: International Work Group for Indigenous Affairs.

Guyot, Sylvain. 2011. "The Eco-Frontier Paradigm: Rethinking the Links between Space, Nature and Politics." *Geopolitics* 16 (3): 675–706. https://doi.org/10.1080/14650045.2010.538878.

Hall, Derek. 2013. *Land.* Cambridge: Polity Press.

Hoefle, Scott W. 2006. "Twisting the Knife. Frontier Violence in the Central Amazon of Brazil." *Journal of Peasant Studies* 33 (3): 445–78. https://doi.org/10.1080/03066150601062993.

Hoffmann, Kasper, and Penelope Anthias. 2021 "The Making of Ethnic Territories: Governmentality and Counter-conducts." *Geoforum* 119: 218–26. https://doi.org/10.1016/j.geoforum.2020.06.027.

Hornsby, Charles. 2012. *Kenya: A History since Independence.* London: I.B. Tauris & Co Ltd.

Hughes, Lotte. 2006. *Moving the Maasai: A Colonial Misadventure.* St Antony's Series. London: Palgrave Macmillan.

Imamura, Masao. 2015. "Rethinking Frontier and Frontier Studies." *Political Geography* 45: 96–97. https://doi.org/10.1016/j.polgeo.2014.09.014.

Kenny, Michael G. 1981. "Mirror in the Forest. The Dorobo Hunter–gatherers as an Image of the Other." *Africa* 51 (1): 477–95. https://doi.org/10.2307/1158950.

Kioko, Eric M. 2016. "Turning Conflict into Coexistence: Cross–cutting Ties and Institutions in the Agro-pastoral Borderlands of Lake Naivasha Basin, Kenya." PhD diss., University of Cologne.

Kioko, Eric M., and Michael Bollig. 2015. "Cross-cutting Ties and Coexistence: Intermarriage, Land Rentals and Changing Land Use Patterns among Maasai and Kikuyu of Maiella and Enoosupukia, Lake Naivasha Basin, Kenya." *Rural Landscapes: Society, Environment, History* 2 (1): 1–16. https://doi.org/10.16993/rl.ad.

Klopp, Jacqueline M. 2001. "Electoral Despotism in Kenya: Land, Patronage and Resistance in the Multi-Party Context." PhD diss., McGill University.

Korf, Benedikt, Tobias Hagmann, and Martin Doevenspeck. 2013. "Geographies of Violence and Sovereignty: The African Frontier Revisited." In *Violence on the Margins. States, Conflict, and Borderlands*, edited by Benedikt Korf and Timothy Raeymaekers, 29–54. New York: Palgrave Macmillan.

Kratz, Corinne A. 1980. "Are the Okiek Really Masai? Or Kipsigis? Or Kikuyu?" *Cahiers d'études africaines* 20 (79): 355–68.

Kröger, Markus, and Anja Nygren. 2020. "Shifting Frontier Dynamics in Latin America." *Journal of Agrarian Change* 20 (3): 364–86. https://doi.org/10.1111/joac.12354.

Kuiper, Gerda. 2017. "The Flowers Are Carrying Us. Agro-industrial Labour and Migrant Workers' Settlements at Lake Naivasha, Kenya." PhD diss., University of Cologne.

Larsen, Peter B. 2015. *Post-frontier Resource Governance: Indigenous Rights, Extraction and Conservation in the Peruvian Amazon*. New York: Palgrave Macmillan.

Lawren, William L. 1968. "Masai and Kikuyu: An Historical Analysis of Culture Transmission." *Journal of African History* 9 (4): 571–83. https://doi.org/10.1017/S0021853700008904X.

Leakey, Louis S.B. 2007 [1977]. *The Southern Kikuyu before 1903. Volume I.* Kenya: Richard Leakey.

Li, Tania M. 2014. *Land's End: Capitalist Relations on an Indigenous Frontier.* Durham: Duke University Press.

Little, Paul E. 2001. *Amazonia: Territorial Struggles on Perennial Frontiers.* Baltimore: John Hopkins University Press.

Matter, Scott. 2004. "'We Have This Land as Our Right': Ethnicity, Politics, and Land Rights Conflict at Enoosupukia, Kenya." MA thesis, McGill University.

Matter, Scott E. 2010. 'Struggles over Belonging: Insecurity, Inequality and the Cultural Politics of Property at Enoosupukia, Kenya." PhD diss., McGill University.

Muriuki, Godfrey. 1974. *A History of the Kikuyu 1500–1900.* Oxford University Press.

Nelson, Robert L. 2011. "Emptiness in the Colonial Gaze: Labor, Property, and Nature." *International Labor and Working-Class History* 79 (1): 161–74. https://doi.org/10.1017/S0147547910000335.

Radcliffe, Sarah A. 2019. "Geography and Indigeneity III: Co-articulation of Colonialism and Capitalism in Indigeneity's Economies." *Progress in Human Geography* 44 (2): 374–88. https://doi.org/10.1177/0309132519827387.

Radcliffe-Brown, Alfred D., and Daryll Forde. 1950. *African Systems of Kinship and Marriage.* London: Oxford University Press.

Rasmussen, Mattias B., Christian Lund. 2018. "Reconfiguring Frontier Spaces: The Territorialization of Resource Control." *World Development* 101: 388–99. https://doi.org/10.1016/j.worlddev.2017.01.018.

Reid, Richard. 2011. *Frontiers of Violence in North-East Africa. Genealogies of Conflict since c. 1800.* Oxford: Oxford University Press.

KNBS. 2019. *Kenya Population and Housing Census, Volume II: Distribution of Population by Administrative Unit.* Nairobi: KNBS.

Republic of Kenya. 2019. *Report by the Ad Hoc Committee on Ng'ati Farm Alternative Dispute Resolution (ADR).*

Schetter, Conrad, and Marie Müller-Koné. 2021. "Frontiers' Violence: The Interplay of State of Exception, Frontier Habitus, and Organized Violence." *Political Geography* 87: 102370.

Schetter, Conrad, and Marie Müller-Koné. 2021. "Frontier – ein Gegenbegriff zur Grenze?" In Handbuch Grenzforschung, edited by Dominik Gerst, Maria Klessmann, and Hannes Krämer, 240–54. Baden-Baden: NOMOS Verlag.

Schlee, Günther. 1997. "Cross-cutting Ties and Interethnic Conflict: The Example of Gabbra, Oromo and Rendille." In *Ethiopia in Broader Perspective: Papers of the 13th International Conference of Ethiopian Studies*, edited by Wolde Gossa Tadessa, Katsuyoshi Fukui, Eisei Kurimoto and Masayoshi Shigeta, 577–96. Kyoto: Shokado.

Schmink, Marianne, and Charles H. Wood. 1992. *Contested Frontiers in Amazonia.* New York: Columbia University Press.

Schmitt, Carl. 1997 [1950]. *Der Nomos der Erde im Völkerrecht des Jus Publicum Europaeum.* Berlin: Dunker & Humboldt.

Spear, Thomas, and Richard Waller. 1993. *Being Maasai: Ethnicity and Identity in East Africa.* Oxford: James Currey.

Tignor, Robert L. 1976. *Colonial Transformation of Kenya: The Kamba, Kikuyu, and Maasai from 1900–1939.* Princeton University Press.

Toulmin, Camilla, Philippe L. Delville, and Samba Traoré. 2002. "Introduction." In *The Dynamics of Resource Tenure in West Africa*, edited by Camilla Toulmin, Philippe L. Delville, and Samba Traoré, 1–24. Oxford: James Currey.

Tsing, Anna L. 2003. "Natural Resources and Capitalist Frontiers." *Economic and Political Weekly* 38 (48): 5100–106.

Turner, Frederick J. 1893. *The Frontier in American History.* New York: Dover Publications.

Waller, Richard. 1993. "Acceptees and Aliens: Kikuyu Settlement in Maasailand." In *Being Maasai: Identity and Ethnicity in East Africa*, edited by Thomas Spear and Richard Waller, 226–48. Oxford: James Currey.

Werthmann, Katja, and Thilo Grätz. 2012. *Mining Frontiers: Anthropological and Historical Perspectives.* Mainzer Beiträge zur Afrikaforschung 32. Köln: Rüdiger Köppe.

CHAPTER 12

Devolution of Governance and the Politics of Fishery at Lake Naivasha, Kenya

Johannes Dittmann and Antony F. Ogolla

1 Introduction[1]

> The capacity of the lake is at its end because management has failed. A few years ago, quite a number of strange things started to happen in the lake.
>
> FISHING-BOAT OWNER, NAIVASHA, AUGUST, 2016

A fishing-boat owner at Lake Naivasha points to recent challenges she faces in her work. As she explained in an interview, since the devolution of the Kenyan government in 2013 local fishery management is being critically compromised. Since then, the fishing communities have experienced a decrease in fish-catch quotas and a resurgence of ethnic tensions among their members. This chapter on the political ecology of fishery at Lake Naivasha deals with the question of how political transformation is related to challenges in local natural-resource governance. It focuses on the influence of decentralising political authority over local fishery management from central government to county governments in Kenya.

Inland fisheries in the Global South are characterised by highly complex social-ecological dynamics that require interdisciplinary approaches to investigate. However, many studies in the field primarily take a natural-science perspective focusing on ecological, physical or demographic factors. Social and political aspects are often not sufficiently recognised in the academic literature on inland fisheries of the Global South (Geheb and Sarch 2002, 1–4; Welcomme et al. 2010, 2888–89; Lynch et al. 2016), though they need to be taken into consideration to understand actor constellations and power relations that determine fisheries management (Njaya, Donda, and Béné 2011; Hara and Raakjær Nielsen 2003; Béné et al. 2009).

1 First submission date: January 15, 2021. Acceptance date: March 13, 2023.

 | DOI:10.1163/9789004695429_013

The fishery at Lake Naivasha is an example of an inland fishery on which academic discourse is dominated by specialised accounts on water quality, vegetation cover and fish stocks (see e.g. Harper, Mavuti, and Muchiri 1990; Becht, Odada, and Higgins 2005; Harper et al. 2011; Aloo et al. 2013; Hickley et al. 2015). There are significantly more studies that explain the ecological significance of individual fish species in very detailed ways (e.g. Britton et al. 2007) than papers on issues of fishery management. Phenomena discussed in academic debates regarding challenges faced by the fishery include population growth, poverty, water abstraction, eutrophication, deforestation and El Niño phenomena (Harper et al. 2011; Hickley et al. 2015; Aloo et al. 2013). Aspects that are less subject to these debates are politics, governance and power relations. Several studies have pointed to the need to integrate a critical social-science perspective into the debates (Hickley et al. 2004; Becht and Harper 2002; Aloo et al. 2013). The importance of recognising social complexities and power relations became most evident during fieldwork with local fishing communities, as they pointed to political driving forces of recent challenges faced by fishery through the devolution of the Kenyan government.

McGuire (1998) suggests political ecology as an interdisciplinary research programme to add social-science perspectives to academic debates on fisheries. Building on literature dealing with inland fisheries in the Global South from a political-ecology perspective (Béné et al. 2009; Tan-Mullins 2007; Menon et al. 2015) our chapter focuses on power relations and social dependencies in human-environment relations. It aims to contribute to a better understanding of the social-ecological complexities of recent challenges arising in the fishery at Lake Naivasha in the context of political transformation. We particularly suggest that attention be paid to a recent increase in political interference through the devolution of the Kenyan government. The chapter deals with the ways in which the decentralisation of political authority affects fishery management, identity politics and vulnerability of fish stocks at Lake Naivasha.

In the following, we will first briefly introduce the research context by describing the organisation of the fishery at Lake Naivasha. We will then revisit the process of devolution of government in Kenya as well as its general meaning for restructuring local power constellations. In its empirical section, it relates these debates to our case study of the fishery at Lake Naivasha, describing how the political circumstances of fishery governance in Naivasha have changed since environmental authority has been transferred from national to county levels. The chapter then proceeds to illuminate the clandestine organisation of informal fishing in Naivasha and how intensified relations to political elites increase the organisation's control over the lake's resources. It continues to show how these patronage networks compromise the governmental control over fishery and the operation of local community-based fishery management. The final part focuses on the progressive ethnicisation of fishery in Naivasha, which is a region marked by a history of ethnic violence. The chapter concludes by advocating political-ecology perspectives when analysing social-ecological dynamics of wetlands in the Global South.

Our data is based on expert interviews in Naivasha Town and ethnographic fieldwork in 2015 and 2016 conducted at the lakeshore and in informal settlements of the area. Unless otherwise indicated, the information in this chapter is based on our own data collection.

2 Organisation of the Fishery at Lake Naivasha

There are four official landing beaches located along the lakeshore which are organised by the fishing communities in Beach Management Units (BMUs) (see Map 12.1). The BMUs are in cooperation with the local Fisheries Department as part of a fishery co-management approach introduced at Lake Naivasha in the early 2000s. Every four years, each BMU elects a new board to represent it. The board consists of a chairperson, a vice-chairperson, a secretary and a treasurer. The chairpersons are very influential as they have the right to admit or exclude

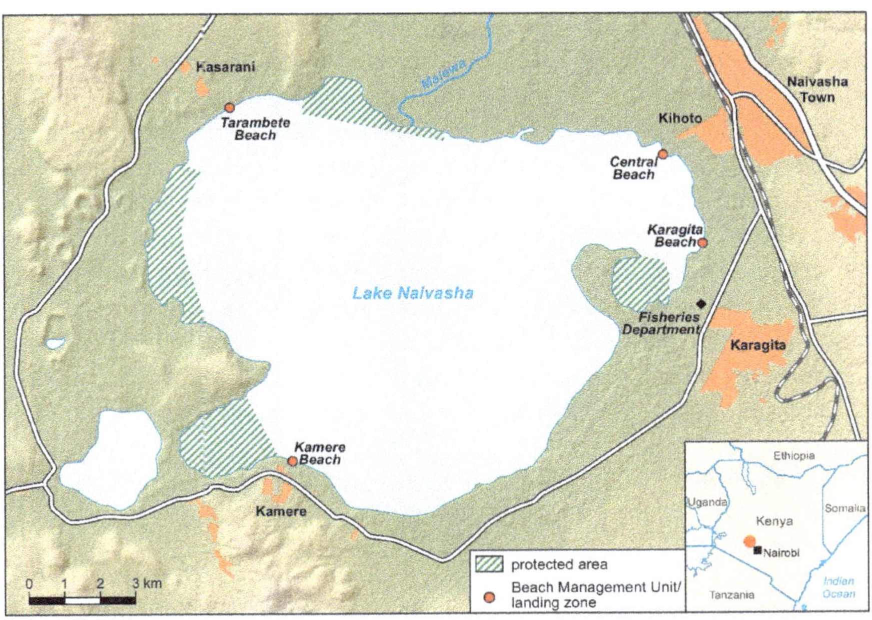

MAP 12.1 Protected areas and fish-landing beaches at Lake Naivasha

FIGURE 12.1 A fisherman (left) hauls in the catch (right)
PHOTO: JOHANNES DITTMANN 2016

boat owners, fishermen and fish traders in their BMU. The mandate of the ten to twelve staff of the Fisheries Department is to issue licences, register catch numbers and provide control services. In the lake, there is a protected shore zone about 100 m wide, which varies depending on the water level, as well as four protected areas. Fishing is not allowed in these areas, to provide a breeding ground for fish. Each boat is allocated a certain zone in the lake, where fishing is allowed with ten nets per boat, which must have a minimum mesh size of 10 x 10 cm (see Figure 12.1).

According to estimates by the Fisheries Department, about 160 boat owners are currently officially licensed who are allowed to own only one fishing boat. For one boat, four people are employed as fishermen, i.e. there are about 640 fishermen working on Lake Naivasha. In addition, there are an estimated 200 fish traders, making the total number of people officially involved in Naivasha's licensed fishery more than 1,000. Fishing boat ownership is the most capital-intensive and profitable activity in the BMU. A boat owner receives on average about KSh 70,000 per week from the sale of fish. Depending on this revenue, fishermen receive their wages, which range from KSh 300 to KSh 600 per person per day.[2] The fish are registered by officials at the BMU handling facilities and then sold to traders. Most fish are sold fresh at markets, restaurants and hotels in Nairobi, Nakuru and Naivasha, as well as in smaller towns in Rift Valley Province and Central Province. According to local fish traders Gikomba and other urban markets in Nairobi are the main outlets for fish from Naivasha.

Informal fishing is classified as illegal by the government. It is estimated that 2,000 to 3,000 people are involved in informal fishing in Naivasha, who,

2 At January 1, 2016, KSh 1 was equivalent to $0,0096 (source for the exchange rate: https://www.oanda.com/currency-converter/en/). Hence, KSh 300 to KSh 600 was equivalent to $2,88 to $5,76 at the time of research.

according to local experts, catch twice to three times as much fish as all licensed fishing.

3 Devolution of the Kenyan Government

Kenya's new constitution 2010 is seen as a major turning point in the country's history. The 2013 general elections brought in political leaders who were to govern under a devolved system of governance. In the same year, the Kenyan government started to implement the country's most crucial political transformation since independence. Devolution decentralised allocative and authoritative resources from the hitherto strictly centralised state to the 47 counties of Kenya. These county governments were assigned extensive powers managed by a directly elected governor and legislative assembly (Basset 2017).

Devolving political power as an inevitable element of good governance became popularised and supported by international development communities in the 1990s (see Ostrom 1990). Devolution constitutes a key element in development visions and political programmes in the Global South. In Kenya, the desire to devolve government was driven by a long history of politicised ethnicity and its dramatic implications for Kenya's population, politics and economy. A centralised state system inherited from colonial times allowed specific ethnic groups to politically dominate the whole country. Ethnic polarisation of political power repeatedly led to nation-wide violence and depression, especially in the wake of elections. Majimboism, a Kenyan-specific kind of federalism based on the linkage of ethnic identity and territoriality, became increasingly popularised. This concept first gained wider support in 1972 from opposition groups that aimed to transform the centralised system of government of Kenya, which was dominated by members of the Kikuyu ethnic community. It re-emerged in the following decades especially during elections and was constantly revived by politicians of the opposition in Kenya (Anderson 2005; Nabwire 2018). Majimboism popularised desires for decentralisation of government but also fuelled an ethnically determined political culture dubbed "our turn to eat" (Branch, Cheeseman, and Gardner 2010) which alludes to prioritising one's own ethnic community and not others (D'Arcy and Cornell 2016, 246).

Patronage, corruption and politicised ethnicity were seen as effects of an over-centralisation of government, and devolution of political power as the logical solution to these issues. Thus, expectations regarding devolution were very high in Kenya, especially after the post-election violence in 2007/2008 which resulted in ethnic massacres, evictions of thousands of people, and a national economic crisis (see Müller-Koné and Kioko, this volume=. Naivasha was one of the hotspots of the post-election violence that, according to Murunga (2011), was mobilised in the area in revenge against members of communities that were seen to have been involved in previous ethnic conflicts. Devolution is believed to overcome the root causes of political conflicts in Kenya. By transferring substantial legislative capacities and accountabilities of budget administration to county governors, the Kenyan government aimed for a more equal distribution of resources to be realised. It was meant to move decision-making closer to the people and make their involvement more pragmatic (Ngigi 2019).

However, the effects of devolution in Kenya are mixed and often not unproblematic, as empirical evidence shows. According to Ngigi (2019) devolution also served to alienate a significant part of the Kenyan society by leaving decision-making to the political elites within the county. He argues that there has been no proper legal framework to guide the way in which citisens participate in the day-to-day affairs of a devolved system (*ibid.*). Further, different studies argue that devolution has not led to an eradication of the root causes of political conflicts but to an exacerbation of corruption (Cheeseman, Lynch, and Willis 2016), an increase of political patronage (Cornell and D'Arcy 2014) and a resurgence of tribalism (Lind 2018; Scott-Villiers 2017).

Most of these studies deal with the implications of devolution for the whole country in general from a meta-political perspective, mainly concerning changes in administration. Less research has been done concerning the implications of Kenyan devolution at the local level, especially in terms of natural-resource management. For our case study on Naivasha, we focus on the implications of devolution, as Béné et al. (2009) emphasise the need to consider the effects of decentralising political power in order to understand local fisheries governance in Sub-Saharan Africa. Drawing on our ethnographic fieldwork in the following we will illustrate the impacts of devolving political authority from national to county level in Kenya on local regimes of natural resource use.

4 Increasing Political Interference in Lake Naivasha Fishery

Prior to the start of the devolution process in 2013, fishery management at Lake Naivasha was under direct mandate of the National Ministry of Agriculture, Livestock, and Fisheries in Nairobi. With the coming in of devolution in 2013, the control over natural resources was transferred to politicians at the county level. Fisheries was one of the sectors whose governance was devolved. The fishery at Lake Naivasha is now under the mandate of the county government of Nakuru. Its management has been sub-divided between two sub-counties. Since then, the BMUs Central Beach, Karagita Beach, and Kamere Beach are under the Fisheries Department of Naivasha Sub-County whereas Tarambete Beach has been assigned to Gilgil Sub-County. The county government of Nakuru today rates the economic significance of the fishery at Lake Naivasha much higher than the national government did before devolution. According to the local Fisheries Department in Naivasha, the county government recognises that the lake's full potential with regard to fishing has not been exploited so far. The increase in political interest is rooted in a stronger sense of ownership towards the lake and its resources:

> Without devolution, nobody would be interested in the lake... Now with devolution the people here feel that they own the waters; that they own the resource at local level. Now the governor has more power than the president towards the lake. He can decide on who to be involved and what to be done.
>
> LOCAL NGO EXPERT, NAIVASHA, SEPTEMBER, 2016

Through devolution, Nakuru County politicians received the authority to govern the fishery in Naivasha directly. The interviewed government representatives agree that devolution has had very positive impacts on local environmental governance. However, a majority of research participants perceive the effects of devolution on the fishery at Lake Naivasha very differently. The transfer of political authority over natural resources from national to county level is associated with increased liabilities in terms of economic development of the region, through job creation or tax increase. In Naivasha this resulted in expanding fishery by lifting the limit on the number of fishing licenses allowed. The county government increasingly seeks to exploit the lake's fish resources to increase revenues:

> The sub-county politicians were told to make their own money so if the governor sees a resource like this he has to use it maximally. He can say now, we are going to increase the number of boats so we can make more money.
>
> LOCAL NGO EXPERT, NAIVASHA, AUGUST, 2016

The main source of county income from fishery is licensing. Since devolution the county government is more dependent on such revenues in order to get recognised by the central government. The Sub-County Commissioner of Naivasha highlights the importance of fishery in this regard as he claims that the county government earns millions of shillings through licensing. The Kenya Marine and Fisheries Research Institute (KMFRI) recommended 43 fishing boats for Lake Naivasha (divided among three BMUs) to ensure sustainable resource use. Since 2013, more than 100 additional boat licenses have been issued. According to the Fisheries Department, there are now officially about 163 boat licenses and more than 1,000 people engaged in the BMUs. In addition, Naivasha Sub-County and Gilgil Sub-County are in competition with each other in maximising the number of fishing licenses:

> You know Lake Naivasha is now divided into Gilgil site and Naivasha constituency. Now there is a fight in the lake among the politicians. They want to show their supremacy in the lake.
>
> FISHING BOAT OWNER, NAIVASHA, AUGUST, 2016

In the BMU of Tarambete Beach alone, which is the only landing beach in Gilgil Sub-County, the number of boat licenses recommended by KMFRI was exceeded, with 55 being granted. The Sub-County Commissioner claims Gilgil's share of fish resources to be equal to Naivasha's.

Further, particular groups in the fishing communities convinced county politicians to lift several restrictions developed by KMFRI. In the early 2000s an annual four-month fishing ban was introduced in Naivasha to allow fish stocks to recover. This ban on the lake was lifted in 2013 to allow year-round fishing. In addition, the allowed number of nets per boat has been increased. According to local experts the most influential boat owners have up to 100 nets cast in the lake instead of the officially allowed ten. What is particularly interesting about the suspension of many restrictions is that it is unofficial:

> After devolution when new politicians went into this county KMFRI's recommendations were thrown away! ... Our technical advice was that the ban should not be lifted and actually legally it has not been lifted. It is only that the people have, how can you say, they have overruled what is supposed to be done.
>
> LOCAL FISHERIES EXPERT, NAIVASHA, SEPTEMBER, 2016

That means that actions that violate the restrictions are deliberately overlooked and no longer sanctioned if they are in the interests of certain political leaders. County politicians circumvent resource use regulations of KMFRI in order to increase county revenues or personal profit, as many fishing community members assert. Furthermore, granting a license is determined by the "commercialisation of bureaucratic relations". How many licenses are allowed for the fishery depends on how many people are willing to pay the negotiated price for a license: "Once you have sufficient money, you just get a license" (fishing boat owner, Naivasha, August, 2016). This kind of rent-seeking by political elites has increased in Kenya in several counties since devolution (D'Arcy and Cornell 2016) and is described by Béné et al. (2009, 1941–42) as a common consequence of rescaling governance in fisheries of Sub-Saharan Africa.

This section has shown that since devolution, the criteria for obtaining a fishing license are less determined by fishing experience or scientific management recommendations than by rent-seeking. The following sections deal with how particular actors benefit from recent political interference and expand their control over fish resources while other actors at the same time face incremental challenges in fishing at Lake Naivasha.

5 Political Patronage Relations in Fishery

The last section has shown the ways in which increased political interference affects the licensed fishery in Naivasha. In what follows we will portray how the pressure on the lake's fish resources are also increased by the intensification of non-licensed fishing activities. Illegal or unlicensed fishing activities are often seen as survival strategies of poor or marginalised groups (Béné 2003). There are fishermen in Naivasha who are unable to access the licensed (legal) fishery sector or other employment opportunities for various reasons (e.g. see Kuiper, this volume). For many, (illegal) fishing in the riparian zone, which requires little equipment and know-how, is a short-term survival strategy. However, the link between poverty and illegal fishing is often simplified in fisheries policy discourses (Kateka 2010; Béné 2003). Especially in Naivasha, unlicensed fishermen should not be generally considered as disadvantaged actors because many of them make more money through fishing than licensed fishermen. Since devolution a locally operating network of unlicensed fishermen, which locals refer to as the "fishing-mafia" because it makes use of methods including bribery, intimidation and murder, increasingly engages in patronage relationships with political elites. These relationships, which do not involve all unlicensed fishermen but a specific informally organised network of them, lead to a situation where fishing activities intensify.

The informal fishery network is probably the most powerful actor in the fishery of Naivasha. It is characterised by a thoroughly organised system of intensive informal resource use, with its own state-independent rules, hierarchies, and networks that operate in the informal settlements around the lake.

Based on its economic and numeric superiority, it is able to assert its power over BMUs, fishery officials, and the police. For a long time, the Kenyan government did not consider the fishery of Naivasha as being of national interest, nor did it introduce effective control. This power vacuum was filled by a clandestine organisation probably formed in the 1980s. Informal settlements in many regions of Kenya are prone to the emergence of clandestine institutions due to a lack of judicial enforcement. Various forms of self-organised and conspiratorial groups, such as the Mungiki, began to form in the 1980s, increasingly developing their own systems of rules and hierarchies that countered the authority of the state (Wamue 2001). In Naivasha, the Mungiki are particularly active in Kihoto and Karagita, where they exert their power beyond state regulation at economic and political levels (Lang and Sakdapolrak 2015).[3] Among other informal sectors, they control unlicensed fishing and have organised it in the most efficient way. The members of the informal fishing network were described by locals as young men who usually act as a collective. All interviewed fisheries experts, licensed fishermen and unlicensed fishermen who are not part of the network consider them unpredictable, armed and violent.

Due to intensive fishing methods, illegal fishing is three times more profitable than licensed fishing. Furthermore, illegally caught fish is sold at lower prices than the legally caught fish sold by BMUs and is therefore favoured by traders and consumers. The chiefs in the organisation play a decisive role in the coordination of fishing, transport and trade with customers in e.g. Nairobi or Nakuru. Described by a BMU chairman as rich tycoons, these chiefs provide the necessary capital and connections to make the informal fishing network such an efficient organisation:

> These rich tycoons use fish poachers to go and fish for them because they have means and ways for networking with fish customers in Nairobi … So the tycoons support the fish poachers. And if you fight the fish poachers you also fight them. They have a very good network.
>
> BMU CHAIRMAN, AUGUST, 2016

3 "Mungiki" (Kikuyu for "mass") is a spiritual movement of mainly young, male Kikuyu, ideologically related to the Mau Mau rebellion of the 1950s and the Kikuyu religion. Today, Mungiki groups operate particularly in the informal settlements of the former Central and Rift Valley Provinces. There, they control economic sectors and decision-making processes through sometimes extreme violence and brutality. Although the Mungiki were criminalised by the Kenyan government, several politicians instrumentalised them during the post-election violence of 2007/2008 (Wamue 2001; Kagwanja 2003; Lang and Sakdapolrak 2015).

The methods the "fishing mafia" uses to enforce its business have given it its name. According to a licensed fisherman a group of four members of the informal fishing network can raise KSh 10,000 in bribes every week with the support of its bosses. In the past, there have been several violent fights between members of the informal fishing network and BMU members or fishery officials. Theft of nets, boats, or hauls often has to be tolerated by BMU members as they are intimidated by groups of young men armed with machetes. An environmental expert who has lived at the lake since the 1970s confirmed:

> There is a mafia amongst them (illegal fishermen). Yes, there is a mafia and you don't mess with them. If you start messing with them they are all armed and a lot of them turn nasty on you. So you just let them be.
>
> LOCAL ENVIRONMENTAL EXPERT, AUGUST, 2016

In an interview a boat owner in Karagita Beach reported revenge attacks in the past that intimidate the members of the BMUs to this day: "If you report them to police, they bribe themselves out and come to your house to deal with you". Some licensed fishermen spoke of threats they had received from members of the informal fishing network:

> Some fear for their lives because they have been threatened by the fish poachers. For instance, I have been threatened four times. I was actively engaged against poaching... But when they realised that I was intruding and interfering with their businesses I had to stop.
>
> FISHERMAN, SEPTEMBER, 2016

When bribery and intimidation do not work, the informal fishing network does not shy away from violence and even murder. While on patrol in the shore zone of the lake, a senior fishery officer was attacked by several men and almost beaten to death. As fishermen at Kamere Beach and Central Beach reported, licensed fishermen who reported unlicensed fishing to the Fisheries Department have on several occasions been murdered by the members of the informal fishing network. The network's readiness to use violence became particularly evident when environmentalist Joan Root, who was campaigning against illegal fishing, was murdered in her home at Lake Naivasha in 2007. Her murder was never officially solved, but one of her former colleagues asserted in an interview that it was a contract killing executed by the informal fishing network:

> She tried to introduce community patrol to the lake to get rid of illegal fishermen. Initially it worked well but with time everything changed

> because she was recruiting and reforming illegal fishermen and using them to arrest other illegal fishermen. In the cause of implementation everything collapsed. And it was one of the reformed poachers who has murdered Joan Root. Local environmentalist, August, 2016

Local police arrested four suspects in 2007, who were released after a short time in jail, officially due to lack of evidence. The case was subsequently not investigated.[4]

Unlicensed fishing is not a phenomenon unique to Lake Naivasha. What is particularly interesting for our case is that its informally organised network in Naivasha consolidated its power through patronage relationships with political elites. The recent political interference in fishery is largely determined by the interest of politicians in attracting votes. By expanding fishery and lifting restrictions, county politicians not only aim to increase revenues; they also strategically follow the demands of the most influential voter groups in the fishery sector. In terms of election campaigns, Naivasha's informal settlements are considered important constituencies in the region because of their large number of voters. Therefore, county politicians aim to collaborate with those actors who are most influential in lobbying and mobilising voter support in these settlements. Since the informal network of unlicensed fishermen is a powerful player here, political actors attend to their demands, which imply tolerating or even endorsing illegal fishing activities:

> Politicians are after votes and Karagita is one of the strongest voter baskets. … Being after votes the politicians just come and do what the voters want. They tend to mobilise from the back scene to work everything to support the illegal fishermen.
>
> LOCAL NGO EXPERT, SEPTEMBER, 2016

As a consequence, two landing beaches, Karagita and Kihoto Beach,[5] were legally established in 2014 and numerous informal fishermen of the network became licensed. The sub-county government saw this measure as a good way to increase revenues from licensing while combating illegal fishing and unemployment. Instead, illegal fishing has increased. By licensing previously unlicensed fishermen their sections at the lake's shoreline became free for newly recruited members of the informal fishing network. Further, it was the newly

4 For more details on Joan Root's murder see her biography by Seal (2010).

5 Kihoto Beach was closed again in the same year due to land conflicts with smallholders. However, it remains an important landing beach for unlicensed fishermen.

licensed fishermen who demanded the lifting of the annual fishing ban and other restrictions by the KMFRI.

Several informants referred to Karagita Beach as the "fish-poacher BMU". While their members have an official license to fish and use the appearance of a BMU, they continue to use illegal fishing methods such as seining (see Figure 12.2), and remain part of the informal fishing network. Among all BMUs at Lake Naivasha, Karagita Beach records the highest fish catches and has the best transport connection. Tellingly, its establishment was only possible by coupling political influence with violent enforcement on the ground. According to local fisheries experts, the land transaction for the new BMU is still an ongoing court case. Local BMU-members claimed that in colonial times, the land on which the Karagita Beach was established was part of a public corridor that led to the lakeshore between two private properties. In fact, the corridor never was a formal access point (see Kuiper, this volume). Prior to 2013, it was used by a private owner as an access road to his property near the lakeshore. After devolution political elites sought a shoreline section for a new BMU and came across that property. Informal fishermen of the network then put pressure on the private owner to leave his property. A mob destroyed the gatehouse and wrote "*Public Beach – Free Entrance*" on its ruins (see Figure 12.3). Shortly after, the shoreline area to which the corridor leads was recognised as the official landing site for the Karagita BMU. Politically these actions were never officially

FIGURE 12.2 (top left) One of the authors holding illegal fishing gear during a meeting with unlicensed fishermen; (right) fishermen doing seine fishing at the shoreline; (bottom left) unlicensed fishermen in the protected area near Crescent Island
PHOTO: JOHANNES DITTMANN 2016

FIGURE 12.3 The gatehouse of the private owner at Karagita Beach, destroyed by a mob mobilised by the informal fishing network
PHOTO: JOHANNES DITTMANN 2016

endorsed; neither were they prevented. In 2015, there was renewed unrest as the private owner reclaimed the land based on new evidence. When he closed the corridor to public use and fishing members of the informal fishing network then set up a roadblock on Moi South Lake Road, preventing any vehicles from passing, including for the transporting of cut flowers to the airport in Nairobi. After security forces failed to bring the situation under control even after two days (*Daily Nation* 2015), Karagita Beach was reopened through political influence without further consideration of the private owner's new evidence. The roadblock and demonstrations then ended and the transport route was reopened. As a local observer summarised:

> When you follow how land transaction has taken place in that area the title owner is a legitimate owner but through politics and mob justice people forced away the owner, and most of these people were fish poachers.
>
> LOCAL NGO EXPERT, NAIVASHA, SEPTEMBER, 2016

This section has shown how an informal network of illegal fishermen has expanded its influence through patronage with local politicians since devolution. The next section points to ways these clientelist relationships corrode fishery governance.

6 Eroding Governance of Fishery and the Blurry Line between Legal and Illegal

The mandate of the Fisheries Department of Lake Naivasha is to regulate and control fishing activities through licensing, registration and services related to monitoring, control and surveillance. Its limited financial resources have presented a challenge for years. After 2013, the Nakuru County government reduced them even more. Often patrol boats are not operational due to lack of fuel or pending repairs. There are only two patrol boats and twelve officers to control fishing across the whole approximately 160 km^2 lake. Prior to devolution, seven patrols were conducted per week; since 2013, they have decreased to two per week. Fish restocking projects have been drastically reduced as well.

The Fisheries Department was described by several experts and licensed BMU members as highly inefficient due to a lack of funding, corruption and close social ties to unlicensed fishermen. During fieldwork we observed that several internal house memos were posted in the department's offices, which addressed its employees' disregard of punctual attendance as well as the theft of office equipment and confiscated fish and fishing nets. Many BMU members assured that fishery officials do not prosecute but collaborate with unlicensed fishermen, take bribes and confiscate captured fish to resell them. Even the senior fishery officer admitted that the work of his department is significantly affected by bribes:

> It is obvious that the officers are corrupt because they earn so little money. They earn less than the fish poachers and the licensed fishermen. They earn up to KSh 18,000 in a month. A group of poachers when caught fishing illegally offers them KSh 20,000, so they take it.
>
> SENIOR FISHERY OFFICER NAIVASHA, AUGUST, 2016[6]

He continued:

> When I go on patrol with my officers, we usually do not arrest any fish poachers. When I send my officers for patrol without me, they come back without success... Information is leaking from the department to the illegal fishermen.
>
> *IBID.*

6 The amounts in shillings were equivalent to \$173 and \$192 respectively, based on the exchange rate mentioned in footnote 2.

Confiscated nets and fish are often resold by fishery officials on the same day. If unlicensed fishermen get arrested by incorruptible officials, they use their patronage relationships to local elites to manipulate the Fisheries Department and avoid convictions. Fishery officials are then urged by their political superiors to release from custody individuals considered criminals before the law, as a BMU chairman confirmed.

Fishery governance in Naivasha has been massively weakened since devolution. Through underfunding and patronage the Fisheries Department is heavily compromised, which means its actions have hardly any effect. Local experts describe the chief fishery officer as powerless since new county politicians are in power:

> You find today the fishery officer here has nothing. He is basically at the wing of the governor. So if the governor decides today we have to increase the number of boats there is nothing much this guy can do because it's the boss talking here.
>
> LOCAL FISHERIES EXPERT, AUGUST, 2016

The senior fishery officer himself said, "*This is an office where you are not free*". He knows that the consistent increase in fishing boats is unsustainable for the lake's fish resources, but he must enforce the will of his political superiors or that of the informal fishing network.

Political interference also affects the cooperation of fishery authorities with BMUs. As the Fisheries Department is being critically underfunded by the county government, the Fish Levy Trust Fund (FLTF) was established, through which BMU members fund fishing patrols by the department. Through this, the power asymmetries between these governmental and civil-society actors are changing as monitoring-, control- and surveillance-services become dependent on the willingness of fishing communities to have their own activities controlled. According to local fisheries experts, some BMU members seek to take advantage of this dependence. Licensed fishermen pay into the FLTF, but exploit their position as financiers of the Fisheries Department by demanding to be authorised to fish with illegal methods themselves.

> How do you fight the hand that feeds you?...They (Fisheries Department) have been trying to arrest them but the problem is that the fishermen are the only ones who are funding the Fisheries Department right now... So it is hard to enforce the laws when the people you are supposed to arrest are the people who are supplying you with fuel and so on.
>
> LOCAL FISHERIES EXPERT, NAIVASHA, AUGUST, 2016

In Naivasha fishery private entrepreneurs pay for an institution which is meant to control them. This is not a rare set-up in neoliberalised natural resource management settings (see Bixler, Dell'Angelo, Mfune and Roba 2015; Nunan, Hara and Onyango 2015; Li 2007a; Li 2007b). The fact that some licensed BMU members use illegal fishing methods or are even associated with the informal fishery network shows that the line between legal and illegal fishermen becomes increasingly blurry.

A recent phenomenon that further confuses the legal status of licensed fishermen is the rise of elite capture by the BMU chairmen. According to BMU members interviewed, the chairmen do not represent the concerns of their members, and abuse their high position. Several fishermen and boat owners confirmed that they are involved in the current increase in numbers of fishing licenses as they have vested interests in admitting more members to their BMU. Whether one may obtain membership depends on the amount of money one is able to pay to the chairperson. Thus, they benefit personally from the expansion of fishing and allow increased pressure to be placed on Naivasha's fish stocks. Furthermore, BMU members and local experts stated that the chairmen collaborate with the informal fishery network. At one of the landing beaches the chairman allows licensed as well as unlicensed fishers to use illegal fishing methods provided he receives 10% of revenues from fish sold. In interviews the BMU chairmen, one in particular, were denoted as the most serious fish poachers in Naivasha: "Karagita Beach is where poaching is at its peak and its chairman is one of the biggest poachers" (local fisheries expert, Naivasha, August, 2016). Instead of combating illegal fishery, BMU chairmen profit by either engaging in illegal fishing activities or accepting bribes from members of the informal fishing network:

> The problem is the individuals who are in charge of the BMUs. They are not interested in curbing poaching because they are benefitting from it themselves.
>
> LOCAL FISH TRADER, NAIVASHA, AUGUST, 2016

Fishing communities consider themselves powerless against the chairmen as they are intimidated or threatened with the revocation of their membership. The Fisheries Department only accepts the chairmen to speak for the fishing community, which leaves almost no chance for the ordinary BMU members to report any complaints to governmental authorities.

Facing a lack of governmental enforcement, an increase of corruption and elite capture, the relationship between licensed BMU members and unlicensed fishermen has become ambivalent. The informal fishing network has a much more effective organisation and fishing method, in terms of profit maximisation, than the BMUs. As unlicensed fishermen mostly catch juvenile fish in the riparian zones, licensed fishermen are increasingly being deprived of their resource. Fish traders prefer to buy illegally caught fish because it is offered more cheaply, forcing licensed boat owners to lower their regular prices. Therefore, more and more licensed fishermen tend to use illegal fishing methods to increase their income. What makes the whole situation of fishery governance even more difficult are social ties between unlicensed fishermen and licensed BMU members based on neighbourhood, friendship and kinship. In these cases, licensed and unlicensed fishermen tolerate each other as they live in the same settlement, just like fishery officials.

This section has shown how political interference since devolution leads to an erosion of local fishery governance and an increasingly blurry line between legal and illegal fishing communities. In the following the chapter will address how the social cohesion within the BMUs is affected through a resurgence of ethnic tensions.

7 Ethnicising Fishing

In Kenya, identity constructions based on the nexus of ethnicity, territoriality, and economic practices are commonplace in articulating and mobilising natural resource use claims (Ajulu 2002; Lynch 2006). In this section we show how devolution results in ethnicisation of fishing in Naivasha.

Since devolution, voices that deny fishermen from Luo and Luhya ethnic communities the right to fish in Lake Naivasha have become louder. Segments of the local Kikuyu population that are involved in fishing argue for this by territorialising ethnicity and referring to narratives about "traditional homelands" of Kenya's different ethnic communities. In this thinking, the region of Naivasha has been claimed to be a Kikuyu homeland since Kenya's independence. The homelands of Luo and Luhya communities however are considered to be in Western Kenya (former Nyanza and Western Province) (Lang and Sakdapolrak 2014). As Luo have dominated fishing in Naivasha since the 1960s (Yongo et al. 2013), several Kikuyu complain that fishery in "their land" is in the hands of "non-residents" and that members of their ethnic group, who rightfully reside in Naivasha, can hardly benefit from the lake's resources. In 2013, the Kikuyu-dominated local government therefore demanded that fishing licenses would be transferred from Luo to Kikuyu fishermen (local politician, Naivasha, August, 2016). It argued for a quota system and claimed 75% of fishing licenses should be owned by Kikuyu and 25% by other tribes. As that was not legally enforceable, 90% of the new licenses issued since devolution were allocated to Kikuyu. Most of them were assigned to the newly established Karagita Beach BMU. As a local expert explained:

> When the Kikuyu realised that there is money in fishing they went into illegal fishing in crowds, so we had many fish poachers. And because they are so many, politically they had bargaining power. So they could go to the governor because they are indigenous people here and say that there are Western people fishing here and themselves they don't have licenses... Actually, the newest beach which is called Karagita was opened to satisfy the Kikuyu because the other three beaches are run mostly by Luo and Luhya. So it was opened because the governor wanted his people also to get something.
>
> LOCAL FISHERIES EXPERT, NAIVASHA, AUGUST, 2016

Kikuyu in Naivasha use their ethnicity as a political resource. The clientelistic relationship between some Kikuyu County politicians and the Kikuyu-dominated informal fishing network is based on reciprocity as well as ethnicity. Old ethnic enmities between Luo and Kikuyu still flare up today, manifesting in a particularly extreme way during the post-election violence of 2007/2008. According to Lang and Sakdapolrak (2014, 2015), these conflicts still have repercussions and shape the lives of people in the region through fear, mistrust, hypervigilance and ethnic encapsulation. Fishing in Naivasha was severely affected by the 2007/2008 post-election violence, as many Luo members of the BMUs were displaced or killed. Only a few returned after the riots. As a result, their fishing licenses became available and were given preferentially to Kikuyu. Since then, inter-ethnic disputes in fishery have intensified. These tensions have heated up in the BMUs since 2013, with fishing becoming increasingly ethnicised in the course of devolution. Devolution not only decentralises administrative duties but also facilitates the ethnicisation of newly empowered local administration. Since then, Luo fear that Kikuyu could take over the leadership of all BMUs because their numbers in fishing are steadily increasing. The disputes manifest in associations with fishing that Luo and Kikuyu attribute to each other or to themselves, with particular reference to the right of ownership regarding fish resources in Lake Naivasha.[7]

In-depth interviews with Luo at the BMU landing beaches revealed that they identify themselves strongly with fishing and fish consumption. Fishing

7 The outlined associations of particular characteristics with ethnic identities are only based on the perspectives of our research participants.

is seen not only as a source of income, but also as a cultural practice that is firmly rooted in their ethnic identity. Many Luo of the fishing communities in Naivasha consider themselves the descendants of the first migrants from Lake Victoria who came to Naivasha in the 1960s to engage in fishing. Since the population previously living in Naivasha hardly used the lake's fish, Luo consider themselves the original fishermen of Lake Naivasha. Further, based solely on their tribal identity and ancestors, they refer to themselves as "traditional" and professional fishermen. Some Luo informants described how they were once forced to work in other economic sectors, but then switched back to fishing to pursue an activity that was "natural" for Luo. Conceiving of their tribal identity as intrinsically related to fishing, Luo claim to be the rightful owners of fish in Lake Naivasha. Kikuyu are deprived of this right:

> Luo are born fishers. For them it is like a cultural event to go fishing. The Kikuyu are now trying to acquire something that is not theirs. They can go back to their farm to grow maize.
>
> FISHING BOAT OWNER, NAIVASHA, AUGUST, 2016

This view is strongly related to the image of Kikuyu as crop farmers and an ethnic identity closely linked to owning land (cf. Rodrigues 1975, 52; Muriuki 1976, 107). With this view, Luo deny them the right to engage in fishing. Further, they construct an image of a Kikuyu fisherman that is detrimental to sustainable fishing. According to Luo BMU members, Kikuyu have an unsustainable perspective on fish resources and support the informal fishery network. Luo however know how fishing should be done sustainably due to ethnic tradition and long experience. One Luo boat owner described the difference in perspectives of Luo and Kikuyu towards fishery as follows:

> You know the Kikuyu, they do something for profit and they want immediate profit and they want big profit. They don't know we need to control the fishing and that there are some places we must not fish. They don't want to know that. A Luo fisherman knows that I should not fish in the breeding area; I should not fish at the shoreline. The Kikuyu know that those places are the ones where there are many fish. So they go and fish there.
>
> FISHING-BOAT OWNER, NAIVASHA, AUGUST, 2016

The interviewed Luo claim that they depend on fishing and have no economic alternatives. In several interviews the lake was described both as an economic and spiritual resource for current and future generations. A Luo fisherman explained the cultural significance of fishery and the lake like this: "You know for us Luo the lake is what is land for the Kikuyu" (fisherman, Naivasha, August, 2016). This comparison was meant to illustrate that for Luo, the lake is not only a place of work, but also has a profound spiritual meaning essential for their ethnic identity formation. It means that Luo have developed a particular connection to Lake Naivasha which in context of politicised ethnicity in Kenya also has political connotations. It also shows that conflicts about the fishery of Lake Naivasha are explosive, similarly to land conflicts in Kenya (cf. Branch, Cheeseman, and Gardner 2010; Boone 2011).

The interviewed Kikuyu fishermen, on the other hand, feel disadvantaged and under-represented in the fishery and therefore seek more decision-making power. Kikuyu fishermen at Central Beach complained that Luo are always favoured by the BMU board: "The Luo we have here are treated as kings" (fisherman, Naivasha, September, 2016). Fishermen at Karagita Beach argued that only Kikuyu should have the right to fish in Naivasha as the region is one of their ancestral homelands. By referring to Naivasha as their homeland, Kikuyu fishermen justify their exclusive claims on the lake's fish resources and deny Luo the right to take advantage of the same.

Tribalism is a prevalent phenomenon in various social institutions in Kenya. The sense of belonging comes to play in politics at the national level, where different ethnic groups align to support their candidate in elections (Boone 2011; Branch, Cheeseman, and Gardner 2010). In multi-ethnic Naivasha the issue is of particular sensitivity and explosiveness, which is why it was one of the hotspots of the 2007/2008 post-election violence. Inter-ethnic tensions in the BMUs are being reinvigorated since devolution through the ethnicising of fishery licensing and the instrumentalising of ethnic identities to justify claims on rights of ownership of Lake Naivasha.

8 Conclusion

Through employing a political-ecology perspective, this study contributes to an understanding of the political driving forces of recent dynamics of Lake Naivasha fishery. It has shown how shifts in the political system of Kenya through devolution have manifested in the deterioration of local fishery governance.

The chapter has argued that to understand the human-environment relationships involved in fishing at Lake Naivasha a political-ecological perspective is required to comprehend its social-ecological complexities. As we pointed out in our study, it is not only ecological and demographic factors but political processes at different scales, such as the devolution of government, that significantly determine the characteristics of fishery of Naivasha.

With a focus on local implications of devolution in Kenya, we outlined how shifting political authority from national to county level can affect natural-resource use. When the Nakuru County government's authority over Lake Naivasha was extended in 2013 and fishery management was divided between Naivasha Sub-County and Gilgil Sub-County, political interference in fishing increased. Rent-seeking by political elites who take advantage of their new arenas of authority results in a drastic increase in numbers of fishing licenses being issued. However, the increased revenues from licensing are not invested in the Fisheries Department. Instead, its meagre financial resources have been further reduced. In order to secure votes for election campaigns, local politicians collaborate with the informal fishing network or Mungiki, who are very influential in Naivasha's informal settlements. As a result, the illegal fishing methods of the network are tolerated, a separate BMU was established for them, and the regulations for sustainable fishing of the national fishery research institute are unofficially abolished. Efficient control is hardly possible because the local Fisheries Department is compromised by corruption, patronage and well-armed unlicensed fishermen. Thus, there is a massive weakening of fishery governance at Lake Naivasha. This is further exacerbated by chairmen, and other members of the BMUs, who profit from expanding and intensifying fishery or who collaborate with the informal fishing network. This increasingly blurs the line between legal and illegal fishermen. Since fishery licensing has also been based on ethnic criteria, the number of Kikuyu BMU members in the sector has increased. Ethnicised licensing makes the previously more dominant Luo feel threatened. While ecological issues get out of hand, ethnicity is used by both Luo and Kikuyu as a political resource to construct identities and justify claims on fish resources of Lake Naivasha. In the aftermath of the 2007/2008 post-election violence this results in disputes over fishing rights and a resurgence of ethnic resentments and hostilities. Further, we argue that the portrayed situation leads to an increase of the vulnerability of the Lake Naivasha ecosystem.

This chapter shows that normative decentralisation models, such as devolution, can fail if their implementation consolidates the power, status and resource claims of political elites. This is particularly significant because devolution was expected to reduce ethnic tensions in Kenya. Hence, we encourage more research on the implications of devolution for other sectors in Naivasha to contribute to a more nuanced understanding of the rescaling of environmental governance in Kenya.

Acknowledgements

We would like to thank all the people involved in the fishery of Lake Naivasha who have given us insights into their lives. We also thank the University of Bonn and the DFG for funding our research.

References

Ajulu, Rok. 2002. "Politicised Ethnicity, Competitive Politics and Conflict in Kenya: A Historical Perspective." *African Studies* 61 (2): 251–68. https://doi.org/10.1080/00020180220000032947.

Aloo, Peninah, Dalmas Oyugi, George Morara, and Margaret Owuor. 2013. "Recent Changes in Fish Communities of the Equatorial Lake Naivasha, Kenya." *International Journal of Fisheries and Aquaculture* 5 (4): 45–54. http://karuspace.karu.ac.ke/handle/20.500.12092/1932.

Anderson, David M. 2005. "Yours in Struggle for Majimbo. Nationalism and the Party Politics of Decolonization in Kenya, 1955–64." *Journal of Contemporary History* 40 (3): 547–64. https://doi.org/10.1177/0022009405054571.

Basset, Ellen M. 2017. "The Challenge of Reforming Land Governance in Kenya under 2010 Constitution." *Journal of Modern African Studies* 55 (4): 537–66. https://doi.org/10.1017/S0022278X17000441.

Becht, Robert, Eric O. Odada, and Sarah Higgins. 2005. "Lake Naivasha. Experiences and Lessons Learned Brief." International Waters Learning Exchange and Resource Network, January 15, 2021. Accessed May, 15, 2023. http://www.worldlakes.org/uploads/17_Lake_Naivasha_27February2006.pdf.

Becht, Robert, and David M. Harper. 2002. "Towards an Understanding of Human Impact upon the Hydrology of Lake Naivasha, Kenya." *Hydrobiologia* 488: 1–11. https://doi.org/10.1023/A:1023318007715.

Béné, Christophe. 2003. "When Fishery Rhymes with Poverty: A First Step beyond the Old Paradigm on Poverty in Small-Scale Fisheries." *World Development* 31 (6): 911–33. https://doi.org/10.1016/S0305-750X(03)00045-7.

Béné, Christophe, Emma Belal, Malloum O. Baba, Solomon Ovie, Aminu Raji, Isaac Malasha, Friday Njaya, Mamane N. Andi, Aaron Russel, and Arthur Neiland. 2009. "Power Struggle, Dispute and Alliance Over Local Resources: Analyzing Democratic Decentralization of Natural Resources through the Lenses of Africa Inland Fisheries." *World Development* 37 (12): 1935–50. https://doi.org/10.1016/j.worlddev.2009.05.003.

Bixler, Patrick, Dell'Angelo, Jampel, Mfune, Orleans and Hassan Roba. 2015. "The Political Ecology of Participatory Conservation: Institutions and Discourse." *Journal of Political Ecology* 22 (1): 164–82. https://doi.org/10.2458/v22i1.21083.

Boone, Catherine. 2011. "Politically Allocated Land Rights and the Geography of Electoral Violence: The Case of Kenya in the 1990s." *Comparative Political Studies* 44 (10): 1311–42. https://doi:10.1177/0010414011407465.

Branch, Daniel, Nic Cheeseman, and Leigh Gardner, eds. 2010. *Our Turn to Eat. Politics in Kenya since 1950*. Berlin: LIT.

Britton, J., Boar, R., Grey, J., Foster, J., Lugonzo, J. and D. M. Harper. 2007. "From Introduction to Fishery Dominance: The Initial Impacts of the Invasive Carp Cyprinus Carpio in Lake Naivasha, Kenya, 1999 to 2006." *Journal of Fish Biology* 71: 239–57. https://doi.org/10.1111/j.1095-8649.2007.01669.x.

Cheeseman, Nic, Gabrielle Lynch, and Justin Willis. 2016. "Decentralisation in Kenya: The Governance of Governors." *Journal of Modern African Studies* 54 (1): 1–35. https://doi.org/10.1017/S0022278X1500097X.

Cornell, Agnes, and Michelle D'Arcy. 2014. "Plus ça change? County-level Politics in Kenya after Devolution." *Journal of Eastern African Studies* 8 (1): 173–91. https://doi.org/10.1080/17531055.2013.869073.

Daily Nation. 2015. "Transport Paralysed as Locals Block Road." January 10, 2015.

D'Arcy, Michelle, and Agnes Cornell. 2016. "Devolution and Corruption in Kenya: Everyone's Turn to Eat?" *African Affairs* 115 (459): 246–73. https://doi.org/10.1093/afraf/adw002.

Geheb, Kim, and Marie-Therese Sarch, eds. 2002. *Africa's Inland Fisheries. The Management Challenge*. Kampala: African Books Collective.

Hara, Mafaniso, and Jesper Raakjaer Nielsen. 2003. "Experiences with Fisheries Co-management in Africa" In *The Fisheries Co-Management Experience*, edited by Douglas Clyde Wilson, Jesper Raakjaer Nielsen, and Poul Degnbol, 81–97. London: Kluwer Academic Publishers.

Harper, David M., Kenneth Mavuti, and Mucai Muchiri. 1990. "Ecology and Management of Lake Naivasha, Kenya, in Relation to Climatic Change 'Alien Species' Introductions, and Agricultural Development." *Environmental Conservation* 17 (4): 328–36. https://doi.org/10.1017/S037689290003277X.

Harper, David M., Edward Morrison, Michael Macharia, and Caroline Upton. 2011. "Lake Naivasha, Kenya: Ecology, Society and Future." *Freshwater Reviews* 4 (2): 89–114. https://doi.org/10.1608/FRJ-4.2.149.

Hickley, Phil, Mucai Muchiri, Rosalind Boar, Robert Britton, Chris Adams, Nicholas Gichuru, and David M. Harper. 2004. "Habitat Degradation and Subsequent Fishery Collapse in Lakes Naivasha and Baringo, Kenya." *Ecohydrology & Hydrobiology* 4 (4): 503–17. http://hdl.handle.net/1834/7302.

Hickley, Phil, Robert Britton, S. Macharia, Mucai Muchiri, and Rosalind Boar. 2015. "The Introduced Species Fishery of Lake Naivasha, Kenya: Ecological Impact vs. Socio-economic Benefits", *Fisheries Management and Ecology*, 22 (4): 326–36. https://doi.org/10.1111/fme.12130.

Kagwanja, Peter. 2003. "Facing Mount Kenya or Facing Mecca? The Mungiki, Ethnic Violence and the Politics of the Moi Succession in Kenya, 1987–2002." *African Affairs* 102 (406): 25–49. https://doi.org/10.1093/oxfordjournals.afraf.a138810,

Kateka, Adolphine G. 2010. "Co-Management Challenges in the Lake Victoria Fisheries: A Context Approach." PhD diss., Stockholm University.

Lang, Britta, and Patrick Sakdapolrak. 2014. "Belonging and Recognition after the Post-Election Violence: A Case Study on Labour Migrants in Naivasha, Kenya." *Erdkunde* 68 (3): 185–96. www.jstor.org/stable/24365231.

Lang, Britta, and Patrick Sakdapolrak. 2015. "Violent Place-Making: How Kenya's Post-Election Violence Transforms a Workers' Settlement at Lake Naivasha." *Political Geography* 45: 67–78. https://doi.org/10.1016/j.polgeo.2014.09.005.

Li, Tania Murray. 2007a. *The Will to Improve: Governmentality, Development, and the Practice of Politics*. Durham: Duke University Press.

Li, Tania Murray. 2007b. "Governmentality." *Anthropologica* 49 (2): 275–81. http://www.jstor.org/stable/25605363.

Lind, Jeremy. 2018. "Devolution, Shifting Centre-periphery Relationships and Conflict in Northern Kenya." *Political Geography* 63: 135–47. https://doi.org/10.1016/j.polgeo.2017.06.004.

Lynch, Gabrielle. 2006. "Negotiating Ethnicity: Identity Politics in Contemporary Kenya." *Review of African Political Economy* 107: 49–65. https://doi.org/10.1080/03056240600671282.

Lynch, Abigail, Steven Cooke, Andrew Deines, Shannon Bower, David Bunnell, Ian Cowx, Vivian Nguyen et al. 2016. "The Social, Economic, and Environmental Importance of Inland Fish and Fisheries." *Environmental Reviews* 24: 115–21. https://doi.org/10.1139/er-2015-0064.

McGuire, Thomas R. 1998. "A Political Ecology of Fisheries." In *Reinventing Fisheries Management*, edited by Tony Pitcher, Paul Hart, and Daniel Pauly, 239–250. Dordrecht: Springer Science and Business Media.

Menon, Ajit, Maarten Bavinck, Johnny Stephen, and R. Manimohan. 2015. "The Political Ecology of Palk Bay Fisheries: Geographies of Capital, Fisher Conflict, Ethnicity and Nation-State." *Antipode* 48 (2): 393–411. https://doi.org/10.1111/anti.12181.

Muriuki, Godfrey. 1976. "The Kikuyu in the Pre-Colonial Period." In *Kenya before 1900. Eight Regional Studies,* edited by Bethwell A. Ogot, 106–38. Nairobi: East African Publishing House.

Murunga, Godwin R. 2011. "Spontaneous or Premeditated? Post-Election Violence in Kenya." Discussion Paper 57. Nordiska Afrikainstitutet, Uppsala. Accessed May 15, 2023. https://www.diva-portal.org/smash/get/diva2:451262/FULLTEXT01.pdf.

Nabwire, Linda M. 2018. "A Critical Analysis of the Concept Majimbo in Kenya's Political Circle." *International Journal of Language & Linguistics* 5 (3): 131–43. https://doi.org/10.30845/ijll.v5n3p11.

Ngigi, Samuel. 2019. “Devolution in Kenya: The Good, the Bad and the Ugly.” *Public Policy and Administration Research* 9 (6): 9–21. https://doi.org/10.7176/PPAR.

Njaya, Friday, Steve Donda and Christophe Béné. 2011. “Analysis of Power in Fisheries Co-management: Experiences from Malawi.” *Society & Natural Resources* 25 (7): 652–66. https://doi.org/10.1080/08941920.2011.627912.

Nunan, F., Hara, M. and Onyango, P. 2015. “Institutions and Co-Mangement in East African Inland and Malawi Fisheries: A Critical Perspective.” *World Development* 70, 203–14. https://doi.org/10.1016/j.worlddev.2015.01.009.

Ostrom, Elinor. 1990. *Governing the Commons: The Evolution of Institutions for Collective Action*, Cambridge: Cambridge University Press.

Rodrigues, Reinhild. 1975. “Unabhängige Religiöse Bewegungen in Kenia unter besonderer Berücksichtigung des Distriktes Gilgil.” PhD diss., University of Freiburg.

Seal, Mark. 2010. *Wildflower*. London: Weidenfeld & Nicolson.

Scott-Villiers, Patta. 2017. “Small Wars in Marsabit County: Devolution and Political Violence in Northern Kenya.” *Conflict, Security & Development* 17 (3): 247–64. https://doi.org/10.1080/14678802.2017.1319696.

Tan-Mullins, May. 2007. “The state and its Agencies in Coastal Resources Management: The Political Ecology of Fisheries Management in Pattani, Southern Thailand.” *Singapore Journal of Tropical Geography* 28 (3): 348–61. https://doi.org/10.1111/j.1467-9493.2007.00314.x.

Wamue, Grace. 2001. “Revisiting our Indigenous Shrines through Mungiki.” *African Affairs* 100 (400): 453–67. https://doi.org/10.1093/afraf/100.400.453.

Welcomme, Robin, Ian Cowx, David Coates, Christophe Béné, Simon Funge-Smith, Ashley Halls, and Kai Lorenzen. 2010. “Inland Capture Fisheries.” *Philosophical Transactions of the Royal Society B* 365: 2881–96. https://doi.org/10.1098/rstb.2010.0168.

Yongo, E. O., George Morara, Jacob Ojuok, Chrisphine Nyamweya, William Ojwang, Mutune Masai, and Chemiaty Wasike. 2013. “Socio-economic Aspects of Fisheries Management in Lake Naivasha.” *African Journal of Tropical Hydrobiology and Fisheries* 13: 27–32.

CHAPTER 13

Conflicting Futures of Geothermal Energy Development in Naivasha: Between State Visions and Community Expectations

Chigozie Nweke-Eze and Christine Adongo

1 Introduction[1]

In "Kenya Vision 2030" – the national development blueprint of Kenya – the state expresses its commitment to the development of geothermal energy resources in the country as a way of increasing electricity generation capacity from sustainable, "green" and resilient sources (GoK 2007). To this end, the Kenyan state, together with other public and private investors, embarked on several geothermal development activities along the Kenyan Rift Valley. The largest centre of geothermal explorations and electricity production from geothermal sources is located in Olkaria, Naivasha Sub-County of Kenya. These infrastructures fit with the state's sustainability narrative (Nweke-Eze and Kioko 2021) and, partly as a result, have attracted huge infrastructure and climate-change-related financing from public and private investors at national, international, and global levels (Klagge and Nweke-Eze 2020). The Kenyan state's vision for the development of geothermal infrastructures goes beyond the provision of electricity for national benefits. The energy infrastructures are developed with the vision of attracting both local and foreign investments into the state-proposed Naivasha Industrial Park, which is to be located near the Olkaria geothermal plant, benefitting from the cheaper power supply than the market price while expanding and creating jobs for the locals in the region (Ngugi 2020). In the development of these infrastructures, the state often acts through dedicated institutions, including the ministries, parastatals, and Special Purpose Vehicles (SPVs), often with financial and technical support from international development financial institutions (Klagge et al. 2020; Klagge and Nweke-Eze 2020).

The development of these geothermal energy infrastructures, however, often defies infrastructure-affected communities' expectations, raising concern over their exacerbation of land grabbing and resource-use conflicts, their

1 First submission date: January 25, 2021. Acceptance date: March 13, 2023.

 DOI:10.1163/9789004695429_014

destabilising of community bonds and identities by dispossession, as well as their adverse effects on the socio-economic livelihoods and biodiversity in the infrastructure-affected communities (Adongo 2015; Hughes and Rogei 2020; Nweke-Eze and Kioko 2021). The infrastructure-affected community members are predominantly *Maasai* – a pastoral community that has lived in the area for decades, subsisting mainly on low-impact pastoralism and living in coexistence with the wildlife in the area (Adongo 2015; Mariita 2002; see also Waller, this volume).

These narratives on conflicting interests and futures are common with such large-scale development infrastructures (Lind, Okenwa, and Scoones 2020). The development of infrastructures in accordance with state visions and for national benefits, however sustainable they may be framed as being, often beg for substantial socio-economic and environmental sacrifices and relegation of expectations at local levels (Nweke-Eze and Kioko 2021; Hughes and Rogei 2020). In contribution to this scholarship, this chapter explores the processes and dynamics of these conflicting futures through the lens of the interplay between the (mainly state-)investors and infrastructure-affected community, in the development of Olkaria geothermal energy infrastructures in Naivasha. The development of these infrastructures in Naivasha is a classic case for the study, as the construction of the infrastructures involved substantial negotiations and promises, which culminated in the relocation and resettlement of four villages in the infrastructure-affected Maasai community.

We begin with a conceptual framework on future-making and the state-community relations thereof, after which, we describe the methodology of the study. In the section that then follows, we describe and explain the case and contexts of the study (Olkaria geothermal plants, state roles, and the infrastructure-affected community). Thereafter, we empirically explore and discuss the Kenyan state visions and how they come into conflict with expectations and realities at the infrastructure-affected community level. In conclusion, we discuss the implications of our findings for state-community relations in future-making in Kenya and elsewhere in the Global South.

2 State-Community Relations in Future-Making

In the era of capitalism in a globalised world, the notions of state and community are increasingly heterogeneous, both as concepts and as objects of study (Trouillot 2002; Scott 1998). The state can either be seen as existing only as a centralised entity at the national level or as a combination of multiple entities with governance presence at both national and local levels – that is, being closer to the community, for more efficient influence, in a process called decentralisation or devolution of powers (Bardhan 2002; Brown, Cloke, and Harrison 2015). Similarly, the notion of the communities is formed by what they are (identities) and how they exist (practices) (Adler 2015), all of which are constantly made and maintained in creative processes – "communities-in-the-making" (Schiemer 2018). However, when the interests of the state and the communities are considered in the context of future-making – that is, the practice of materialising futures – the state and community appear more as two clear-cut and homogeneous entities, each bonded by common interests. It is this constellation of the state as one entity and community as another that gives both sides force in "making their cases" in future-making.

Following this understanding, we therefore contend that future-making is made up of and shaped by competition between visions of the state and the expectations of the community (also see Appadurai 2013; Granjou, Walker, and Salazar 2017; Müller-Mahn, Moure, and Gebreyes 2020; Bollig 2021). These futures play a central role in social life and are acted upon in the present (Berkhout 2006; Ruivenkamp and Rip 2011; Mathews and Barnes 2016; Urry 2016). Future-making is anticipated (Groves 2017), prepared for (Anderson 2010), and is often contested and saturated with different interests (Brown, Rappert, and Webster 2000; Beckert 2016). The state envisions certain practices and delivers them to foster "development" (Scott 1998; Jasanoff and Kim 2015). Likewise, the community has expectations from the state and aspires to improved socio-economic conditions (Meadowcroft 2010; Appadurai 2013).

The interaction of these futures is premised on the state's complex, dynamic and interdependent relationship with the community in the course of future-making (Nilsson et al. 2011; Klagge et al. 2020). The state's nature and role in the community and the extent to which it 'makes the future', also depends on its relationship with non-state actors in the community, who also have certain degrees of power and influence, such as tribal leaders, religious authorities, and civil-society organisations (Meadowcroft 2010; Hufen and Koppenjan 2015). This process of state formation and interaction with the community is often non-linear, involving tensions, conflicts and compromises (Brown, Rappert, and Webster 2000, Sovacool and Cooper 2013; Nweke-Eze and Kioko 2021).

With each range of non-state actors wielding a certain amount of power and influence, and contesting with the state for their futures, negotiations become important in managing the inter-actor relations in the course of future-making (Marginson, Keune, and Bohle 2014; Klagge et al. 2020). These also bring about innovative ways of negotiating, with non-state actors gaining more momentum fuelled by globalisation trends. Such trends, including major strides in the flow of information, communication, and transportation, have also created avenues for international actors and organisations of various forms, to exert influence on state functions and future development (Alexandre et al. 2012; Mawdsley 2012). This is, in part, evident in the emergence of several international political and economic organisations, such as the World Bank, the International Monetary Fund (IMF), and the African Union.

Notwithstanding the powerful influences of non-state actors in the community, the state often still gains the upper hand in charting the course of the future through its far-reaching powers, functions, and tools. Several studies (Fairhead, Leach, and Scoones 2012; Benjaminsen and Bryceson 2012; Hughes and Rogei 2020; Nweke-Eze and Kioko 2021) focusing on resource appropriation and the correlated community-level consequences often allude to the disjuncture and unbalanced interplay between state visions and community expectations. The state pushes its agenda toward implementation through its function of resource distribution and its power to draw up, repeal and enforce rules, laws, and regulations (Enns 2016). In addition, the state also uses certain community-influencing tools – for example, the media – to persuade, mobilise and enforce its agenda (Migdal 2009 and Beckert 2016).

3 Methodology

This study is based on robust data collected through extensive fieldwork conducted by the two authors. The first author (C.N.-E) carried out several key-informant interviews in Nairobi, Nakuru, and Naivasha (2018–2021) as well as made several fieldwork visits to the geothermal infrastructure project sites and surrounding infrastructure-affected communities in Olkaria (2018–2021). The second author (C.A.) carried out successive periods of ethnographic fieldwork (2014–2020), which comprised two household surveys (2014 and 2015) in the infrastructure-host community. The findings were augmented with documents, communication correspondence, and archival data analyses gathered by the two authors.

The key informants work at government agencies and parastatals at different levels (Ministry of Energy (MoE), Ministry of Environment and Natural Resources, Kenya Wildlife Service (KWS), National Land Commission (NLC), the National Treasury (NT), Rural Electrification and Renewable Energy Corporation (REREC), county commissions and county governments). They also include the infrastructure developers such as KenGen and the Geothermal Development Company, and staff members in development finance institutions (EIB, KfW, AFD, World Bank), involved in the development of the Olkaria geothermal infrastructures. In addition, independent and energy-related consulting firms and consultants (Geohydro limited, EED Advisory) as well as a geothermal research institution, Geothermal Training and Research Institute (GETRI), were interviewed.

The ethnographic methodology featured a series of successive fieldwork periods in Olkaria that began in 2014. Data were collected through key-informant interviews, informal and focus-group discussions with the Project-affected Persons (PAPs), and participant observation in mediation sessions between the investors and representatives of the infrastructure-affected community. The first phase of the study was conducted in four villages, namely Olonongot, Olosinyat, Olomayana Ndogo, and Emanyatta,[2] while the second phase was conducted in RAP-land[3] – the resettlement site that lies between Suswa Oloiruwa plains and Olkaria V stations, 14 km from the aforementioned villages. Two household surveys (N=350) were also conducted: one in early February 2014 before the resettlement and the other, a year later (February 2015) post resettlement, with the same households. Personal correspondence and informal conversations have also been useful in providing additional information to the subject of study.

The study also analyzed secondary data largely comprising minutes from meetings between the developers and the community, Environmental and Social Impact Assessments, and other reports such as those from funders (EIB, JICA, World Bank) such as the RAP, mediation documents, and responses to claims by the PAPs. Through community records, the second author also kept track of correspondences (emails, letters, texts) between the community and the developers and funders.

4 Olkaria Geothermal Infrastructures, State Roles, and Affected Community

The Olkaria geothermal infrastructures are located about 120 km north of Kenya's capital Nairobi, in the Hell's Gate Ward of Olkaria, Naivasha Sub-County. Most of the infrastructures are situated within the Hell's Gate NP, a gazetted protected area that covers about 68 km^2 (see Figure 13.1). The geothermal field within the Greater Olkaria Geothermal Area (GOGA) is subdivided into five blocks, namely: Olkaria East (Olkaria I), Olkaria Northeast (Olkaria II), Olkaria

2 The village of Emanyatta is also called Cultural Centre, because it had been established as a cultural centre for facilitating international cultural exchange (field data, 2014).

3 RAP stands for Resettlement Action Plan.

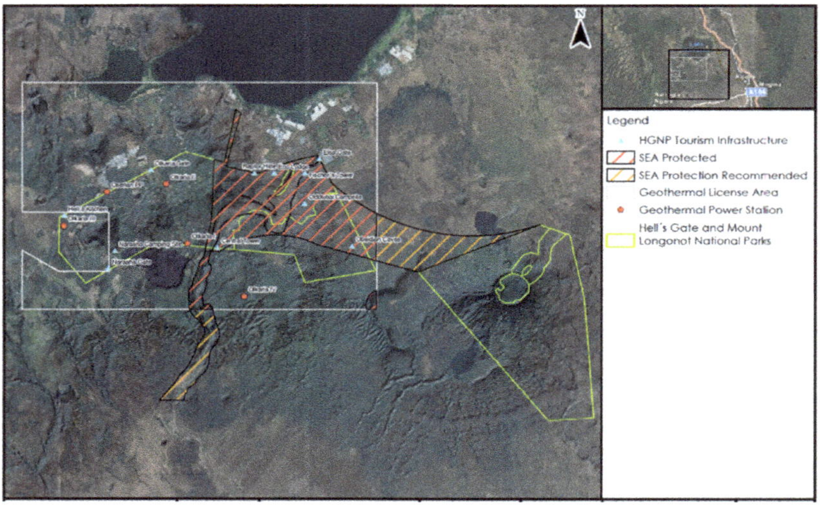

MAP 13.1 Olkaria geothermal infrastructure area
SOURCE: UNPUBLISHED AKIIRA ESIA REPORT, 2016

FIGURE 13.1 Aerial view of Olkaria II geothermal infrastructures
SOURCE: KENGEN ARCHIVES

West (Olakria III), and Olkaria Domes (Olkaria IV, V and VI) (see Table 13.1 for further elaboration).

4.1 State Roles

Most of the Olkaria geothermal infrastructures are owned by the Kenya Electricity Generation Company (KenGen) – a 70% state-owned entity (Table 13.1).

TABLE 13.1 Olkaria geothermal energy infrastructures in Naivasha

Power stations	Capacity (MW)	Ownership	Status/Year of commissioning of finished projects
Olkaria I (Units 1, 2 and 3)	45	KenGen	Finished in 1981, 1982 and 1985 respectively
Olkaria II	105	KenGen	Finished in 2003
Olkaria I (Unit 4 and 5)	140	KenGen	Finished in 2015
Olkaria I (Unit 6)	83.3	Public-private partnership between KenGen and four private companies	Ongoing
Olkaria III	139	Independent Power Producer – Ormat Technologies Inc (OrPower 4)	Finished in 2000
Olkaria IV	140	KenGen	Finished in 2014
Olkaria V	165.4	KenGen	Finished in 2019
Eburru Wellhead	25	Eburru (private company) and KenGen	Various times
Olkaria Wellhead	81	KenGen	Various times
Olkaria VI	140	Public-private partnership between KenGen and five private companies	Ongoing

SOURCES: FIELD DATA, 2018–2020 AND VARIOUS GEOTHERMAL INFRASTRUCTURE DEVELOPERS' WEBSITES, 2019–2020

In total portfolio, KenGen currently owns seven larger geothermal power stations – Olkaria I, Olkaria II, Olkaria IV, Olkaria V, and Olkaria I units 4 and 5, in addition to the wellheads generation plants in Eburu and Olkaria,[4] bringing the entire KenGen geothermal capacity portfolio to 818 MW.[5] Ormat Technologies Inc. owns Olkaria III, with 150 MW installed capacity. In 2008, the 100% state-owned Geothermal Development Company (GDC) was formed as a SPV charged with accelerating the development of geothermal fields by taking up

4 Geothermal wellhead technologies allow for early generation of electricity from single wells before the construction of large power plants.

5 KenGen and MoE interviews, 2019.

the riskiest aspects of the development (exploration and drilling). From 2009 to 2010, the state-owned SPV undertook drilling of 59 wells in Olkaria I and IV, with a total yield of 412MW.[6] In 2014, the company launched its blueprint development strategy, envisioning to install 12 more rigs and to drill at least 1,200 wells in total by 2030.[7] The GDC is currently prospecting for more geothermal development in other areas of Olkaria, in Menengai, Baringo-Silali, Suswa, and Bogoria.[8]

The development of the Olkaria geothermal infrastructures is not only driven by national state commitments and favourable policies; the Kenyan state also received substantial financial and technical assistance from international development financial institutions (DFIs) and technical assistance agencies in Europe, the US, and North Asia. These international institutions include the European EIB, the French AFD, the German KfW, the Japanese JICA, and the World Bank. These institutions played an important catalytic role in the development of the Olkaria energy infrastructure by providing financing, in their various forms (loans, grants, and mezzanine) as well as technical assistance. The DFIs often blended their financing with climate-related financing (mainly grants) from specialised funds such as the Green Environmental Facility (GEF). With technical assistance from DFIs, KenGen also leveraged carbon credits from traded carbon, under the Clean Development Mechanism (CDM) to augment geothermal financing. Table 13.2 summarises the DFIs and their specific roles in the development of one or more of the Olkaria infrastructures.

TABLE 13.2 DFIs and roles in the development of Olkaria infrastructures

DFIs	Roles and activities
EIB	Loans, technical assistance in due diligence
AFD	Loans, project procurement, technical assistance in the resettlement of project affected persons (PAPs) and in the implementation of environmental and social safeguards
KfW	Loans
JICA	Loans, technical assistance in field exploration
World Bank	Loans, technical assistance in the resettlement of PAPs

DATA SOURCES: FIELD DATA, 2018–2020; SCHADE (2017)

6 GDC interviews, 2019.
7 GDC interview, 2019.
8 GDC and KenGen interviews 2018–2019.

4.2 *The infrastructure-Affected Community*

Although the wider Naivasha area is largely cosmopolitan with a diversity of livelihood options such as horticulture, fishing, and irrigation agriculture, the pastoralist Maasai largely inhabit the Olkaria ward, with a population of about 64,507(KNBS 2019). From oral history, the Maasai in the area claim to have migrated from Kinangop, specifically during the forced migration to the southern reserves established by the British colonial rule, after the Maasai treaties with the British in 1904 and 1911 (Rutten 1992; Waller, this volume). While some Maasai continued southwards, others chose to settle nearer to Lake Naivasha. After Kenya's independence, around 1964, these Maasai who settled closer to the lake would again be pushed further, as a result of land grabbing by the first independent state government, which had made Nakuru the headquarters of the Rift Valley province. Other forced migrations would again occur due to the establishment of the Hell's Gate NP as well as Olkaria geothermal explorations in the area.[9] The two enterprises attracted immigrant workers, who ended up establishing their own village (Olomayana Ndogo), in Olkaria, with the consent of the local Maasai. With time, these immigrant workers had somewhat become an extension of the Maasai through intermarriages. Thus, while the Maasai remain the majority inhabitants, the Turkana, Rendille, Pokot, Borana, and Luo represent the minority inhabitants. This diversity, as we show in the following section, has created tensions in the resettlement and compensation process. While the Maasai largely depend on livestock, practising seasonal transhumance, part of the population, especially the immigrant workers, depend on casual labour from the Hell's Gate NP and the geothermal industry. Some Maasai also supplement pastoral activities with tourism-based livelihood such as community tour guiding and beaded jewellery trade (see Figures 13.2 and 13.3).

5 State Visions, Community Expectations, and Conflicting Futures in Geothermal Infrastructure Development in Naivasha

5.1 *Geothermal Infrastructures in Naivasha as State Development Vision*

The Olkaria geothermal energy infrastructure development was envisioned by the Kenyan state as a means of achieving sustainable socio-economic growth and development through the provision of electricity to stimulate industrial investments, create jobs, and increase tourism and government revenue

9 According to informants recounting their history in the Olkaria area (field data, 2014)

FIGURE 13.2 Example of Maasai socio-economic livelihood in Naivasha
PHOTO: FIELD DATA 2016–2017

FIGURE 13.3 Example of Maasai socio-economic livelihood in Naivasha
PHOTO: DR. BENOIT HAZARD

potentials, all with minimal negative environmental and socio-economic impacts.[10] Olkaria geothermal electricity has sufficient base-load to securely provide for household and industrial electricity needs in the country, as its reliance on hydropower continues to diminish due to drought.[11] Some of this generated electricity is envisioned to be sold to large-scale industries and flower farms for the planned Naivasha Industrial Park – a Special Economic Zone (SEZ) near the geothermal operations at Olkaria.[12] Investors in the industrial park are to be offered a lower power tariff of KSh 5/ kWh to set up their businesses as approved by the country's regulatory agency, the Energy and Petroleum Regulatory Authority (EPRA).[13] This electricity tariff deal is roughly half of the market price for other commercial consumers, who are currently paying KSh 10–12/kWh at peak hours.[14] KenGen (the partly state-owned company, owning most of the geothermal plants in the area) has already been able to sell not only electricity to flower firms in the area but also its steam.[15] Oserian Development Company, for instance, has been using geothermal heat from KenGen for its rose-growing business over 50 hectares since 2003.[16] Furthermore, the initially planned East African railway network, a rail connection between the Kenyan harbour city of Mombasa with the nation's capital, Nairobi, is now planned to be extended to connect to the market town of Naivasha in proximity to the Olkaria geothermal field.[17] This was deliberately planned to further enable manufacturing facilities in the region to benefit not only from easier access to cheap geothermal power, but also from easier access to markets by rail transport to the port of Mombasa in the western part of the country.[18]

The state also envisioned that the development of the Olkaria Geothermal Project would create direct and indirect jobs for skilled, semi-skilled, and non-skilled workers in the project host community and beyond.[19] The construction of the projects was expected to create job opportunities for highly skilled workers, including geologists, geochemists, geophysicists, engineers (electrical, civil, drilling, and mechanical among others), environmentalists, community

10 MoE, KenGen and county governments interviews, 2019.
11 MoE interview, 2019.
12 KenGen interviews, 2019.
13 EPRA interview, 2019.
14 EPRA and MoE interviews, 2019. At January 1, 2019, KSh 1 was equivalent to $0,00976 (source for the exchange rate: https://www.oanda.com/currency-converter/en/).
15 KenGen interviews, project site visits, 2019.
16 KenGen interviews, 2019.
17 KenGen and MoE interviews, 2019.
18 KenGen and MoE interviews, 2019.
19 KenGen, NT, and MoE interviews, 2019–2020.

liaisons, and human resource personnel.[20] The infrastructure host community and their neighbours were also to benefit from the need for unskilled and semi-skilled labour during the construction of the geothermal projects. Such job opportunities include construction of access roads, clearing of projects sites, driving, masonry, carpentry, loading/off-loading of project equipment, provision of security for the project area, among others.[21] Other indirect employment opportunities, more in the form of self-employment, were also expected to emerge due to the construction of the projects. They include petty trading, restaurant trade, housing construction, and provision of accommodation, for the benefit of the workers as well as other members of the community.[22]

Furthermore, due to the strategic location of Olkaria geothermal plants within the Hell's Gate NP, it was envisioned that it will contribute to boosting the park's tourism potentials.[23] Other than animals within the park, the state expected that most of the tourists would be eager to see the geothermal plants and to visit the Geothermal Spa and Recreational Centre – a direct-use feature of geothermal energy, as part of sightseeing activities during touristic tours.[24] Other than enhancing the touristic value of the NP,[25] Olkaria geothermal energy infrastructure itself was also expected to earn the state some additional revenue through taxes and fees.[26] The energy infrastructures were also expected to generate operating license fees, Value Added Tax for most of the procured project construction equipment, as well as corporation tax at 30% of net income.[27] After the completion of the projects, the state and KenGen also look forward to earning additional revenue through the sale of carbon credits, officially called Carbon Emission Reductions (CERs), as well as to meet the state's climate-change mitigation commitments contained in the Paris Agreement.[28]

20 KenGen and independent consultants' interviews, 2019.

21 KenGen, county commissions and county governments interviews, 2018–2020.

22 KenGen, independent consultants, county commissions and county governments interviews, 2018–2019.

23 Tourism is one of the Kenyan state's major sources of national revenue and foreign exchange (NT interviews, 2019).

24 KWS and KenGen interviews, 2019.

25 This takeover by geothermal energy production as the focal point of tourism in Hell's Gate NP has generated mixed feedbacks among tourists. Some see it as having added sightseeing value, while others miss the earlier completely natural environment, complaining that geothermal development has eroded the natural beauty of the park (field data, 2019).

26 NT and KenGen interviews, 2019.

27 NT and KenGen interviews, 2019.

28 NT and KenGen interviews, 2019.

5.2 *Conflicting Interface of State Visions with Community Expectations, and Realities of Living with Geothermal Infrastructures*

In this section, we discuss the expectations of the community, its conflicts with the state-level visions, and the ensuing and persistent local consequences and realities of the Olkaria geothermal infrastructure development.[29] Although the Olkaria geothermal project has also had ecological impacts, especially in the Hell's Gate NP, we focus on the experiences and perceptions of the human inhabitants – the Maasai.[30] As we will find, and as opposed to their own initial expectations, the Maasai continue to struggle to coexist with the state's development of Olkaria geothermal infrastructures, against the backdrop of historical continuities of land grabbing, dispossession, resource alienation, and human rights violation from the colonial to the post-independence era (Rutten 1992; Fratkin and Sher-Mei-Wu 1997; Hughes 2008; Sena 2015).

The expectations of the predominantly Maasai community from the development of Olkaria evolved over time.[31] In the early stage of project-infrastructure development, there was no resistance from the communities on the planned development and expansion of geothermal infrastructures in Olkaria.[32] One reason for this is that the land for the energy infrastructure had long been gazetted as a national park under the management of the Kenyan Wildlife Service (KWS).[33] KenGen acquired the land from KWS, for a repurposed usage for geothermal development, governed by a Memorandum of Understanding.[34] Although the communities did not resist the development of the energy infrastructure at this stage, they implicitly expected that their construction would enable the provision of better amenities such as water and health dispensaries as well as better socio-economic opportunities, including market access, within their communities.[35] These expectations were premised on the assumption that the presence of geothermal energy in their community would not disrupt their existing livelihood and cultural heritage.[36] In the initial stages, these concerns were expressed informally to visiting project developers,

29 The data and views presented here span 7 years (2014 to 2020). A recent survey by Hughes and Rogei (2020) reaffirms the persistence of these sentiments and the overall situation of the Maasai at RAP-land.

30 Although there are other ethnic minorities in the area, we use Maasai, who are the majority, to refer to the inhabitants unless specified otherwise.

31 Field data, 2014–2020.

32 Field data, 2014.

33 Field data, 2014; KWS interviews, 2019.

34 KenGen interviews, 2019.

35 Field data, 2014.

36 Field data, 2014.

but not formally established, as the community were still ignorant of their benefit-sharing rights from local resources as contained in the Kenyan law.[37]

However, as Olkaria projects development advanced to cover more areas in the Hell's Gate NP, with risks of air pollution and land acquisition increasingly becoming apparent, expectations of benefits from the infrastructure development grew among members of the community. These expectations grew partly through sensitisation by civil-society organisations as well as due to the insistence from international DFIs that the developers adhere to sustainability principles.[38] Soon after, more serious and formal talks about compensations for pollution, land acquisition, and other socio-economic benefits arose among the state, DFIs, and the community. This led to the creation of RAP committees and the signing of a Memorandum of Understanding for the resettlement of the PAPs and socio-economic benefits for the infrastructure host and affected community.[39] These relocation and resettlement plans were against the initial expectations of the communities, though. However, the state, together with its partner DFIs, promised to resettle the PAPs in areas with better living conditions, to provide opportunities for the restoration of their socio-economic livelihoods and to provide infrastructures and scholarships to the PAPs and surrounding communities.[40] In the following paragraphs, we discuss how most of these promises and agreements were not met, at least according to the rising expectations of the affected Maasai community, and how the socio-economic conditions of the community worsened due to the geothermal infrastructure development in the area (Kimani 2020).

After several attempts to relocate the predominantly Maasai inhabitants from the Olkaria geothermal complex,[41] KenGen successfully achieved this aim, displacing the inhabitants of the first four villages in August 2014. At the time, KenGen, in accordance with its Environmental and Social Impact Assessments (ESIAs), cited health hazards (JICA 2015) as the motive for such displacement. Coincidentally, at the same time, the Akiira geothermal company was in the process of establishing its first plant adjacent to the resettlement location, causing the Maasai to question the validity of KenGen's justification

37 Field data, 2014.

38 Independent consultants and DFI Interviews, 2019.

39 KenGen and DFIs interviews, 2019.

40 Field data 2014–2015, and KenGen interviews, 2019.

41 FGD with Maasai of Emanyatta (Cultural Centre) February 2014: they narrate two incidents when their village was set ablaze in an attempt to evict them. The latest prior to resettlement occurred on July 26, 2013.

for the displacement.[42] "What is different about Akiira plant? Is it not the same steam, the same effluents, the same noise? Why take us there then?" they pondered.[43]

Despite expressing their concerns, the displaced PAPs found themselves in 'RAP-land", lost and aggrieved.[44] To them, the land that KenGen had evicted them from had always belonged to them, regardless of who had the legal title deed.[45] Following the relocation, the communities were also dissatisfied with the value of compensation and the relocation processes contained in the proposed resettlement plan of KenGen and their financiers.[46] They had not been prepared to move into the "*wilderness*".[47] To compound matters, the violence with which they had been evicted provoked distress for many families, especially those with young children[48] (see also Kimani 2020).

Further, the positioning of the RAP-land between geothermal consesssions and the NP in GOGA has not favoured the Maasai,[49] and this is set to worsen with tentative plans for land acquisition for further geothermal development.[50] Consequently, the Maasai are subject to restrictions in terms of grazing and access to water sources. The RAP-land prevents them from advancing their pastoral activities.[51] Six years after the resettlement, the Maasai still lack water

42 Akiira is a private geothermal developer at Olkaria owned by a consortium of American investors. Power Africa – a major funder of Akiira.

43 Interviews and FGD between February and June 2014.

44 3 weeks fieldwork at RAP-land and 1 day after displacement.

45 The Maasai have been living on this land without a legal title deed since independence due to historical continuities of land injustice and corruption that dispossessed the Maasai from their lands, see Lotte Hughes (2006) and Marcel Rutten (1992).

46 A house and moving allowance of KSh 35,000 (approximately $350) (Field data, 2014). Many community members also stated that they did not receive the allowance (Field data, 2014).

47 Interviews in RAP-land 2014, just after resettlement. The community members claimed that many hyenas that threatened their livestock inhabit the RAP-land and that they did not feel safe (Field data, 2014).

48 Interview with a mother of three children (Field data, 2014): the morning following night evictions – I found her preparing tea for her three children under a tree next to her demolished house. "They gave me KSh 1,000 ($10) and told me to leave", she stated.

49 We discuss Maasai because of their predominance in Olkaria. There are however other (minority) ethnic groups present, represented by Samburu, Borana, Turkana, and Rendille. From interviews with them, we found they had settled in Olkaria in search of job opportunities, marriage, or had come to visit their relatives at a young age but ended up not returning home.

50 Threats of degazettement (interview with Nature Kenya, 5th February 2015, personal communication with a board member of the LNRA.

51 Field data, 2014-2019.

and certain social amenities, still feel poorer, and struggle to adapt to their "new" homes and lifestyle.[52]

Despite having been increasingly sedentarised over the past decades (Fratkin and Mearns 2003; Adano and Witsenburg 2005), the Maasai in general, including those in the Olkaria area, continue to depend upon expansive land to support their pastoral livelihoods. They graze strategically, moving between areas of pasture and water depending on the seasons. The Maasai of Olkaria, for example, used the Orbatata not only as a dry-season grazing area and watering ground for their livestock,[53] but also to access pasture in the Suswa lowlands during wet seasons.[54] Loss of land coupled with restricted access to pasture in the NP and around geothermal plants has thus curtailed their transhumance and livelihoods as pastoralists.[55] Even though at RAP-land these pastoralists have access to 1,700ha, they claim that the vegetation there is unpalatable to livestock.[56] Besides, the deep gulleys, loose soil, and severe soil erosion in RAP-land render grazing impossible.[57] Several of the pastoralists attested to have lost their livestock in the gulleys, which forced some of them to keep their livestock with relatives living outside of the Olkaria area.[58] The Maasai in RAP-land recount painfully how they have lost the gorge (Orbatata), a culturally and historically significant area. They feel that by being denied access to their cultural sites, they have lost a part of their culture. For instance, one elder told of how they used to conduct child-naming ceremonies under a special rock in the gorge, something they have been unable to do since being resettled at RAP-land.[59] Besides this, the entire *Orbatata* not only constitutes caves and hideouts that the Maasai say had been used for their meat-eating ceremonies (*Olpul*) and as hideouts during the Mau Mau uprising, but also contains springs and grass-banks crucial for their livelihood as pastoralists as well as red and white ochre sites that the Maasai use during ceremonies.

52 Field survey results (2014-2015).

53 Orbatata is the Maa reference to the Orjorowa gorge that connects Olkaria to the Suswa plains.

54 We followed the herders from February to July 2014, to determine livestock routes and grazing areas.

55 FGDs with PAPs 2014, RAP document, shows that the Maasai lost 4,200 hectares.

56 Our observation of RAP-land ecology suggests that the predominant vegetation is mostly *Taconanthus conpharatus* (leleshwa in Maa language).

57 Interviews 2014, RAP-land.

58 Interview with five livestock owners, September 10, 2014.

59 FGD, elders, February 11, 2014.

Presently, the effluents from geothermal wells heavily pollute some of the springs, rendering it unhealthy for community members and livestock.[60]

During fieldwork at RAP-land following the resettlement,[61] we observed families in congested houses sleeping on the cold cement floors and sitting on stones.[62] Unlike in their previous settlements where the bed and seating areas were inbuilt and made from readily available natural resources (sticks and earth), in these new RAP houses, there was the need to purchase furniture – which many of the settlers could not afford. Many attest to being poorer because not only have they had to sell their livestock in order to finance a new lifestyle, but they have also lost jobs.[63] Even without the casual employment from the geothermal industry, in their previous villages the Maasai had their livestock, grazing areas, beaded-ornaments businesses, and tour-guiding activities at the edge of the Orjorowa gorge "*Orbatata*". Having been displaced more than 25 km away, the trek to the Gorge and back to RAP-land became expensive and impossible,[64] especially for the women[65] and the old.[66]

Traditionally, the husband gets his own "*Enkaji*"[67] while the wives share their *Enkaji* with their children and spend the night with the man in his space, depending on their arrangement in case several wives are involved.[68] Circumcised boys prefer to have their own independent *Enkaji* away from their parents. This presents the dilemma facing several families in the RAP houses[69] meant for a nuclear family with one or two children, in an urban set-up. For our72-year-old polygamous informant with his ten children, among them adolescent sons and daughters, for example, this was a nightmare.[70] The physical fence around each housing unit, fitted with gates, further enforced seclusion. During

60 One of the interviewees described how their cows started dying after drinking from the springs (interview with a PAP in Emanyatta, February 15, 2014).

61 Fieldwork started on the day of resettlement.

62 For example, a disabled single mother of 4, living with her mother and brother, who had been in prison at the time when censuses were conducted to determine who gets resettled; her brother did not get a house as a result (Field data, 2014).

63 They were poor in comparison to the rest – they largely depended on tourism (FGD, Maasai market, 13 September, 2014), with women depending on Maasai market and tour guiding (especially those from Cultural Centre).

64 They spent at least KSh 600 on transport – their daily average income is KSh 200 (FGD with tour guides, 1 October 2014).

65 FGD with women at Maasai market gorge, September 13, 2014. Also see a short fieldwork documentary on the RAP-land by EHESS (2019).

66 Field observation, 2014-2019.

67 Maasai traditional hut.

68 FGD with the PAPs 10 October 2014, at the RAP-land.

69 Comprising a living room, two bedrooms, and a kitchen.

70 He and others finally resorted to building his *manyatta* in the compound.

an interview with a 43-year-old father of one,[71] he narrated how he had been alone with his wife during the emergency: "I shouted but no one came because they could not hear me – the houses are far apart. If it were it back in the village [referring to their previous settlement], people could have heard me and come to my aid. … By the time I got back from seeking help, I found the child dead". The Maasai also claim that RAP-land has also promoted individualism, contrary to the Maasai culture. For instance, during a focus-group discussion (FGD) with the women of RAP-land, on 6 February 2017, they claimed: "The fencing deters people from visiting freely; it is like our minds have developed fencing also… In the village [referring to their previous settlement] it was not a problem to borrow sugar or salt just next door".

The binary categorisation of the PAPs as deserving or non-deserving further deepened the wedge between the Maasai and non-Maasai, with the latter accusing the Maasai of unjustly locking them out of the entire process.[72] These non-Maasai persons had been classified as tenants. One of our informants – a Turkana, who came to Olkaria at the age of fourteen – has had three children whose fathers are Maasai; and having given birth to all her children in Olkaria, she felt that she deserved to partake in the resettlement process.[73] In response to such claims, one chairperson told us:

> Yes, they are pastoralists like us; in fact, they are Maasai in a way, so they are part of us. We have lived together for many years, but they have land where they came from because their ancestors owned land in those places.

The elders on the other hand are confused about whether to call themselves Maasai elders or members of the Community Advisory Council (CAC). In RAP-land, their functions have been usurped by the chairpersons,[74] who have become more powerful as a result of their roles as brokers. Specifically, discussions and negotiations have shifted from how to apportion and manage grazing and water, or whose calves access the grass banks at the *Orbatata*, to who gets the next casual employment and tenders from the geothermal industry. Reasons for conflicts also shifted from whose animals have trespassed in the

71 Whose second child had died shortly after being born at his home.

72 Interviews with individual non-Maasai 12 October 2014, RAP-land and FGD with the "non-deserving PAPs"; Emanyatta (June 6, 2014).

73 Interview with Narumbe; she was given KSh 300,000 ($3,000) as compensation, used the money and came back to Olkaria market to sell beads.

74 The rest of the PAPs accuse these chairpersons of being brokers, getting richer from tenders and bribes from KenGen (field data, 2014).

communal grazing areas reserved for dry-season grazing to whose sons have got employment in the geothermal industry.

6 Discussion and Conclusion

According to Appadurai (2013), the major contestation for future-making is not between dreams and realities, but between dreams and other dreams – so that future-making is essentially the product of contestations between dreamscapes. This contribution engaged with this understanding of future-making in the context of the interface between state and community dreamscapes in the materialisation of large-scale geothermal energy infrastructures in Naivasha. Our study shows how state-level futures diverge from those of the communities, alluding to similar observations by Lind Okenwa, and Scoones (2020), Mukeu and Langat (2016), Fairhead, Leach, and Scoones (2012), Benjaminsen and Bryceson (2012), and other studies focusing on resource appropriation and infrastructure development in the Global South. Our findings also highlight the unbalanced interaction between state-level and community-level dreams, leading to exclusion, alienation, and in some cases conflicts at the community level. As we show, these dreams are divergent because of equally divergent futures and views of nature and resources, their appropriation and management (see also Tilly and Cameron-Daum 2017). Whilst the state views its exploitation of geothermal resources as development projects in accordance with its development visions of bringing about positive change in the nation as a whole, these visions conflict with the infrastructure-affected community's alternative use of these resources in accordance with their views and expectations regarding socio-economic and cultural wellbeing.

We also show how, notwithstanding these conflicting futures, the state, using its dominant power, enables resource availability for its visions, while attempting to compensate for the loss of dreams at community levels (see also Koissaba 2018). These compensation and reparation attempts, as we show, are non-optimal, as they do not fully compensate for the socio-economic and cultural losses, especially for the relocated PAPs. The community complains that even though electricity is generated within what they consider "their" land, 90% remain without electricity, as the electricity cost for them is unaffordable.[75] The questions, therefore, remain: whose socio-economic development aspirations is prioritised, and why? Should the socio-economic livelihoods

75 FGD with PAPs (9th May 2018).

of the infrastructure-host communities be traded off for national and/or global good?

The practice of "near-forceful" dispossession of land for development projects, without sufficient regard for the socio-economic wellbeing of its occupants, as shown in our study, reflects the Kenyan state's negative perceptions of pastoral communities and pastoralism since colonial times (Cavanagh, Weldemichel, and Benjaminsen 2020; Hughes and Rogei 2020; Hazard and Adongo 2015). In Kenya, pastoralism has always been seen at the state level as a degrading practice, an unrewarding activity, and a waste of space (Hazard and Adongo, 2015). With such views, pastoral areas thus remained marginalised, susceptible to land grabbing for "development" projects (in conservation and more recently in renewable-energy infrastructures) for "general national interests". In the case of the development of Olkaria geothermal energy infrastructure in Naivasha, the Kenyan state maintains that the development of the infrastructures will put lands, which are often the subject of conflict,[76] to "better" and more profitable use.[77] They also see it as a way of "opening up" the culture of the traditional communities.

> With the anticipation of coming of new people from varied cultures into the project area, it is projected that there will be cultural exchange which will lead to the adoption of new ways of life and shedding off of traditional ways of living that have stagnated development in the area.[78]

The above stated prejudicial comment clearly perpetuates the image of Maasai, their culture, and livelihoods as backward and insinuates that they "need" to be encapsulated. In reaction to these reproaches, the infrastructure-affected community, with help of civil-society organisations, often filed complaints to the county government and other non-national local authorities. These attempts have, however, been unfruitful,[79] as the vision of the national state traverses and influences other non-national state authorities in the country. When the communities contest the development of the infrastructures, the assumption of the state is often that the locals are against the project, and thus it asserts more of its powers as the central authority and mobilises tools to quell such

76 There have been several confronting and conflictual attempts by the "original owners" of these lands to evict the Maasai.

77 KenGen and GDC interviews, 2019.

78 GDC interviews, 2019.

79 Interview with 39-year-old PAP four days after resettlement, August 2014: "*The DC came and told me I could not win against the government and that if I loved my life, I should move out*".

contestation. In contrast to this assumption, however, our findings show that the infrastructure-affected communities do not attest to being against the project – all the community members interviewed or engaged in discussion expressed pride in the fact that resources from their lands would serve the entire nation.[80] They do however contest the modalities and processes of dispossession that the development of the infrastructures have caused and their adverse effects on their livelihoods.[81]

The Environmental and Social Impact Assessment (ESIAs) reports are important instruments through which the state and its supporting international development and financial institutions attempt to ensure the application of a balanced sustainability triad in the development of such large-scale "green" infrastructures in Kenya (Nweke-Eze and Kioko 2021; Klagge et al. 2020). In the case of Olkaria geothermal infrastructure development, its ESIA reports (JICA 2015) assert that the development of the infrastructures will provide affected and host communities with immense benefits through CSR practices, including employment opportunities, electricity and social-amenities provision, and international tourism. Our study and others,[82] however, find that these CSR practices have given rise to new forms of exclusion often within the local community itself. For instance, the provision of casual jobs has given rise to a clamour for tenders, and disputes arising from allegations of corruption in jobs allocation among community members.[83] In addition, these infrastructures are not provided with the long-term socio-economic benefits of the communities in mind. They are chiefly a necessary means for the infrastructure developers (predominantly constituting the state) to access and extract the resources. This explains why broken-down infrastructures are often not repaired, at least not promptly.[84] Furthermore, the geothermal industry requires highly skilled professionals[85] – which the local communities do not have.[86] As a result, the employment opportunities left for the local communities are intermittent, poorly remunerated and casual.[87] A culmination of these factors gives rise to the infrastructure-affected community's view of these CSR

80 Field data, 2014-2019.
81 Field data, 2014-2019.
82 For example, Hughes and Rogei, 2020.
83 Field data, 2019.
84 Field data, 2014-2019.
85 KenGen Interviews, 2019.
86 Less than 1% employed are local Maasai (Hughes and Rogei, 2020).
87 These are not necessarily from Olkaria – but most of them involve working as security guards. (field data, 2018; also see Hughes and Rogei, 2020).

practices as an enticement to persuade them into accepting the development of the infrastructures in their community.

Overall, our analyses highlight the growing imbalance between dreamscape contestations in future-making. Our study shows that the state often superimposes its visions, while neglecting the expectations of infrastructure-affected communities in the course of future-making. Community involvement and participation in the planning of large-scale infrastructure projects is still insufficient, despite several attempts made to close this gap through the production of ESIAs and the institution of laws such as the Environment Management and Coordination Act of 1999, which require community consultations prior to and during project development. We hope that the findings of our study will improve inter-stakeholder relations understanding and inform procedures and standards for successful community engagements in the development of renewable energy projects in the peripheries.

Acknowledgements

The fieldwork for this research was made possible through funding by the DFG under the Collaborative Research Centre *Future Rural Africa: Future-Making and Social-Ecological Transformation*, and by the EU-funded Marie Curie Innovative Training Network *Resilience in East African Landscapes* (*REAL*). We are thankful to them.

References

Adano, Wario R., and Karen Witsenburg. 2005. "Once Nomads Settle." In *As Pastoralists Settle*, edited by Elliot Fratkin, and Eric Abella Roth, 105–36. Boston: Springer.

Adler, Paul S. 2015. "Community and Innovation: from Tönnies to Marx." *Organization Studies* 36 (4): 445–71. https://doi.org/10.1177/0170840614561566.

Adongo, Christine. 2015. "Y a-t-il des énergies vertes? Le cas de la géothermie au Kenya." In *Déprogrammer l'apocalypse: le passe, la présent et l'avenir du climat: Les leçons pour éviter le pire*, edited by Raymond Woessner, 31–42. Paris: Livre Capes.

Alexandre, Marc, Alys Willman, Ghazia Aslam, Michelle Rebosio, and Kanishka Balasuriya. 2012. *Societal Dynamics and Fragility: Engaging Societies in Responding to Fragile Situations*. Washington DC: World Bank.

Anderson, Ben. 2010. "Preemption, Precaution, Preparedness: Anticipatory Action and Future Geographies." *Progress in Human Geography* 34 (6): 777–98. https://doi.org/10.1177/0309132510362600.

Appadurai, Arjun. 2013. *The Future as Cultural Fact: Essays on the Global Condition.* London: Verso.

Bardhan, Pranab. 2002. "Decentralization of Governance and Development." *Journal of Economic Perspectives* 16 (4): 185–205. https://doi.org/10.1257/089533002320951037.

Beckert, Jens. 2016. *Imagined Futures: Fictional Expectations and Capitalist Dynamics.* Cambridge, MA: Harvard University Press.

Benjaminsen, Tor. A, and Ian Bryceson. 2012. "Conservation, Green/Blue Grabbing, and Accumulation by Dispossession in Tanzania." *The Journal of Peasant Studies* 39 (2): 335–55. https://doi.org/10.1080/03066150.2012.667405.

Berkhout, Frans. 2006. "Normative Expectations in Systems Innovation." *Technology Analysis and Strategic Management* 18 (3–4): 299–311. https://doi.org/10.1080/09537320600777010.

Bollig, Michael. 2021. "Materiality, Inequality, and Future-making as Focal Points of Future Engagement of Economic Anthropology with Climate Change." *Economic Anthropology* 8 (1): 180–82. https://doi.org/10.1002/sea2.12199.

Botelho, Andre. 2014. "Political Sociology: State–society Relations." *Current Sociology Review* 62 (6): 868–85. https://doi.org/10.1177/0011392114533213.

Brown, Ed., Jon Cloke, and John Harrison. 2015. *Governance, Decentralization, and Energy: A Critical Review of the Key Issues.* Working Paper 1. Renewable Energy and Decentralization (READ).

Brown, Nik, Brian Rappert, and Andrew Webster. 2000. *Contested Futures: A Sociology of Prospective Techno-science.* Aldershot, UK: Ashgate.

Cavanagh, Connor J., Teklehaymanot Weldemichel, and Tor A. Benjaminsen. 2020. "Gentrifying the African Landscape: The Performance and Powers of For-profit Conservation on Southern Kenya's Conservancy Frontier." *Annals of the American Association of Geographers* 110 (5): 1594–612. https://doi.org/10.1080/24694452.2020.1723398.

EHESS. 2017. *"RAPland", in Anthropo-Scènes.* [Video]. Canal-U. Accessed on October 31, 2023. https://www.canal-u.tv/79247.

Enns, Charis. 2016. "Experiments in Governance and Citizenship in Kenya's Resource Frontier." PhD diss., University of Waterloo.

Fairhead, James, Melissa Leach, and Ian Scoones. 2012. "Green Grabbing: A New Appropriation of Nature?" *Journal of Peasant Studies* 39 (2): 237–61. https://doi.org/10.1080/03066150.2012.671770.

Fratkin, Elliot M., and Tiffany Sher-Mei-Wu. 1997. "Maasai and Barabaig Herders Struggle for Land Rights in Kenya and Tanzania." *Cultural Survival Quarterly* 21 (3): 55–61.

Fratkin, Elliot, and Robin Mearns. 2003. "Sustainability and Pastoral Livelihoods: Lessons from East African Maasai and Mongolia." *Human Organization* 62 (2): 112–22.

Government of Kenya (GoK). 2007. *Kenya Vision 2030: The Popular Version.* Nairobi: The Republic of Kenya.

Granjou, Celine, Jeremy Walker, and Juan F. Salazar. 2017. "The Politics of Anticipation: on Knowing and Governing Environmental Futures." *Futures* 92 (September): 5–11. https://doi.org/10.1016/j.futures.2017.05.007.

Groves, Christopher. 2017. "Emptying the Future: On the Environmental Politics of Anticipation." *Futures* 92 (September): 29–38. https://doi.org/10.1016/j.futures.2016.06.003.

Hazard, Benoît, and Christine Adongo. 2015. "'Green Grabbing' Pastoralism and Environmental Dynamics in Northern Kenya. An Assessment of Conservation Models and Practices in Marsabit County." *Cahier de l'Afrique de l'Est* 50: 40–62.

Hufen, Hans, and Joop Koppenjan. 2015. "Local Renewable Energy Cooperatives: Revolution in Disguise?" *Energy, Sustainability and Society* 5 (18). https://doi.org/10.1186/s13705-015-0046-8.

Hughes, Lotte, and Daniel Rogei. 2020. "Feeling the Heat: Responses to Geothermal Development in Kenya's Rift Valley." *Journal of Eastern African Studies* 14 (2): 165–84. https://doi.org/10.1080/17531055.2020.1716292.

Hughes, Lotte. 2006. *Moving the Maasai: A Colonial Misadventure*. Basingstoke: Palgrave Macmillan.

Hughes, Lotte. 2008. "Mining the Maasai Reserve: The Story of Magadi." *Journal of Eastern African Studies* 2 (1): 134–64. https://doi.org/10.1080/17531050701847326.

Japan International Cooperation Agency (JICA). 2015. *Preparatory Survey on Second Olkaria Geothermal Power Project*. Nairobi: Final Report.

Jasanoff, Sheila, and Sang-Hyun Kim. 2015. *Dreamscapes of Modernity: Sociotechnical Imaginaries and the Fabrication of Power*. Chicago: University of Chicago Press.

Kimani, Joyce. 2020. "150 Nomadic Families Leave their Manyattas for Life in Stone Houses." *Nation*, June 29. Accessed June 30, 2021. https://nation.africa/kenya/counties/150-nomadic-families-leave-their-manyattas-for-life-in-stone-houses-1017520.

Klagge, Britta, and Chigozie Nweke-Eze. 2020. "Financing Large-scale Renewable-energy Projects in Kenya: Investor Types, International Connections, and Financialization." *Geografiska Annaler: Series B, Human Geography* 102 (1): 61–83. https://doi.org/10.1080/04353684.2020.1729662.

Klagge, Britta, Clemens Greiner, David Greven, and Chigozie Nweke-Eze. 2020. "Cross-Scale Linkages of Centralized Electricity Generation: Geothermal Development and Investor-Community Relations in Kenya's Semi-arid North." *Politics and Governance* 8 (3): 211–22. https://doi.org/10.17645/pag.v8i3.2981.

KNBS. 2019. *The 2019 Kenya Population and Housing Census*. Nairobi: KNBS.

Koissaba, Ben. R.O. 2018. *The Olkaria Projects: a Case Study of Geothermal Energy and Indigenous Communities in Kenya*. Brussels: Heinrich Boell Stiftung.

Lind, Jeremy, Doris Okenwa, and Ian Scoones, eds. 2020. *Land, Investments, and Politics: Reconfiguring Africa's Pastoral Drylands*. Woodbridge: James Currey.

Marginson, Paul, Maarten Keune, and Dorothee Bohle. 2014. "Negotiating the Effects of Uncertainty? The Governance Capacity of Collective Bargaining under Pressure." *European Review of Labour and Research* 20 (1): 37–51. https://doi.org/10.1177/1024258913514356.

Mariita. Nicholas O. 2002. "The Impact of Large Renewable Energy Development on the Poor: Environmental and Socio-Economic Impact of a Geothermal Power Plant on a Poor Rural Community in Kenya." *Energy Policy* 30: 1119–28. https://doi.org/10.1016/S0301-4215(02)00063-0.

Mathews, Andrew S., and Jessica Barnes. 2016. "Prognosis: Visions of Environmental Futures." *Journal of the Royal Anthropological Institute* 22 (April): 9–26. https://doi.org/10.1111/1467-9655.12391.

Mawdsley, Emma. 2012. "The Changing Geographies of Foreign Aid and Development Cooperation: Contributions from Gift Theory." *Transactions of the Institute of British Geographers* 37 (September): 256–272. https://doi.org/10.1111/j.1475-5661.2011.00467.x.

Meadowcroft, James. 2010. "Who Is in Charge Here? Governance for Sustainable Development in a Complex World." *Journal of Environmental Policy and Planning* 9 (3–4): 299–314. https://doi.org/10.1080/15239080701631544.

Migdal, Joel. S. 2009. *Boundaries and Belonging: States and Societies in the Struggle to Shape Identities and Local Practices*. Cambridge: Cambridge University Press.

Mukeu, Pharis, and Reuben Langat. 2016. *Olkaria (Kenya) Geothermal Project Case Study*. Naivasha: GRC Transactions.

Müller-Mahn, Detlef, Mar Moure, and Million Gebreyes. 2020. "Climate Change, the Politics of Anticipation and Future Riskscapes in Africa." *Cambridge Journal of Regions, Economy and Society* 13 (2): 343–62. https://doi.org/10.1093/cjres/rsaa013.

Mwangi, Martin. 2007. *Geothermal Development in Protected Areas: Kenyan Experience*. San Salvador, El Salvador: UNU-GTP.

Nilsson, Mans, Lars. J. Nilsson, Roger Hildingsson, Johannes Stripple, and Per Ove Eikeland. 2011. "The Missing Link: Bringing Institutions and Politics into Energy Future Studies." *Futures* 43: 1117–28. https://doi.org/10.1016/j.futures.2011.07.010.

Ngugi, Brian. 2020. "KenGen Eyes Industrialists for Direct Electricity Sales." *Business Day Africa,* March 11. https://www.businessdailyafrica.com/markets/marketnews/KenGen-eyes-industrialists-for-direct-electricity-sales/3815534-5487498-view-asAMP-e6ndgyz/index.html.

Nweke-Eze, Chigozie, and Eric M. Kioko. 2021. "'But We Cannot Do it All': Investors' Sustainability Dilemmas and Strategic Selectivity in Geothermal Energy Development in Kenya." In *Sustainable Development in Africa*, edited by Walter Leal Filho, Rudi Pretorius, and Luiza Olim de Sousa, 385–404. Cham: Springer.

Ruivenkamp, Martin, and Arie Rip. 2011. "Entanglement of Imaging and Imagining of Nanotechnology" *Nanoethics* 5 (2): 185–93. https://doi.org/10.1007/s11569-011-0122-2.

Rutten, Marinum. M. E. M. 1992. *Selling Wealth to Buy Poverty: The Process of the Individualization of Land Ownership among the Maasai Pastoralists of Kajiado District, Kenya, 1890–1990*. Saarbrücken: Breitenbach.

Schade, Jeanette. 2017. *Kenya "Olkaria IV" Case Study Report: Human Rights Analysis of the Resettlement Process*. Bielefeld: COMCAD. Accessed October 20, 2023. https://www.ssoar.info/ssoar/handle/document/51409.

Schiemer, Benjamin. 2018. *Problematizing Communities in Creative Processes: What They Are, What They Do, and How They are Practiced Organized Creativity*. Berlin: Discussion Paper No. 1/1.

Scott, James C. 1998. *Seeing Like a State: How Certain Schemes to Improve the Human Condition Have Failed*. New Haven, CT: Yale University Press.

Sena, Kanyinke. 2015. *Renewable Energy Projects and the Rights of Marginalized/Indigenous Communities in Kenya*. Copenhagen: IWGIA Report 21.

Sovacool, Benjamin K., and Christopher J. Cooper. 2013. *The Governance of Energy Megaprojects: Politics, Hubris and Energy Security*. Cheltenham and Northampton, MA: Edward Elgar.

Tilly, Christopher, and Kate Cameron-Daum. 2017. *An Anthropology of Landscape: The Extraordinary in the Ordinary*. London: UCL Press.

Trouillot, Michel-Rolph. 2002. "The Anthropology of the State in the Age of Globalization: Close Encounters of the Deceptive Kind." *Current Anthropology* 42 (1): 125–38. https://doi.org/10.1086/318437.

Urry, John. 2016. *What is the Future?* Cambridge: Polity Press.

PART 5

The Future of the Lake Naivasha Basin

CHAPTER 14

COVID-19 and the Cut-Flower Industry in Naivasha: Risk, Uncertainty and Preparing for the Future

Eric M. Kicko, Kristin Schmit, Annalia Gminder, Selina Emmanuel, Anne Achieng and Silas Wanjala

1 Introduction: The Cost of a Pandemic[1]

> We have to discard millions of stems at farms, it is disheartening … We have made some critical decisions of limiting the number of the work-force at farms, cutting back plants to reduce the production of flowers as there were no markets for it. We try to control costs as much as possible to manage the day-to-day operations.
>
> FLOWER-FARM MANAGER, MARCH, 2020

> The flower sector is losing approximately KES 250M (USD 2.5 million) per day and is estimated to lose half of its value, KES 60 billion (USD 6 billion), by end of 2020 … Kenya's flower exports have so far recorded a more than 50% drop in exports with indications that production is currently at less than 10% and facing the risk of total collapse.
>
> DELOITTE 2020

These sentiments describe the situation in Naivasha, one of the world's largest suppliers of cut flowers, following the outcome of the global rush to contain the spread of COVID-19. Globally, the public-health measures aimed at controlling the spread of the pandemic included suspension of international and domestic flights, social distancing and restrictions on gathering and meetings, stay-at-home orders including dawn to dusk curfews, and public transport regulations. In Kenya, these containment measures included closure of all learning institutions for a year, closure of businesses like bars and hotels, and a night-time curfew from 07:00 p.m. to 04:00 a.m.

The immediate economic effects of these containment measures for the agro-industry in Naivasha included significant reduction of air freight services, limited cargo capacity accompanied by high freight costs, and a drastic reduction of demand for cut flowers in Europe and other markets. This massive

1 First submission date: October 8, 2022. Acceptance date: March 13, 2023.

 | DOI:10.1163/9789004695429_015

disruption of flow of goods and people led to the collapse of the global supply chain particularly between March and April 2020 (Carlsson-Szlezak, Reeves, and Swartz 2020). In early 2019, prior to the pandemic, the Kenya National Bureau of Statistics recorded that the industry was exporting approximately 140,000 metric tonnes of produce valued at KSh 120 billion ($1.2 billion), with the European market accounting for up to 70% of the export (KNBS 2020). In March 2020, flower exports dropped by more than 50%, accounting for a loss of approximately KSh 250 million ($2.5 million) daily (Deloitte 2020).[2] Supplies of farm inputs such as fertilisers, chemicals and energy also dropped.

With regard to social effects of the containment measures, the welfare of industry labourers is important. In 2019, before the pandemic, the industry employed approximately 150,000 workers nationwide, with 65–75% being women, thousands of whom were seasonal and casual workers (UNDP 2020, based on information from the Kenya Flower Council). Naivasha accounted for approximately 40,000 workers in the flower farms as of 2017 (Kuiper and Gemählich 2017). According to UNDP (2020), out of the 150,000 workers, companies sent 70,000 away on unpaid compulsory leave, while seasonal workers left with no anticipation of further work. The disruptions greatly sabotaged the livelihoods and well-being of labourers and particularly women workers, whose struggles were compounded by soaring food prices, unpaid house-rent bills, and lack of food for children including those sent home after the closure of schools (Hivos 2020).

This chapter focuses on how the flower industry deals with risks and uncertainty, as well as emerging ways in which it is preparing for the future. Specifically, we reveal new strategies that the industry is utilising to reduce economic shocks arising from surprises like the COVID-19 pandemic. Primary data was obtained from in-depth interviews with purposively sampled flower-farm managers in Naivasha in 2022 while secondary sources provided data on the situation of flower farms at the height of the COVID-19 containment measures. In the next section, we explore the nexus between risk, uncertainty and preparing for the future to underscore the lessons and opportunities that emerge from risk and uncertainty, and how these inform actions in the present aimed at preparing for different futures. The Naivasha industry places Kenya as the leader in global export of cut flowers from Africa. Unlike tea and coffee, for which Kenya is also a major player, cut flowers are often considered luxury commodities and could hence become more vulnerable in terms of value compared to the foods.

2 Currency conversion as indicated in the source.

2 Risk, Uncertainty and Preparing for the Future

The agro-industrial establishment at Lake Naivasha has been historically defined by unpredictability, referring to effects of climate change, instability of the political set-up, and changing land ownership, which contribute elements of risk and uncertainty. The term *risk* is understood as "unpredictable variation in some ecological or economic variable (e.g. variation in rainfall, prices, etc.), and an outcome is viewed as riskier if it has greater variance" (Cashdan 2019, 2).[3] Scoones and Stirling (2020, 3) observe: "Risk is where we know what the possible outcomes are and can estimate their probabilities". Therefore, risk, or the probability of an undesirable event happening, can be assessed, measured, and quantified, thus giving actors a frame of reference from which decisions and actions may be taken.

Risk defines the agro-industrial investments and the social-ecological system on which they are set up in Naivasha. Flooding and drying of the lake due to natural and anthropogenic factors present a risk to investors and labourers. For example, in 2008/2009, the lake witnessed rapid water-level decline, while between 2011 and 2013, flooding submerged some flower farms. The flooding continued in 2020, a problem also characteristic of other Rift Valley lakes as a result of varying rainfall and unexpected subterranean exchanges between aquifers.

Markets and value chains also present a risk to the industry because they are highly dependent on turbulent global economies, which have dealt a blow to the industry in the past, such as the 2008–2010 global economic crisis, which reduced the demand for cut flowers. Moreover, the effect of the Brexit has led to considerable market upsets and renegotiations of market access between Kenya and the UK which continue to affect the flower industry in Kenya. The sustainability of the value chain and markets is also dependent on efficient linkages and networks, which, obviously, may be disrupted. For example, investors deal in highly perishable goods and their produce is dependent on the efficiency of technology and transport infrastructure. However, technology and transportation can fail and hence the efficiency of linkages and networks is not guaranteed. Moreover, the market and value-chain structure are also informed by the perceived social value of cut flowers, which is fundamentally different for European consumers (e.g. flowers as a form of communication, a gift for friends and family, or a luxury good that serves to brighten living rooms) than for Kenyan producers (e.g. flowers as a source of income). The value that

3 For other approaches to "risk", see Douglas (1992); Boholm (2015).

society attaches to cut flowers is subject to change and particularly so when confronted with a crisis like the COVID-19 pandemic.

Politicised conflicts pose a risk for the agro-industrial establishments because conflicts or violence have a direct effect on the labour force, which is a critical node of the entire enterprise. Naivasha is prone to social-political instabilities that, for example, led to the disturbance occasioned by the 1993 as well as the 2007/2008 post-election violence. Such disruptions mean that labourers flee for their safety and hence do not attend work in the farms. The risks associated with the labour force, including possible waterborne diseases and exposure to agro-industrial chemicals, have a direct effect on flower farms because such risks limit workers' ability to discharge their duties in the farms.

The industry investors, managers and labourers perceive these risks; they are able to make judgements and evaluations of the hazards that they (or their facilities, or their environment) are or might be exposed to (Rohrmann 2008). They make rational calculations of risk-taking and are often involved in strategising regarding how to control or reduce the outcome of these risks. We explore these strategies in the next section.

Our main argument here is that the industry actors are experienced risk-takers, but COVID-19 was unprecedented; no one expected that markets would be completely down the following day, or that operations would come to a complete halt. This leads us to the concept of uncertainty.

As applied in the study context, uncertainty refers to things that people simply do not know in advance: the inability to predict events or their outcome (Cooper and Pratten 2015). Unlike risk, where possible events or outcomes may be calculated, uncertainty means that people do not know what is going to happen, and hence cannot know what to prepare for. For example, the industry did not know about and could not have prepared for the clouds of volcanic ash across Europe that grounded flights in 2010, nor could it have prepared for the COVID-19 shock. However, from May 2020, the industry began to recover and, one year on, flower and horticulture farms had fully recovered and bounced back from COVID-19 (Murage 2022), indicating full operations and complete reinstatement of the labour force.

Against this backdrop of risks, uncertainties, and the recovery process, the important question is how stakeholders act in the present to prepare the ground for processes that shape future conditions. This leads us to briefly explore the concept of the "future" and how actors prepare for it.

Appadurai (2013) argues that we cannot research the future because the future is not here, but what we can research is how human beings try to make sense of the future – their imaginations, anticipations, expectations, hopes, anxieties, fears, speculations, and uncertainties. Therefore, agency becomes the empirical entry point. Making sense of and acting upon the future is what Van Wolputte, Greiner, and Bollig (2022, 1) call *futuring* – that is, "planning, imagining, making, building, expecting, avoiding, policing, preventing, writing, speculating, and anticipating whatever is becoming." Futuring is "the way individuals and societies think about and act upon their futures" (Müller-Mahn and Kioko 2022, 28). In the context of the cut-flower industry at Naivasha, *futuring* entails deliberate planning and action by flower farms (either singly or jointly) for possible and probable futures.

Flower farms engage in *futuring* through three overlapping practices that Appadurai (2013) describes: (i) anticipation, which is based on calculations and modelling and from which we can draw conclusions; (ii) imagination, which is based on influencing the future through creative action; and (iii) aspiration, which is hope. From these practices, Appadurai distinguishes two kinds of futures that are central to the flower industry at Naivasha: The *future of probabilities*, which relates to anticipation and risk, that is, something one can calculate and is close to the real world; and the *future of possibilities*, which is based on aspirations, dreams, imaginations of shared futures. Furthermore, following the COVID-19 shock, *futuring* for the flower farms involves preparing for unforeseeable surprises (uncertainties). For the agro-industrial developments at Naivasha, these two kinds of futures, despite questions of how realistic they are, inform actions and decisions in the present. In the following, we briefly delve into the question of how risks and uncertainties inform actions and create opportunities for the cut-flower industry at Lake Naivasha.

3 Preparing for Probable and Possible Futures

Risk and uncertainty inform ways of acting upon the future. By assessing and coming to an understanding of what is coming, people can build strategies that aim to control or reduce risks. This is the notion of risk-taking, which characterises the agro-industrial investments in the Lake Naivasha region. What is unique about the experiences of the COVID-19 pandemic, as we will show, is that some strategies and actions that emerged prior to the pandemic proved their worth at the height of the pandemic and hence were intensified. For example, existing and newly established social networks between investors proved essential for joint assessment and risk-taking. Consequently, investors often come up with similar strategies to reduce or control risks, and hence prepare for imagined or expected futures. However, the similarity of strategies for preparing for the future could lead to the possibility of producing more risks, as shown in the discussion section.

Uncertainty also becomes a resource for preparing for the future. Cooper and Pratten (2015) note that uncertainty, which is often used in its negative and constraining sense referring to a lack of absolute knowledge, the inability to predict the outcome of events or to establish facts about phenomena and connections with assurance, is rather productive and provides us with opportunities. They argue that "uncertainty is a social resource and can be used to negotiate insecurity, conduct and create relationships, and act as a source for imagining the future with the hopes, and fears this entails" (Cooper and Pratten 2015, 2). Uncertainty becomes "the basis of curiosity and exploration; it can call forth considered action to change both the situation and the self" (ibid., drawing from Whyte 2009, 213–214).

In the context of the flower investments in Naivasha, uncertainty therefore opens a window of opportunities meant to cushion the industry against unforeseeable happenings. In the following, we briefly discuss seven (7) opportunities or strategies that global-industry players, linked to the cut-flower sector in Naivasha, have employed to prepare for risk and uncertain futures. For the purposes of anonymity, we refrain from associating the direct quotes used here with our informants or their companies.

3.1 *Intensified Diversification of Produce and Markets*

COVID-19 became more of a catalyst for existing risk-coping / future-making activities such as diversification of produce and markets. To account for COVID-19-related risks and challenges, representatives of the flower farms interviewed relied on some form of diversification of produce, markets and market segments.[4] A diversified portfolio consisting of horticultural goods for consumption, including vegetables and fruits, ensured revenue in a time when demand for luxury goods such as cut flowers was unstable. One farm for instance had previously recognised the potential of avocados as a crop in high demand and began cultivating these prior to the pandemic, further demonstrating the importance of foresight in capitalising on opportunities. The farm manager told us that selling avocados can provide an additional income of about $11 per managed square metre annually and gives many farms the opportunity to

4 Seven (7) farms were included in the study and in-depth interviews conducted with farm representatives including a managing director, two production managers, a marketing manager, two human-resources managers, and a sales manager. The farms represented a mixed picture in terms of ownership: one farm owned by Kenyans of Indian origin; one farm fully owned by Kenyans, though with headquarters in the Netherlands; one farm owned by Kenyans; one farm foreign-owned, and three farms with unclear ownership. In terms of size, farms included in the study ranged from smaller farms (35–70 ha) to medium/large-sized farms (200–500 ha).

make money. In many farms, activities which originally started as sideline and small-scale ventures were accelerated due to COVID-19 and expanded into an important additional source of revenue.

As an added benefit, diversifying product portfolios also helped flower growers unlock domestic markets and additional market segments. Two managers highlighted the opportunity for diversifying to seedlings, i.e., of vegetables and herbs, due to their importance for both large- and small-scale, non-commercial growers. As one manager put it, "when people were told to stay back at home, they had nothing much to do except farming" (human resources manager, interview, September 23, 2022). As such, for propagators specialising in domestic markets, targeting multiple segments and offering a range of products, the pandemic has had an inverted outcome, allowing them to considerably increase sales. However, turning to an internal market and to set up a value chain that is not directed toward the international market may not account for the volumes of supply that often go to international outlets. Attendance to the internal market was a reaction intended to maintain some sales when exports were not possible.

While viewed as being of secondary importance, as COVID-19 affected the nature of exporting as a whole, geographic diversification, i.e., marketing to multiple European countries and to Australia and South Africa for instance, helped to spread risks and reduce dependence on one single market. Exports to Scandinavia, where government restrictions were less stringent, allowed one farm to make up for pandemic-related losses in other parts of Europe. Nevertheless, as emphasised by one manager, the time it takes to adapt both internal and external processes to the various likings and preferences of different markets proves challenging, and thus must be pursued as a long-term strategy.

3.2 *Mode of Sales*

During the pandemic, direct sales to retailers of cut flowers largely outnumbered sales at auctions, which were affected by lockdown-related closures (Andreas Gemählich, Kioko personal communication, July 18, 2022). While only 25% of flower farms at Lake Naivasha are currently engaging in direct sales (Button 2020), COVID-19 has proven that the large initial effort in establishing direct links with buyers has proven to be highly beneficial in reducing systemic risks, associated with the complexity of cut-flower supply chains. As one farm manager explained, "we got by scot-free, simply because we were in the right segment; we already had direct marketing and the access to those markets" (farm manager, interview, September 19, 2022). Direct sales to supermarkets in particular were perceived as a great asset during the pandemic. When florists in most European countries were forced to close due to lockdown policies, supermarkets remained open for essential services even in spring 2020, at the most difficult time for the industry. In addition to continued sources of revenue, a direct-sale model also enabled flower farms to plan ahead, as they were able to count on fixed prices for their goods without variances caused by seasonal and crisis-induced fluctuations. This perceived stability carried over to a section of the workforce, as employees were not hired on a seasonal basis, but on fixed-term contracts throughout the whole year (Andreas Gemählich, Kioko personal communication, July 18, 2022), although one cannot generalise this effect across all farms during and after the pandemic.

3.3 *Economic Networks and Partnerships*

In order to respond to global crises with turbulent effects, most industry players emphasised the importance of not working in silos, but rather to work "together as a team" (marketing manager, interview, September 20, 2022), referring to cooperation with other local, national and international actors. In response to COVID-19, the emergence and strengthening of networks and partnerships among key industry players has been experienced to be vital in the sharing and managing of risks. Such cooperation can be seen across various scales:

a As the floriculture sector is Kenya's second-highest export earner and a major source of foreign currency reserves (Central Bank of Kenya 2021), *the government* is an important stakeholder in the national cut-flower industry. In response to the pandemic, flower farms around Lake Naivasha experienced varied levels of support by governmental institutions. A staff member of the Kenya Flower Council (KFC), one of the local umbrella organisations, reiterated during an interview that "the government [has been] fully supportive" through provision of safety nets for investors while also offering tax reductions to certain farms in the first 6 months of the pandemic. Such support constituted a major asset in upholding competitive advantages of the industry, especially during disruptive crises. However, some flower-farm managers noted that the support from the government was quite low and that "the government could do more" (production manager, interview, September 20, 2022). They wished for more support especially in relation to entering new markets and the disruption of air-freight services during the peak of the COVID-19 pandemic. One farm manager at Lake Naivasha contrasted their situation to Ethiopia, Kenya's biggest competitor in the cut-flower segment on the African continent. Due to strong support and subsidisation by the Ethiopian government, air-freight services were maintained at the peak of the pandemic and industry players in Ethiopia "did not really feel the impact of COVID-19" (marketing manager, interview, September 20, 2022). In terms of shifting to alternative transportation services (i.e., sea freight), some farm managers noted that KFC and governments of foreign markets (e.g. the governments of the Netherlands and the United Kingdom) have offered more assistance than the Kenyan government. Complaints by flower farms about the lack of support by the Kenyan government is not new, but this became more poignant during the pandemic as farms pushed the Kenya Revenue Authority for tax refunds to ensure struggling farms remain afloat, as well as assistance with alternative transportation services.[5]

b To unlock stronger assistance from the Kenyan government, *umbrella organisations of the industry* have dispensed additional efforts into lobbying and "work[ing] as one" (KFC official, interview, September 17, 2022), especially in the context of prioritising COVID-19 vaccination among flower-farm workers, whom they considered to be essential service providers. Collaboration has extended beyond the cut-flower sector to include manufacturing and tourist associations. They are not only helping to highlight gaps, but are also advising the government on how to fix them. Networking and collaborating with such organisations have enabled the latter to act as a voice for the whole industry. In this way, they have successfully advocated for improved economic and trade terms and pushed for VAT refunds and the lifting of the air-freight ban.

c Maintaining *horizontal and vertical networks* has also turned out to be a way of strategising in response to risk and uncertainty. Many industry players reported that efforts have been made by all stakeholders in the value chain. For instance, although flower growers around Lake Naivasha compete for the same markets and buyers, flower growers helped to shore up the capacities of other growers struggling to maintain operations by buying up produce. "Those who had access in the market helped others" (Gemählich, interview, July 18, 2022). This also applied to plant breeders, who have been supporting their clients (i.e., flower growers) by allowing them longer credit terms. In an interlinked industry, actors along the supply chain rely on each other, as highlighted by one of interviewed breeders: "if they [growers] are affected, we [breeders] are affected in a bigger way" (sales manager, interview, September 22, 2022). In order to prevent such ripple effects that affect the whole supply chain, partnerships were not only created among directly linked industry players (such as growers

5 Eventually, the Kenya Revenue Authority paid flower farms KSh 7 billion, accounting for 70% of the tax refunds they had demanded (Muchangi 2020).

and breeders), but also among growers and actors in the segment of transport and logistics, with the latter providing technical expertise regarding sea freight for instance.

d Flower farms around Lake Naivasha have also experienced support beyond the local level by *buyers*. Many flower farms engaging in direct sales had pre-established close and long-term relationships based on trust (see Gemählich 2022). Such producer-client relationships that are built on trust ensured continued access to markets during the pandemic. Buyers also provided support in kind. For instance, funds were created "to support the workers in the form of sustenance, materials [and] masks" (production manager, interview, September 20, 2022), particularly from the retail markets, which prioritised the health and well-being of the workforce. Simultaneously, buyers in Europe "have taken quite some losses" (marketing manager, interview, September 20, 2022) for the flower farms in Naivasha, as they bought the produce, despite uncertain levels of market demand during the crisis. Producer-client relationships seemed to strengthen during the COVID-19 crisis, with some clients offering direct assistance to the farms. Flower-farm managers acknowledged the fact that this had also positively affected the livelihood of flower-farm workers, which would be put at risk if international buyers were to reduce their orders. In this respect, "consumption [can be considered] as a form of support" (KFC official, interview September 17, 2022). Hence, retaining long-term clients is also a strategy to confront risks, as it guarantees continuous supply and secures the livelihood of the local flower-farm workforce. One industry player further underlined the importance of customer relations beyond financial support, as these helped them to take necessary steps, plan ahead and keep up with the dynamics of the COVID-19 pandemic.

e *Certification programmes* such as Fairtrade have been stepping in as supporting actors for the industry as well. Industry players praised Fairtrade for coming in "very strongly to provide food staples" (marketing manager, interview, September 20, 2022) and making health and hygienic equipment available at the farms. Likewise, special Fairtrade committees, created by the certification programme at the farm level, were considered a strong influencing factor when it came to the welfare of flower-farm workers during the pandemic. Notably, there was a form of relaxation on the certification requirements from the side of Fairtrade because the focus was on getting through the COVID-19 crisis; as one informant put it, "[they have] not been running as much as [they] used to" (human resources manager, interview, September 22, 2022). While this was good for the farms, it was not good for the workers.

3.4 *Innovation and New Technologies*

As emphasised by one manager, the COVID-19 pandemic was unique in that it affected both production and markets. Flower growers had to grapple with increased costs of production on the one hand, and rising costs of living affecting consumers and the demand for cut flowers on the other – both of which have recently been exacerbated by the war in Ukraine and its impacts on energy, fuel, and fertiliser prices. Experiences of the last few years, according to one marketing manager, "have necessitated a very high rate of innovation" for the industry in Naivasha. These innovations are reflected in a number of ways:

a Organic farming and renewable energy: To counteract rising production costs, while simultaneously responding to growing consumer demand for organically farmed cut flowers, farms have increasingly invested in organic and sustainable farming methods. According to one manager, alternative sources of fertilisers and composting, which were previously used at a smaller scale, are "[now] a fully blown activity" (marketing manager, interview, September 20, 2022). The KFC estimates that 70–80% of Kenyan cut-flower farms already produce organically,[6] investing in biological pest control and shifting from pesticides to mulching, for instance. In addition to further investing in hydroponics or rainwater harvesting, flower farms also increasingly draw energy from sustainable sources (i.e., solar energy). This shift to solar power is also highly motivated by flower farms' striving for independence from the Kenyan Power System and its sole power distributor. While transition to green energy is not new, recent developments have intensified green-energy uptake among players in the cut-flower industry in Naivasha.

b Sea freight: Prior to the pandemic, cut floewers were almost exclusively transported via air freight. The cancellation of both passenger and cargo flights, in addition to their diversion to the transport of more lucrative goods such as vaccines, forced flower farms to consider shipment as an alternative transport mode. Farm managers voiced mixed experiences. While some have invested in sea freight and profited from high success

6 The estimates may be subjective and are only verifiable following a keen evaluation of the inputs used at the firm level over a given period of time. Furthermore, organic farming follows specific laid-down guidelines and certification criteria (e.g. Demeter Certification) that also affect neighbouring farms.

rates of up to 80% linked to lowered transport costs, higher loading capacities, and a reduced carbon footprint, others remain sceptical. Especially the high perishability of cut flowers was perceived as a major challenge to the use of sea freight, with prompt transportation and logistics being key in ensuring the timely delivery of quality produce. The transportation via sea freight from Kenya to Europe for instance can take up to four weeks and requires a high level of resilience, which only some varieties of flowers are able to offer (farm manager, interview, September 19, 2022). That there is little guarantee that shipments arrive on time especially strains farms engaging in direct sales, as they are dependent on punctual delivery and client satisfaction. It also means that payments are made when buyers receive the shipment and ascertain the quality, as opposed to air freight, in which prompt delivery and payments have a competitive advantage. Nevertheless, with fuel and energy costs remaining high and air freight being considered the "biggest struggle for [the] business right now" (marketing manager, interview, September 20, 2022), the (re)consideration of sea freight has gained traction, further motivating growers to invest in extensive infrastructure, knowledge, and capacities needed to adapt entire production processes and supply chains.

While these strategies were initially considered due to the pandemic, increasing investment in new technologies and innovations is also motivated by potential cost savings in the long run and the shielding from externalities that are bound to have negative effects on the operations at flower farms in Naivasha. Growing competition from Ethiopia, where freight costs are lower, workforce is cheaper and sustainable energy such as solar power is abundant, puts additional pressure on the Kenyan floriculture sector. As the costs of living and production continue to rise, industry players are thus obliged to "be innovative and creative" (KFC official, interview, September 17, 2022). Some of the farms have incorporated value-adding activities, such as exporting ready-made flower bouquets, while others have focused on creating efficiencies internally through smarter irrigation or reduced packaging, for instance. Two informants summarised the outcome of these innovations:

> As much as [the pandemic] had a negative impact in the beginning, we came away stronger.
>
> SALES MANAGER, INTERVIEW, SEPTEMBER 22, 2022

> We have lessons that we learned to help us move faster to create a working formula.
>
> PRODUCTION MANAGER, INTERVIEW, SEPTEMBER 20, 2022

3.5 *Investing in Labour Well-Being*

> All in all, [due to] the power of togetherness [...], we were able to get through the crisis as a farm, by talking to each other, by consoling each other, and by comforting each other.
>
> HUMAN RESOURCES MANAGER, INTERVIEW, SEPTEMBER 23, 2022

Taking care of their employees and building a trustful and respectful labour relationship enabled farms to better master the difficulties of the COVID-19 pandemic, simultaneously maintaining the quality of their produce. Doing so has thus had multiple benefits to several players in the industry, including workers, farm managers, investors, buyers and customers. As shown below, farm managers made efforts to guarantee the health of the workers. Maintaining safety on the farms allowed them to continue production and thus to secure jobs at the farms. This has created a mutually beneficial situation, while also improving trust and respect between managers and employees.

The COVID-19 pandemic and related sanitary measures have necessitated the reinvention of processes at the farm level to simultaneously guarantee the health of workers and the continuity of production. Flower farms took precautionary measures, including the distribution of face masks and hand sanitisers, and the introduction of social-distancing measures in packing halls, buses and at water points. Our interviews did not reveal if farms advocated for the vaccination of their staff, although the Kenya Flower Council strongly advocated to prioritise farm workers in vaccination plans. By implementing these measures, most of the interviewed farms around Lake Naivasha did not experience severe effects on the health of their workers.

A major difference in the farms' approaches has been whether they decided to temporarily lay off employees during the lockdown in spring 2020. While some flower-farm managers showed relief that they did not have to lay off workers or shift to part-time employment of their workers, others believed that it was imperative to introduce shift work at their farms, "doing only what was necessary to keep the farm alive" (marketing manager, interview, September 20, 2022). Low levels of orders as well as the protection of workers' health figure among the main reasons for this policy. Most of the flower farms interviewed at Lake Naivasha took responsibility for their employees, which in turn was appreciated by the latter, as they could sustain their source of livelihood thanks to these efforts.

During COVID-19, certification labels such as Fairtrade also voiced their concern for the well-being of flower-farm workers. The Fairtrade premium, for instance, is reinvested in programmes benefiting the local community, such as health infrastructure and subsidised staple foods. These activities have cushioned flower-farm workers and the community from severe effects of the pandemic, in the long-term contributing to the reduction of staff turnover. In turn, this offers greater planning certainty and stability for flower farms and empowers them in the face of future uncertainties.

Thus, advantages that emerge from investments in the employees' well-being are gained by all industry players, even beyond the pandemic: 1) By the workers, who have more secure jobs with improved labour conditions; 2) by the farm managers and investors, who employ skilled, loyal and healthier workers; and 3) by buyers and consumers, who have a quality guarantee for the end products thanks to continued reliance on the experienced labourers.

3.6 *Social Networks*

The global cut-flower industry can no longer only be conceptualised as a value chain. Instead, existing networks between industry players are increasingly shifting towards horizontally better-integrated social networks with interests beyond the economic. During the COVID-19 crisis, social networks especially served to "give everybody confidence" (marketing manager, interview, September 20, 2022) and to ease fears about the prevailing uncertainty. Due to an increasing importance of social capital, investors are in closer contact, some living within the same space in Naivasha (e.g. Flower Park) as well as in the Netherlands (Gemählich 2022). These social aspects of partnerships and networks play a pivotal role in understanding not only the dynamics of the value chain, but also other related factors such as risk-taking, joint actions, and sharing information, skills and technology. Against the backdrop of turbulent global economies in the aftermath of the COVID-19 crisis, social networks have also been a response in themselves, preparing industry players for future possible crises. Social networks contribute to joint strategising and planning for economic performance in difficult times. They also provide an opportunity for exchange of technology and new strategies aimed to confront competition from other flower-producing countries in the Global South like Ethiopia.

The industry is known to be risky business – anything can happen at any time. Hence, the maintenance of such a social network is not only crucial in brainstorming response mechanisms, but also vital to prepare for the uncertainties more generally – through joint risk assessments and calculations as well as shared strategies of risk-taking. Especially national and international umbrella organisations of the industry have come into play as important actors within these social networks, agreeing on a common frame of reference for decisions and actions. They have provided and shared important information on containment measures and travel restrictions, which were put in place to contain the spread of the virus, but also provided expertise on quality assurance, sea freight and pest control (e.g. Flower Watch). Most importantly, they offer a platform for social networks, facilitating stronger cohesion and solidarity among industry players and shaping greater agency. Additionally, events such as the Naivasha Horticulture Fair, which were suspended during the pandemic but have since resumed, contribute to establishing and strengthening social networks beyond the object of showcasing products and services from stakeholders in the horticultural industry.

3.7 *Leadership and Financial Capacity*

Leadership, financial capacity and resulting agency are observed as key factors for managing the risks – of COVID-19 and beyond – and can be linked to the three above-mentioned practices of future making (cf. Appadurai 2013).

Continuous cash flow and financial buffers from previous business operations, as well as the absence of loans or investments in bigger projects, enabled interviewed farms to successfully weather the crisis. For instance, producing for the British market has been a stroke of "great luck" (farm manager, interview, September 19, 2022) to one farm, as, due to high revenues during Valentine's Day and Mother's Day, they were cushioned from the financially disruptive effects of the first lockdown in March 2020. In contrast, farms heavily relying on daily turnovers to cover running costs had to lay off workers or, even worse, close their facilities. Sufficient financial capacity can thus be linked to the future-making practice of *anticipation,* enabling farm management to make decisions that are largely informed by calculations or modelling of risks and risk-taking (e.g. by not shutting down operations). Whereas many farms completely shut down production and laid off workers, for others, not following the grain and succumbing to initial shock turned out to be both a risk-taking manoeuvre and a successful strategy.

Financial capacity thus also lays the foundation for decisions and actions and therefore also leadership. Previously mentioned strategies that farm managers adopted to respond to COVID-19 related risks – such as the diversification of produce and markets, operations via direct sales, investments in partnerships and networks, and innovation – can all be traced back to decisions and actions informed by leadership. Actions such as investing in the welfare of workers or "build[ing] teams better internally" (sales manager, interview, September 22, 2022) are a function of leadership as well, and have the potential to shape industry players' preparedness for crises and uncertain futures.

Leadership, however, can be linked not only to this future-making practice of *imagination,* but also to the practice of *aspiration,* because the farms' leadership can ultimately draw upon lessons learned from the COVID-19 pandemic. For instance, for those that made rash decisions and shut down their operations, "the COVID-19 period was [still] an eye-opener" (human resources manager, interview, September 23, 2022). As an outcome of felt impacts, they learned that there is not necessarily a "need to take [immediate] drastic measures" (sales manager, interview, September 22, 2022) but rather that they should draw lessons from the experience about how to better deal with global crises in the future. It is worth noting that these lessons learned, for instance about networking or customer retention, go far beyond the COVID-19 crisis. One of the interviewed farm managers highlighted that hygiene practices, which were introduced at the height of the COVID-19 pandemic, were maintained afterwards. This was positively received by workers, simultaneously having a positive effect on the workers' health and lowering the risks for production. On a more general note, one of the managers reiterated that they "learned that life can change overnight" (human resources manager, interview, September 22, 2022), and hence sound leadership and decision-making become an important prerequisite for the future.

4 Discussion and Conclusion

Several related points are worth noting with reference to the foregoing conceptual and empirical discussion: the future as volatile and unpredictable; the robustness of the industry in Naivasha; and the potential of riskscapes as industry players adopt similar risk-taking mechanisms or responses. Taking these aspects into account allows us to broaden the perspective on Naivasha and its industry players.

What emerged from the conversations with cut-flower-industry players in Naivasha was that the notion of the future itself is highly volatile. Whereas for some farms, future making consists in planning one day at a time, other farms see the future in terms of a time span of three months, reflecting the time needed to plant and grow a flower destined for export. Others yet, often simultaneously, look to explore long-term ventures such as the diversification of their operations. Hence, uncertainty not only spawns choices, but also forms the foundation on which notions of different and sometimes colliding futures are built. The volatility and unpredictability of the future informs deliberate actions on the side of the industry players in Naivasha aimed at increasing the industry's robustness. For the industry, robustness aims to avoid potential collapse when faced with possible perturbations.

The strategies to increase the robustness of Naivasha's agro-industry define its sustainability. Robustness, a concept well developed in engineering to refer to the maintenance of system performance either when subjected to external, unpredictable perturbations, or when there is uncertainty about the values of internal design parameters (Carlson and Doyle 2002), is increasingly applied in the social sciences, for instance, to analyse resource regimes and value chains under different market conditions. In the context of the flower and horticulture farms in Naivasha, robustness reflects the industry's dependence on diverse strategies and networks across scales. Innovations, inventions, strategic positioning, joint risk-taking and social capital compound these multiscalar strategies and networks. Put together, these make the system more versatile, which over time increases robustness. While our interviews were limited to allow generalisations, we did not encounter information about flower farms that either collapsed or sold their businesses under the weight of the effects of the pandemic. While we cannot claim that the flower industry is robust, we argue that actors are increasingly devising strategies that strengthen their ability to cope with and withstand risks and uncertainties.

Furthermore, we are aware that the development of robustness is a process, and just like resilience, it is affected by both internal and external factors, and, that robust systems may also collapse. Just as is the case with governance of social-ecological systems (see Janssen and Anderies 2007) robustness involves trade-offs, which for the agro-industrial hub of Naivasha includes trade-offs between risk and productivity and between short-term and long-term goals, but also between strategic positioning and joint action in preparing for uncertain futures. Notably, some factors that increase robustness, e.g. diversifying risk, may in themselves produce more risks and potentially lead to collapse. Hence, the concept of riskscapes is vital here.

Riskscapes can be understood as socially produced spatio-temporal phenomena, which combine the material and practice components of risk and relate them to space (Müller-Mahn, Everts, and Stephan 2018). The authors argue that riskscapes are socially produced, as risks are not conceived in isolation, but through common perceptions, communication and collective action at societal level. They are spatio-temporal phenomena, because they play out in space and time. In the context of Naivasha, industry players talk to each other, calculate risks and often come up with similar strategies to manage expected risks and to prepare for uncertainty. For example, diversifying production and markets comes through joint risk assessment and actions aimed to "control" uncertain futures through similar strategies. However, instead of managing risks and "controlling" uncertainty, actors may end up producing new incalculable and unexpected risks because of the similarity of their strategies, and hence translate uncertainty into risks. For example, many farms are diversifying towards fruit (especially avocado) and vegetable production. Despite the possibility of increased competition, such diversification amounts to increased natural-resource use including land and water. All these accumulated risks may lead to a collapse, and although no one can say this with certainty, the future of the floriculture industry in Naivasha and beyond is open.

Acknowledgements

The fieldwork for this research was made possible through the generous support of the DFG under the Collaborative Research Centre *Future Rural Africa: Future-Making and Social-Ecological Transformation*. Data collection was realised though a field school in Naivasha with mixed groups of German and Kenyan master students (among them the co-authors). Eric Kioko wishes to thank all the co-authors and the DFG for the support.

References

Appadurai, Arjun. 2013. *The Future as Cultural Fact: Essays on the Global Condition*. London: Verso.

Button, Kenneth. 2020. "The Economics of Africa's Floriculture Air-Cargo Supply Chain." *Journal of Transport Geography* 86: 102789. https://doi.org/10.1016/j.jtrangeo.2020.102789.

Boholm, Åsa. 2015. *Anthropology and Risk*. London: Routledge. https://doi.org/10.4324/9781315797793.

Carlson, J. M, and John Doyle. 2002. "Complexity and Robustness." *PNAS* 99 (1): 2538–45. https://doi.org/10.1073/pnas.012582499.

Carlsson-Szlezak, Philipp, Martin Reeves, and Paul Swartz. 2020. "Understanding the Economic Shock of Coronavirus." *Harvard Business Review*, March 27, 2020. Accessed September 9, 2023. https://hbr.org/2020/03/understanding-the-economic-shock-of-coronavirus.

Cashdan, Elizabeth. 2019. *Risk and Uncertainty in Tribal and Peasant Economies*. London: Routledge.

Central Bank of Kenya. 2021. *Monetary Policy Committee Flower Farms Survey*. January 2021. Nairobi: Central Bank of Kenya. Accessed September 9, 2023. https://www.centralbank.go.ke/uploads/market_perception_surveys/848246841_MPC%20Survey%20of%20Flower%20Farms,%20January%202021.pdf

Cooper, Elizabeth, and David Pratten. 2015. *Ethnographies of Uncertainty in Africa.* London: Palgrave Macmillan.

Deloitte. 2020. *Economic Impact of the COVID-19 Pandemic on East African Economies: Summary of Government Intervention Measures and Deloitte Insights.* May 2020. Nairobi: Deloitte. Accessed September 9, 2023. https://www2.deloitte.com/content/dam/Deloitte/tz/Documents/finance/Economic_Impact_Covid-19_Pandemic_on_EastAfrican_Economies.pdf.

Douglas, Mary. 1992. *Risk and Blame.* London: Routledge. https://doi.org/10.4324/9780203430866.

Gemählich, Andreas. 2022. *The Kenyan Cut Flower Industry & Global Market Dynamics.* Suffolk: Boydell and Brewer. https://doi.org/10.1017/9781800105607.

Hivos. 2020. *Impact of Covid-19 on Women Workers in the Horticulture Sector in Kenya.* April 2020. Den Haag: Hivos People Unlimited. Accessed September 9, 2023. https://hivos.org/assets/2021/08/A-rapid-assessment-on-the-prevalence-of-sexual-harassment-in-Kenya-s-informal-sector.pdf.

Janssen, Marco A, and John M. Anderies. 2007. "Robustness Trade-offs in Social-Ecological Systems." *International Journal of the Commons* 1 (1): 43–65.

KNBS. 2020. *Economic Survey 2020.* Nairobi: Kenya National Bureau of Statistics. Accessed September 9, 2023. https://www.knbs.or.ke/?wpdmpro=economic-survey-2020.

Kuiper, Gerda, and Andreas Gemählich. 2017. "Sustainability and Depoliticisation: Certifications in the Cut-Flower Industry at Lake Naivasha, Kenya." *Africa Spectrum* 52 (3): 31–53. https://doi.org/10.1177/000203971705200302.

Mbembe, Achille, and Sarah Nuttall. 2004. "Writing the World from an African Metropolis." *Public Culture* 16 (3): 349–72. https://doi.org/10.1215/08992363-16-3-347.

Müller-Mahn, Detlef, Jonathan Everts, and Christiane Stephan. 2018. "Riskscapes Revisited – Exploring the Relationship between Risk, Space and Practice." *Erdkunde* 72 (3): 197–213. https://doi.org/10.3112/erdkunde.2018.02.09.

Müller-Mahn, Detlef, and Eric Kioko. 2022. "COVID-19, Disrupted Futures, and Challenges for African Studies." In *African Futures,* edited by Clemens Greiner, Steven van Wolputte, and Michael Bollig, 28–35. Leiden: Brill.

Muchangi, John. 2020. "Flower Farms Receive Sh7 Billion Tax Refund. 80 per Cent of Workers Resume, Hivos Sends Food to Help 5,000." *The Star,* August 6, 2020. https://www.the-star.co.ke/business/kenya/2020-08-06-flower-farms-receive-sh7-billion-tax-refund/.

Murage, George. 2022. "Flower Farmers Recover from Covid-19. Cargo Easily accessing EU Market." *The Star,* April 22, 2022. https://www.the-star.co.ke/print/business-page-fourteen/2022-04-22-flower-farmers-recover-from-covid-19/.

Rohrmann, Bernd. 2008. "Risk Perception, Risk Attitude, Risk Communication, Risk Management: A Conceptual Appraisal." Paper Presented in the 15th International

Emergency Management Society Annual Conference, June 2008. Accessed September 9, 2023. https://www.tiems.info/dmdocuments/events/TIEMS_2008_Bernd_Rohrmann_Keynote.pdf.

Scoones, Ian, and Andy Stirling. 2020. *The Politics of Uncertainty. Challenges of Transformation.* London: Routledge. https://doi.org/10.4324/9781003023845.

UNDP. 2020. *Articulating the Pathways of the Socio-Economic Impact of the Coronavirus (COVID-19) Pandemic on the Kenyan Economy.* Policy Brief no. 4, April 2020. Accessed September 9, 2023. https://data.unhcr.org/en/documents/details/78194.

Van Wolputte, Steven, Clemens Greiner, and Michael Bollig. 2022. "Futuring Africa: An Introduction." In *African Futures*, edited by Clemens Greiner, Steven van Wolputte, and Michael Bollig, 1–16. Leiden: Brill.

Whyte, Susan Reynolds. 2009. "Epilogue." In *Dealing with Uncertainty in Contemporary African Lives,* edited by Liv Haram, and C. Bawa Yamba, 213–16. Stockholm: Nordiska Afrikainstitutet.

CHAPTER 15

Megaprojects and the Transformation of Lake Naivasha Basin: Aspirations, Uncertainties and Social-Ecological Implications

Eric M. Kioko and Silas Wanjala

1 Introduction: Naivasha – Caught between Aspirations, Visions and Megaprojects[1]

Aspirations and visions are key elements in Africa's national development plans. Finnemore and Jurkovich (2020) describe aspiration as a "social and political phenomenon which implies determination and the desire to attain goals that could be difficult to achieve within a given time". Aspirations are reflected in the promises contained in development plans, which are based on visions of economic, social, and environmental prosperity. Irrespective of the persistence of calamities such as climatic perturbations, economic meltdown and collapse, and the lingering COVID-19 pandemic, the governmental tendency to envision prosperous African futures is unwavering. Central to the achievement of these visions and aspirations are large infrastructure projects, commonly referred to as megaprojects.

In Kenya, as elsewhere in Africa, megaprojects feature a combination of ports, pipelines, roads, railways, economic zones, energy investments, agricultural schemes, industrial parks, and new modern cities, among other features. Most of these projects are implemented within "development corridors", defined by the Development Corridors Partnership (n.d.) as "geographical area[s] identified as a priority for investment to catalyze economic growth and development". Terming them "dreamscapes of modernity", Müller-Mahn (2019, following Jasanoff and Kim, 2015) views development corridors as dominant blueprints for spatial development because of the specific way in which they express, perform, and implement desirable futures.

In Kenya, the Lamu Port-South Sudan-Ethiopia-Transport (LAPSSET) Corridor Project is the most conspicuous of all grand projects. LAPSSET is an amalgamation of numerous projects including a port, standard-gauge railway,

1 First submission date: May 12, 2021. Acceptance date: March 13, 2023.

 | DOI:10.1163/9789004695429_016

highways, crude oil pipelines, international airports, and resort cities, among others. These projects are contained in Kenya's Vision 2030 (n.d.a), which "aims to transform Kenya into a newly industrialising, middle-income country providing a high quality of life to all its citizens by 2030 in a clean and secure environment". Out of the many promises of Vision 2030, as of March 4, 2021 the current political regime identified four priority development areas as listed on the home page dubbed "The Big 4 Agenda": manufacturing, affordable housing, universal health coverage and food security (see also Kirigia, this volume).

Megaprojects do not just appear from nowhere – they are products of political processes that involve negotiations and justification between different actors and interests. As Finnemore and Jurkovich (2020) note, "political structures are built on grand aspirations". Some of these aspirations may not materialise, while others may collapse midway (see Müller-Mahn, Mkutu, and Kioko 2021), but, either way, they serve certain interests – they mobilise political support, serve as "magnets" to attract development grants, and reproduce future plans.

Naivasha and other smaller towns in Kenya are caught up in these megaprojects and within the visions and aspirations of desirable futures. Currently, the Lake Naivasha basin serves as a nerve centre for market-driven agro-industrial plantations featuring flower farms and vegetable farms. Other activities include tourism (hotels and lodges), geothermal development (see Nweke-Eze and Adongo, this volume), wildlife conservation (Hell's Gate and Mount Longonot National Parks, private game sanctuaries and Lake Naivasha's status as Ramsar site), and smallholder cultivation in the wider basin, to mention a few. Combined, the basin has witnessed tremendous competition between agricultural intensification and conservation at least since the 1980s. Competition for water and land is also a key factor driving conflict between smallholder farmers and the big horticultural players as well as upstream and downstream actors. Consequently, some have questioned the management and sustainability of the social-ecological systems that constitute the basin (Hickley et al. 2004; Boar and Harper 2002; Becht and Harper 2002).

On top of the activities that characterise the basin, and despite concerns over its sustainability, this nerve centre is already witnessing a new dawn characterised by ambitious space- and energy-consuming megaprojects. Naivasha Town became included into the LAPPSET corridor network because it was already strategically located, and had been ever since the colonial railway passed there, thereby revealing what Enns and Bersaglio (2019, 102) term "the coloniality of new mega-infrastructure projects in East Africa", suggesting the "newest wave of transport infrastructure sweeping across East Africa [Kenya and Tanzania] is reproducing colonial legacies". Despite this longue durée of some mega-infrastructure projects in Kenya, and in Naivasha particularly, they are politically "sold" as new and as having the potential to bring modernity and prosperous futures.

In December 2019, President Kenyatta launched the Chinese-built Nairobi-Naivasha Standard Gauge Railway (SGR), marking a successful completion of Phase 2 of the proposed Mombasa-Malaba railway. In addition, the state in liaison with the Nakuru County government (the county in which Naivasha is located) proposed the development of a waterfront on the shores of lake Naivasha, construction of a mega-dam at the Malewa River, expansion of the Nairobi-Nakuru highway into a dual carriageway, development of Naivasha as an industrial city with a Special Economic Zone (SEZ) and an inland Container Depot joining the SGR, and expansion of geothermal activities, as well as the expansion and marketing of ecotourism and conferencing facilities at Naivasha.

This chapter is primarily based on an analysis of literature including grey literature (e.g. reports, newspaper articles) as well as the blueprint plans for the proposed megaprojects at the Naivasha basin. It also draws from the authors' in-depth knowledge and experiences due to their connection to the study area and their interest in these developments. The chapter discusses five megaprojects and their social-ecological impacts on an already strained ecosystem. These are: the Naivasha SEZ; the mega-dam at Malewa River; geothermal expansion; Naivasha Inland Container Depot; and the expansion of transport, ICT and housing infrastructure.

Despite the anticipation, hopes, and desired futures that these megaprojects elicit, we argue that their conceptualisation, planning, and (possible) implementation gives little consideration to possibilities and probabilities of risks, surprise, shocks and stress. Impact assessments for these megaprojects only mention the expected negative social-ecological impacts without a clear strategy for how to mitigate and address them. In addition, the megaprojects create uncertainties that are linked to resource competition, fears, and conflicts pitting different actors and land-use systems against each other. We conclude that visions and aspirations of desirable futures inform decisions and actions at the present, which have unforeseen consequences and possible externalities beyond their intentions.

In the following, we briefly explore the condition of the Lake Naivasha basin against the backdrop of social-ecological changes since the 1800s, and thereafter examine the proposed megaprojects, their conditions of implementation and the linked social-ecological uncertainties. In the conclusion, we revisit the concepts of "uncertainty" and "future-making" in relation to the Lake Naivasha basin.

2 Lake Naivasha Basin: Past, Present, and Desired Future

Human occupation of Kenya's Rift Valley, including the Naivasha region, stretches deep into precolonial history. Richard Waller (this volume) notes that Maa-speakers (and perhaps other groups) were already in occupation of the Nakuru-Naivasha corridor before 1800, and by 1820, the basic ethno-linguistic map of the central Rift was becoming established, although ongoing processes of expansion, conflict, incorporation and acculturation would simplify and to some extent obscure its original complexity. At the time, the lake and its ecological structure can be said to have been relatively pristine with limited human interference and modifications, as Joseph Thomson described in 1887, painting a picture of grassy plains surrounding the lake as well as wildlife, such as zebras (Kuiper 2019, 36). However, this "pristine" basin would drastically change following colonialism.

According to Harper et al. (2011, 92), "the land around the lake was subdivided by the government early in colonial history (early twentieth century) and sold to settlers. All riparian edge land was in private ownership, except for a few land corridors, which were retained in government control for watering livestock and oxen". Rinderpest epidemic and outbreak of smallpox between 1891 and 1892 made the Rift Valley groups vulnerable to British colonisation, allowing settlers to invade and take control of community land, whose production structure was largely subsistence-based, and in place establish large-scale ranching and farming. Colonialism also brought about the development of infrastructure such as the Meter Gauge Railway which reached Naivasha in 1900, and from which present visions of the SGR are derived.

The railway, according to Richard Waller (this volume), "provided administrative convenience to the imperial authority and, more importantly, Eliot, Governor since February 1901, hoped to attract white settlers who would pay rent and create traffic for the railway because the volume of African production seemed insufficient". Apart from the railway, seeds of some of the current megaprojects began to be planted from the mid-1900s. For instance, exploration of geothermal resources in Kenya started in the 1950s with mainly geological investigations in the region between Olkaria and Lake Bogoria in the north rift (Mariita 2009). The first geothermal plant started in the early 1980s, coinciding with the establishment of Mount Longonot NP and Hell's Gate NP, the immigration of labour force, population explosion, and the expansion of unplanned settlement areas in Naivasha between 1980 and the 1990s (Thorn, Alpha, and Mwangi 2019).

Notably, independence heralded land claims and the search for settlement, which resulted in the restructuring, adjudication and allocation of land, starting with the famous settlement schemes, which targeted communities whom colonial authorities had dispossessed of their land (see Wasserman 1976). From the 1970s, the Naivasha basin received immigrants from central Kenya whose relatives and friends had found settlement within the basin, as exemplified by the occupation of the Maiella Estate on the fringes of the Lake (see Kioko 2016). More migrants from virtually all parts of Kenya arrived at Naivasha following the expansion of the cut-flower industry in the 1990s and early 2000s. Kuiper (2019, 40) described Naivasha as having turned into a "small island of wage labour". Ogutu et al. (2017, 2) note that the human population in the Nakuru County (an area of 7,510 km^2) "increased by more than 300% from 1969 to 1,187,039 in 1999 and by a further 74% to 1,603,325 by 2009, with the greater percentage of populations settling in Naivasha and Nakuru towns. Average human population density was 41/km^2 in 1969, 158/km^2 in 1999 and 214 people/km^2 in 2009, with higher densities concentrated around the urban centres and areas of intensive agriculture".

Since 2010, and following the promulgation of the new Constitution that paved way for devolution, agriculture became a devolved function, making it a priority sector for Nakuru County (in which Naivasha is located), as envisioned in the Nakuru County Integrated Development Plans for the periods 2013–2017 and 2018–2022 (Republic of Kenya 2013). These plans come amidst rapid expansion of the tourism sector and facilities, urbanisation, disappearance of wildlife corridors due to increased fencing and human-wildlife conflict, increased land subdivision, and accelerated geothermal power exploration and supply beyond Naivasha, as well as increased water abstraction (see Thorn, Alpha, and Mwangi 2019). The launch of Kenya's Vision 2030 in June 2008 particularly informed changes relating to and associated with the establishment of mega-infrastructure projects. Being the most important geothermal site in the country and the main transit town west of Nairobi, Naivasha became a critical space for the establishment of some of the projects envisioned in the country's development plans. Although the COVID-19 pandemic drastically reduced operations in Naivasha's key economic sectors (including the flower and horticulture industry) and slowed the progress of implementation of some megaprojects, construction of the Nairobi-Naivasha SGR and the Naivasha dry port was completed and operations launched in 2020/2021.

What do future desires and expectations of Lake Naivasha basin look like? Beckert (2016, 35) argues that "perceptions or expectations of the future are reflected in the way actors associate certain future results with the course of action they are contemplating, connecting numerous outcomes with different possible decisions". The planned megaprojects at Naivasha mean different things to different actors, and these meanings inform their expectations, desires, fears, and hopes. For the state and political elite, megaprojects promise to spur economic growth – what Larkin (2013) refers to as the "politics of infrastructure". Conservationists fear for the quality of natural resources following capitalist investments. At the local level, the completion of these projects is not necessarily the driver of change – it is rather the possibility of their implementation (the planned phase) that creates huge expectations, informing people's actions and behaviours. The idea therefore is not necessarily to look at the infrastructures themselves but also people's expectation of them – not necessarily the "expectations of modernity" (following Ferguson 1999), but rather expectations of infrastructures, valorisation, and connectedness. Therefore, some of the aspirations and desires associated with the (possible) implementation of the megaprojects at Naivasha include more job opportunities due to the influx of investors leading to increased immigration, and increased value of land, leading to continued subdivision and privatisation (see Thorn, Alpha, and Mwangi 2019).

However, conspicuously missing from the prognosis for the Lake Naivasha basin are elements of risk, surprise, and uncertainty. Uncertainty, according to Beckert's thinking, "is paramount in situations associated with change, novelty, and crisis". Beckert (2016, 43) notes that "the complexity and interdependencies of parameters, the unforeseeability of the reactions of relevant third parties, and the nonlinearity of economic processes make the (probabilistic) calculation of outcomes of decisions impossible". Indeed, it is one thing to aspire and imagine a desired future, and it is another to actualise that future. In addition, a future that some aspire to can imply a risk or uncertainty to others.

In the following we describe five megaprojects envisioned for the Lake Naivasha region, focusing on their conditions of implementation and the linked social-ecological uncertainties.

3 Megaprojects and Uncertainties at the Lake Naivasha Basin

3.1 *The Naivasha Special Economic Zone*

Just like the rest of Sub-Saharan Africa, Kenya is in a rush to replicate East Asia's economic success that is modelled around export-oriented manufacturing through SEZs. The establishment of the Special Economic Zones Authority (SEZA) in Kenya through an Act of Parliament (Special Economic Zones Act No. 16 of 2015) is a key catalyst of these proposed developments. SEZA is responsible for "designing, approving, establishing, developing, operating, promoting and regulating an SEZ" (Kenya Investment Authority, n.d.). So far, the state has gazetted 10 SEZs out of which 7 are private and 3 are public. SEZs fall under the manufacturing sector of Vision 2030 and the Big 4 Agenda as the key drivers for achieving this vision. Farole and Moberg (2017, 1) define SEZs as "demarcated geographical areas within a country's national boundaries [in which] rules of business are different, and generally more liberal, from those that prevail in the national territory".

SEZs confer four advantages to investors relative to what they enjoy in the domestic economy: infrastructure including serviced land and utilities; special customs regimes such as duty-free import and export; improved regulatory and administrative regimes; and a range of fiscal incentives such as reduction or elimination of corporate taxes (Farole and Moberg 2017, 3). Kenya's SEZ policy (2015) defines an SEZ as a "designated geographical area that includes a Customs Controlled Area, with liberal economic laws and on-site administrative regulations, management and services, where the benefits to licensed firms apply; and wherein goods introduced into the SEZ Customs Controlled Area are exempted from customs duties and taxes" (SEZA 2020).

In July 2019, the state secretary in charge of Industry, Trade and Cooperatives declared 1,000 acres in the Mai-Mahiu area south of the Naivasha basin a SEZ.[2] According to the Naivasha SEZ Master Plan, "the most prioritised industrial sectors to be located in this zone include cotton, textile and apparel, footwear and leather, motor vehicle assembly, steel industries, agro-processing (grain and milling, dairy products, meat processing, horticulture and floriculture), construction materials, ICT and electronics, pharmaceuticals and chemical products, and bulk storage facilities" (SEZA 2020, 13). Moreover, the choice of the location is based on agglomeration economics including the "the ability to support heavy industry/manufacturing, and noxious industries with huge volumes of waste; linkage to national/ international communication and transport networks; and accessibility to labour" (SEZA 2020, 35).

With inspiration from the state-led Naivasha SEZ, the private sector is also keen to set up industries in the Lake Naivasha basin. Oserian Two Lakes Industrial Park – a 150-hectare sustainable industrial space development has taken lead on the southern banks of Lake Naivasha. The industrial park is linked to the Oserian flower farm and "offers a strong commercial proposition and safe landing spot to textiles and apparel, agri-processing, light industrial, Fast-Moving Consumer Goods, waste-to-value and logistics services businesses" (Savo Project Developers, n.d.). This indicates that decisions and actions of private actors and investors are influenced by government plans. However, actual implementation of this project has not begun. In addition, as of July 21,

2 Kenya Gazette, Notice No. 7,650, August 16, 2019.

2021, the Kenya eOpportunities Platform listed that Kenya's electricity giant, KenGen, is also planning to set up a 453ha Industrial Park at its geothermal power-generation hub at Olkaria-Naivasha, although the plan was still at feasibility phase.

Despite the hype and rush that SEZs receive in Sub-Saharan Africa, a 2017 World Bank report paints a gloomy picture of the outcome of these developments. According to the report, rather than catalysing economic development, most zones' performance has resembled their national average; SEZ may increase unemployment due to immigration; zone growth is difficult to sustain over time; and fiscal or non-fiscal incentives have little bearing on zone performance (World Bank 2017). To date, none of the SEZs and their associated infrastructure have been realised: plans remain on paper. SEZs and related industrial developments are also linked to Kenya's deepening debt crisis, which has seen an increase in the cost of living due to tax increase across board. In April 2021, Members of Parliament pushed for the suspension of projects as an austerity measure to ease the debt burden (Anyanzwa 2021).

Suspension of these projects may mean the "suspension" of people's aspirations and hopes, but, for some, the possibility of their implementation is equally important because it influences their decisions and actions based on their own hopes and aspirations. Consequently, our observation at the sites of the proposed projects revealed ongoing land subdivision, immigration, and expansion of low- and medium-income housing in readiness for the possible implementation of the SEZ and associated developments.

3.2 *Mega-Dam at Malewa River*

Plans to construct Malewa dam scheme and Turasha dam scheme at the Lake Naivasha basin are traceable at least to the 1980s and 1990s. A 1990 feasibility study for the construction of a dam in the Malewa River system by the Japan International Corporation Agency (JICA) for the then Ministry of Water Development identifies the sites and conditions for the development (JICA 1990; see also Nyanchaga 2016, 56–57). Locations for planned construction of both dam sites were set approximately 10 km upstream from the confluence of the Malewa and Turasha rivers. The justification for the dams was mainly to supply domestic water to the growing populations of Gilgil, Naivasha, and Ol Kalou towns and their environs, a problem that JICA projected to hit the area towards the mid-1990s (JICA 1990).

Turasha dam (see Map 5.1) was constructed in 1992 to supply water to Kabarak in Nakuru County, home to the former President Moi. This out-of-basin transfer accounts for 20% of the total river-water abstraction in the basin (De Jong 2011), which certainly reduces river flow to Lake Naivasha. Illegal abstraction of water from the rivers and the lake is another major challenge. Based on a water-abstraction survey, De Jong (2011) found out that only 8% of all abstractions had valid permits (see also Pacini, Giberti, and Harper, this volume).

Thirty years on, the plan to implement the Malewa dam scheme has remained on paper. However, since 2010, there has been a renewed push to construct dams across Kenya, including the Malewa dam. These plans are linked to the Vision 2030 water and sanitation sector on rural water supply. The sector aims to "construct and rehabilitate 150 rural water schemes annually, drill an average of 70 boreholes annually in areas lacking adequate surface water and construct 160 small dams/pans" in Arid and Semi-Arid Lands (Kenya Vision 2030, n.d.b). In the National Water Master Plan 2030, it is noted that dams were planned as necessary to supply relatively large demands such as domestic, industrial and irrigation water, thereby playing an important role in meeting the large water demands projected for 2030 (JICA 2013, 25). The World Bank and other international agencies are at the centre of financing dam construction in Kenya under the Kenya Water Security and Climate Resilience Project (World Bank, n.d.).

Notably, the push to construct the KSh 35 billion[3] Malewa dam comes amidst demands for water supply to the Naivasha SEZ, which hosts an industrial park and dry port, with the government hoping that the dam will ease the future water-related pressure for both industrial and domestic uses (Mwangi and Wangui 2019). However, construction of the Malewa dam has been hit with a lot of controversies and criticisms, particularly from "environmentalists, flower-farm owners, fishermen, hoteliers and others [who] fear the dam will lower water levels in the lake, hurt the ecosystem and, in effect, 'kill' the lake" (Ngotho and Murage 2020). Malewa River supplies 80% of water to Lake Naivasha and thus abstraction of the water is expected to have serious implications, particularly downstream. The water discharge of the river is also quite irregular, with variabilities already reported in the 1980s and 1990s (see Odongo et al. 2015).

The lack of consensus between the Ministry of Water and stakeholders in the basin on the construction of Malewa dam has seen a stalling of the project and a series of negotiations between the state and relevant stakeholders. During the stalemate, the Lake Naivasha Water Resource Users Association (LANAWRUA) cautioned the World Bank against funding the construction

3 At January 1, 2019, KSh 1 was equivalent to $0,00976 (source for the exchange rate: https://www.oanda.com/currency-converter/en/). Hence, KSh 35 million was equivalent to roughly $341,600 at the time of planning.

of the dam (Murage 2020). Driven by its strategic role in Naivasha's flower sector, the Dutch government offered to fund an impact study for the dam to ascertain the extent of impacts on the downstream agro-industrial hub Therefore, despite the promise to increase resilience against future risks (water scarcity, food insecurity), the proposed construction of Malewa dam has generated new risks and uncertainties for local populations as well as foreign investments already in the present, with water conflicts already witnessed in the basin.

Dams, such as the proposed Malewa scheme, generally play an ambiguous role in future-making, because they are on the one hand justified as promising solutions to the anticipated future problems of climate change and population growth, but on the other hand they create new uncertainties for local populations, especially when they do not materialise as originally planned. After thirty years of the proposed Malewa dam existing on paper, and notwithstanding the hopes attached to it, plans to construct the dam face enormous criticisms and constraints today.

For instance, over the years, the population the dam was meant to serve has increased sharply, meaning that there is more thirst to quench; the basin has witnessed intensification of agriculture (large-scale and small-scale) and therefore water for irrigation is at a higher demand; and new demands for land linked to future-making projects of Vision 2030 dominate the basin. Moreover, over the thirty years, the basin has received new actors (investors, migrants,

FIGURE 15.1 Olkaria geothermal plant with Lake Naivasha at the background
PHOTO: CHIGOZIE NWEKE-EZE 2019

the land-hungry, lobbyists, institutions and organisations) who express their interests in the basin and decide on the relevance of the dam against these interests. Moreover, Turasha dam does not seem to serve the purpose for which it was built – that of supplying water within Naivasha as earlier stipulated in the feasibility study of 1990.

3.3 *Expansion of Geothermal Energy Plants at Olkaria*

Nweke-Eze and Adongo (this volume) provide a detailed description of geothermal exploration in Olkaria with a focus on state-investor-community relationships. Our focus here is to examine the environmental and social-economic effects of this megaproject. Geothermal exploration in the Naivasha region started in the 1950s, and since the early 1980s, when the first geothermal energy plant was set up in Olkaria, expansion plans have been underway, and remain so to date.

The Hell's Gate NP is situated within the rapidly expanding geothermal plants in Olkaria, a fact that has elicited serious environmental-management concerns related to wildlife conservation. Such expansions of exploration sites and the accompanying energy infrastructure and operations contribute to wildlife-habitat destruction in the immediate areas of drilling or operations. Kubo (2003, 72) notes "the most important environmental impacts brought about by geothermal development include surface disturbance during drilling whereby a lot of civil works take place such as access roads for drilling, pipeline routes and well pads [which] can lead to serious degradation of the landscape". Such impacts include interference with wildlife corridors and food resources. Noise from geothermal operations and traffic certainly affect wildlife behaviour (see also Blickley and Patricell 2010). Further, apart from the effects of discharge of hot wastewater associated with geothermal exploration, which can cause scorching of the surrounding vegetation, geothermal plants at Olkaria are dependent on water abstraction from Lake Naivasha for construction, reinjection, and domestic use by staff, thereby contributing to the pressure on the Lake (Kubo 2003).

In addition to these environmental effects of expansion of geothermal activities, several studies have identified a number of socio-economic issues associated with geothermal development, particularly among the Maasai pastoral community who have lived in the Olkaria area since precolonial times (see Nweke-Eze and Kioko 2021; Mariita 2002). Mariita (2002) found that the living standards of most of the community members were very low, the majority of them struggling to meet their basic daily needs for food, clothing, healthcare and housing. He further observed that only a handful of community members were employed as watchmen, cleaners, drivers, and office messengers at

KenGen and OrPower companies, and therefore the companies had failed to fully empower poor villagers through employment. Other concerns that community members raised some twenty years ago included: the increasing dust levels and smell the project could bring following expansion; a possible rise in respiratory diseases like asthma, eye problems, cold and flu; displacement/resettlement away from present homes, reduction in grazing land for livestock, reduction in family size due to gradual decrease of land size, an increase in miscarriages or children being born with deformities or learning difficulties following project expansion, and erosion of cultural values (Mariita 2002, 1125).

Two decades later, most of these concerns have actually manifested in the Olkaria area. Nweke-Eze and Kioko (2021) show that four neighbouring pastoral communities, namely Cultural Centre, OloNongot, OloSinyat, and OloMayiana, have been adversely affected by the expansion of geothermal activities. Relocation of these communities from the geothermal drilling sites has contributed to profound social-economic and cultural effects. Contestations of the relocation quickly turned into a human-rights concern, because it caused drastic interference with the pastoral activities and small-scale trading by women in the Cultural Centre as well as disintegration associated with scattering of previously tightly bound kraals (see Nweke-Eze and Adongo, this volume; Nweke-Eze and Kioko 2021).

Traditionally, Maasai developed the ingenious technology of tapping steam from underground and cooling it in order to obtain drinking water. With the expansion of geothermal power generation, some of these water sources are either targeted for geothermal expansion or completely replaced with drilling points. In addition, the fringes of the geothermal area continue to witness an influx of land-hungry people who have purchased land within the Suswa and Kedong regions along the SGR lines and close to the exploration area. The majority of the new land buyers engage in land speculation hoping that geothermal operations will eventually incorporate their land parcels in order to benefit from compensation, or the possibility to sell their parcels for more money when more people arrive in the future.

3.4 *Naivasha Inland Container Depot*

The Naivasha Inland Depot falls under Kenya's Vision 2030 and the government's objective to improve port efficiency and facilitate trans-shipment of cargo at the port of Mombasa. Other related megaprojects include construction of a second container terminal and provision of new handling facilities at the Mombasa Port, development of Dongo Kundu Free Trade Port, and construction of the Nairobi Inland Container Port. In December 2019, the dry port at Naivasha built by the China Construction Corporation was opened for business, aiming to bring services closer to customers and reduce port congestion at the coast. The Inland Container Depot (ICD) at Naivasha acts as the main trans-shipment point between the SGR and road for freight traffic that is destined for Naivasha and western Kenya, Uganda, Southern Sudan, the Democratic Republic of Congo, Rwanda and Burundi.

According to the 2020 Environmental and Social Impact Assessment (ESIA) report for the Naivasha ICD, the direct benefits associated with the depot include local provision of jobs, appreciation of land value, and construction of housing and other facilities within the area. Indeed, workers left jobless following the collapse of Karuturi Flower Farm hoped to find jobs at the construction of the ICD, an example of how "ordinary" people adapted their aspirations based on the proposed plans (Gerda Kuiper, Kioko personal communication, July 15, 2021). The report also identifies some anticipated environmental impacts including land-use changes featuring commercial, unplanned housing developments; vegetation loss; implications for drainage; and pollution (air, noise, waste, etc.) (Kenya Railways Authority 2020).

Commissioning of the Nairobi and Naivasha ICDs had a drastic effect on related job losses at the coastal facilities. Following a state directive that cargo passing through the Mombasa port must be transported via the SGR and cleared at the two ICDs, protests erupted at the coastal facility because of an immediate loss of about 3,000 jobs for Container Freight Stations, fuel stations and drivers and a further threat to more than 8,000 jobs (Chome 2019). The protests were feared to escalate tensions between coastal and upcountry people as some believed that the directive was politically intended to impoverish the already economically poor Mombasa County and its residents for failing to support Uhuru Kenyatta's Presidency in preceding elections.

Further, the Naivasha ICD has been hit with conflicts and lack of infrastructure that could potentially affect the success of the megaproject. Moreover, the ICD has exacerbated regional economic struggles, with Uganda backing out from the compulsory requirement to have its cargo and that of neighbouring landlocked countries transported via the SGR (Kitimo 2020a), as Kenya hoped to attract funds to pay the Chinese SGR debt. Moreover, according to a local news agency, poor road connectivity, inefficient clearing systems and limited space for cargo verification could hamper cargo uptake at the depot (Mwita 2020).

The scramble for land around the Naivasha ICD and industrial park resulted in massive land subdivision led by brokers, land conflicts related to irregular allocation of land and issuance of title deeds, and fears of mushrooming of unplanned and informal settlements that could jeopardise the security and economy of the new establishments (Murage 2019). These conflicts south of

Naivasha add a new layer to an already precarious social situation in the lake region occasioned by land shortage and related conflicts, some of which have taken on an ethnic dimension, with competing claims going back to the 1960s.

3.5 *Transport, ICT and Housing Infrastructure*

Construction of the SGR has gone hand in hand with construction of commuter stations, road network expansion, and the rehabilitation of the old Metre Gauge Railway line.[4] The construction of a commuter railway station at Mai Mahiu, south of Naivasha, is meant to facilitate transportation from Naivasha to Nairobi, with some residents of Naivasha hoping to use the SGR to go to work in Nairobi and to ferry their supplies, as well as movement of workers from Nairobi and its environs to the Naivasha industrial park, ICD and SEZ. The planned upgrading of the Rironi–Nakuru–Mau Summit Highway is hoped to influence the growth and transformation of the economy by facilitating easier transportation of raw materials and finished goods as well as movement of people. The expectations of income from the growing population in Naivasha have prompted growth in the urban area with the construction of new business centers, hotels and residential homes near the busy roads.

Other planned infrastructure includes the ongoing laying of fibre-optic cable to the SEZ, construction of internal link roads, construction of internal water- and power-reticulation systems, and setting up a power substation starting with 11KV, then 66KV and later increasing to 220KV by 2024. The ambitious Affordable Housing Project within the President's Big 4 Agenda is also taking shape at Naivasha. Under a partnership between the World Bank, United Kingdom and the Kenya government, construction of 2,400 affordable homes was set to begin in February 2020 with the aim of easing the housing pressure in Naivasha (Matara 2020), but construction was delayed following the COVID-19 pandemic. However, given the current population of Naivasha, which stands at over one million (Chief of the area, Kioko personal communication, August 17, 2021), and the expected population influx associated with the megaprojects, the 2,400 houses would hardly reduce the pressure on the Naivasha housing market, particularly in the lower segments. Indeed, the nonexistence of a mega-housing project puts Naivasha and the ambitious developments in a very awkward position with regard to where the people will live. Taking advantage of the housing gap, some local entrepreneurs and residents of Naivasha are converting their landholdings into low- and medium-income housing in anticipation of rental-income returns (see Figure 15.2).

---

4 As of April 21, 2021 the Kenya Railway's website indicated that the revitalisations of the Metre Gauge Railway line to Western Region had kicked off.

FIGURE 15.2 Ongoing construction of an already occupied business facility at Naivasha, indicating the growing demand for business space
PHOTO: ERIC M. KIOKO 2021

4 Discussion and Conclusion

Based on existing conditions driven by hopes, aspirations and visions, the question of the future of the Lake Naivasha basin becomes more important than ever before. The visions of and aspirations toward desirable futures in the basin have informed decisions and actions in the present, with reference to the megaprojects, which have profound impacts in terms of land-use change and elicit uncertainties particularly at the local level. Kenya's Vision 2030 seems to prioritise economic growth with the hope that all developments will be achieved by means that safeguard the environment and enhance social justice. However, based on this discussion, tensions are already evident with reference to balancing between the three aims of sustainability. Implementation of the megaprojects at Naivasha could significantly grow the economy of the region and the state at large, but the cost of doing so raises serious concerns in the present and for the future.

The discussed megaprojects share similar characteristics. First, they are arenas of negotiation for strategic groups (following Bierschenk 2008), mainly state and development partners who draft and inspire these projects, but agency at the local level is very restricted. These plans therefore embellish the enforced globalisation of Naivasha. The discussed megaprojects bring together various actors across scales (mainly national and international) who pursue their strategic interests. For example, expansion of the geothermal projects in Olkaria involves several actors, namely the Government of Kenya, line ministries and parastatals; KenGen and its consultants; development financial institutions (including EIB, AFD, KfW, JICA, and the World Bank). Local communities and wildlife-conservation groups have limited if any say in what happens.

Pursuit of strategic interests inevitably leads to forms of appropriation and contestations. For example, the development financial institutions, themselves seemingly held within their own competition for relevance in local development, supply loans and technical assistance and may not necessarily have a direct "eye" on the ground. Consequently, they come into conflict with local affected groups because they are viewed as exacerbating social injustices by the very virtue of supporting these investments. State actors use such promises of "big and fast" to win political mileage and to strategically position themselves for their future political careers. Confused about the future, some local people "engage in relentless determination to negotiate conditions of turbulence to introduce order and predictability into their lives" (Mbembe and Nuttall 2004, 349). They do so through land subdivision and speculation in anticipation of the arrival of land-hungry investors, but certainly oblivious of the fact that they may "sell their wealth to buy poverty" as Marcel Rutten (1992) described in a separate case.

Secondly, the megaprojects already elicit fears and uncertainty over the outcomes of their (possible) implementation or completion. For the locally affected persons, uncertainty is "the lived experience of a pervasive sense of vulnerability, anxiety, hope, and possibility mediated through the material assemblages that underpin, saturate, and sustain everyday life" (Cooper and Pratten 2015, 1). Across all megaprojects, uncertainty is also reflected in the ecological, environmental, socio-economic and political facets. However, uncertainty, according to Cooper and Pratten (2015) may "serve productive purposes going beyond the often used negative and constraining sense that refers to a lack of absolute knowledge, the inability to predict the outcome of events or to establish facts about phenomena and connections with assurance". The authors argue that "uncertainty as a social resource can be used to negotiate insecurity, conduct and create relationships, and act as a source for imagining the future with the hopes and fears this entails". Simply, uncertainty can inform positive decision-making and action in the present that could address fears.

In the context of this discussion, however, the concept reveals structural weaknesses where, for instance, possible undesirable futures and externalities are easily decipherable but are rarely acted upon to manage or reduce possible severe outcomes. For example, the freshness of Lake Naivasha's water relies on dilution episodes from the river floods and sub-surface floodplain inflows. Thus, damming of the rivers (abstraction) has a high potential of increasing salinity and alkalinity levels as evaporation increases at the lake. Nevertheless, the push to supply water to the Naivasha SEZ and to geothermal plants outweighs concerns about the sustainability of the basin system.

Thirdly, the megaprojects have an inherent component of failure. Based on the conditions of their implementation, one may argue that failure is actually embedded in the design of such projects and therefore megaprojects may only serve anticipatory roles by channelling the thoughts of the masses in a given direction. None of the megaprojects discussed here have achieved full completion – most of them are not ready for operation until associated infrastructure (road network, energy supply, water supply, waste disposal and management, etc.) is completed.

The case of dam construction in Kenya offers a good example. Based on the Parliamentary Committee on Environment and Natural Resources tasked with evaluating the status of dams in Kenya in 2019, only four dams were found to be in existence (Thika dam, Chemsusu dam, Kiserian dam, and Sesimua dam) while six dams were in the process of construction at the time, and one had stalled (Itare dam). The slow pace of dam construction, according to the committee, is attributable to inadequate financial resources, inefficient and costly financing models, and unsettled resettlement action-plan issues. The aims to construct and rehabilitate 150 rural water schemes annually, drill an average of 70 boreholes annually in areas lacking adequate surface water and construct 160 small dams/pans in arid and semi-arid lands, as outlined in the Vision 2030 sanitation and water sector, is therefore a commitment that may be impossible to realise.

Finnemore and Jurkovich (2020) question why states and other political actors routinely commit to attaining goals they know they cannot reach, including the SDGs. Müller-Mahn, Mkutu, and Kioko (2021) give four explanations for the failure of megaprojects, which have the potential to affect some of the discussed megaprojects: (1) the complexity of large infrastructure projects, which make them difficult to manage and control; (2) "high modernism" associated with political regimes and the state – the logics of state-led development; (3) the vulnerability of megaprojects in the context of social and economic change; and (4) their potential to upset local conditions by interfering with land rights, ecology and social conditions, thereby creating conflicts.

Even where successful completion of megaprojects has been realised, the desired outcomes simply do not occur, or take longer to materialise. Several example of megaprojects in Kenya can be pointed out: the completed Isiolo airport has remained an "empty airport" with no foreseeable operations in the near future; despite the Chemsusu dam being completed in 2014, it took the state five years to connect it to the Last Mile energy supply, an indication of the postponement of usage of the facility; and despite completion of the Naivasha ICD, its potential has not been fully realised and, notwithstanding the capital investment, the state has turned to offering free storage for containers as a marketing strategy for the facility, which has witnessed an annoying slow start (Kitimo 2020b). Ironically, the "Naivasha ICD remains inaccessible by road" (Kariuki 2021). The SGR has also been marred with controversies despite the promises of a prosperous economy, including complains of "secrecy, poor planning, corruption, lack of commercial viability, [and] excessive indebtedness to China" (Shiundu 2019).

Similar cases of slow-paced development, disconnected implementation, corruption in allocation of tenders, conflicts between locals and developers, extension of completion deadlines and stalling of projects constitute an aspect of failure across most megaprojects. However, irrespective of whether a megaproject succeeds or fails, some of the social-economic and ecological changes at the local level may be irreversible. Based on these conditions of perceived and real social-ecological changes driven by the planned megaprojects in the already stressed ecosystem of Lake Naivasha, it is difficult to predict with any certainty the trajectory the basin will take in the future.

Acknowledgements

The fieldwork for this research was made possible through the generous support of the DFG under the Collaborative Research Centre *Future Rural Africa: Future-Making and Social-Ecological Transformation*. We are grateful for this support.

References

Anyanzwa, James. 2021. "Kenya's MPs Push for Suspension of Projects in Austerity Drive." *The East African*, April 6, 2021. https://www.theeastafrican.co.ke/tea/business/kenya-mps-push-for-suspension-of-projects-3350760.

Becht, Robert and David Harper. 2002. "Towards an Understanding of Human Impact Upon the Hydrology of Lake Naivasha, Kenya." *Hydrobiologia* 488: 1–11. https://doi.org/10.1023/A:1023318007715.

Beckert, Jean. 2016. *Imagined Futures: Fictional Expectations and Capitalist Dynamics*. Cambridge, MA: Harvard University Press.

Bierschenk, Thomas. 2008. "Development Projects as Arenas of Negotiation for Strategic Groups: A Case Study from Benin." *Sociologia Ruralis* 28 (2–3): 146–60. https://doi.org/10.1111/j.1467-9523.1988.tb01035.x.

Blickley, Jessica L. and Gail L. Patricelli. 2010. "Impacts of Anthropogenic Noise on Wildlife: Research Priorities for the Development of Standards and Mitigation." *Journal of International Wildlife Law & Policy* 13 (4): 274–92. https://doi.org/10.1080/13880292.2010.524564.

Boar, Rosalind, and David M. Harper. 2002. "Magnetic susceptibilities of Lake Sediment and Soils on the Shoreline of Lake Naivasha, Kenya." *Hydrobiologia* 488: 81–88. https://doi.org/10.1023/A:1023357910441.

Chome, Ngala. 2019. "Elite Feuds: Are the SGR Protesters in Mombasa Pawns in a Larger Battle for Control of Port Services?" *The Elephant*, November 9, 2019. https://www.theelephant.info/op-eds/2019/11/09/elite-feuds-are-the-sgr-protesters-in-mombasa-pawns-in-a-larger-battle-for-control-of-port-services/.

Cooper, Elizabeth and David Pratten. 2015. *Ethnographies of Uncertainty in Africa*. London: Palgrave Macmillan.

De Jong, Thomas. 2011. *Water Abstraction Survey in Lake Naivasha Basin, Kenya*. Internship Report, Wageningen University, February 2011.

Development Corridors Partnership. n.d. "What Is a Development Corridor?" Accessed August 16, 2023. https://developmentcorridors.org/development-corridors/.

Enns, Charis and Brock Bersaglio. 2019. "On the Coloniality of 'New' Mega-Infrastructure Projects in East Africa." *Antipode* 52 (1): 101–23. https://doi.org/10.1111/anti.12582.

Farole, Thomas and Lotta Moberg. 2017. "Special Economic Zones in Africa: Political Economy Challenges and Solutions." In *The Practice of Industrial Policy: Government—Business Coordination in Africa and East Asia*, edited by John Page and Finn Tarp, 234–54. Oxford: Oxford University Press.

Ferguson, James. 1999. *Expectations of Modernity: Myths and Meanings of Urban Life on the Zambian Copperbelt*. Oakland: University of California Press.

Finnemore, Martha and Michelle Jurkovich. 2020. "The Politics of Aspiration." *International Studies Quarterly* 64 (4): 759–69. https://doi.org/10.1093/isq/sqaa052.

Gitonga, Anthony. 2021. "Dutch Govt to Fund Impact Study on Sh35b Malewa Dam." *The Standard*, April 17, 2021. https://www.standardmedia.co.ke/health/rift-valley/article/2001409916/dutch-govt-to-fund-impact-study-on-sh35b-malewa-dam.

Harper, David, Edward H.J. Morrison, Michael M. Macharia, Kenneth M. Mavuti, and Caroline Upton. 2011. "Lake Naivasha, Kenya: Ecology, Society and Future." *Freshwater Reviews* 4 (2): 89–114. https://doi.org/10.1608/FRJ-4.2.149.

Hickley, Phil, Muchai Muchiri, Rosalind Boar, Robert Britton, Chris Adams, Nicholas Gichuru, and David Harper. 2004. "Habitat Degradation and Subsequent Fishery

Collapse in Lakes Naivasha and Baringo, Kenya." *Ecohydrology & Hydrobiology* 4 (4): 503–17.

JICA. 2013. "The Project on the Development of the National Water Master Plan 2030." Final Report, vol. 1. Republic of Kenya. Accessed September 14, 2023. https://open jicareport.jica.go.jp/pdf/12146353_01.pdf.

JICA. 1990. "Study on Construction of Dam in Malewa River System for Greater Nakuru Water Supply Project." Final report, vol. 3. Republic of Kenya. https://openjicareport .jica.go.jp/pdf/10879187_02.pdf.

Kariuki, James. 2021. "New Proposals to Unlock Potential of Naivasha Depot." *Business Daily*, January 13, 2021. https://www.businessdailyafrica.com/bd/corporate/shipping -logistics/new-proposals-naivasha-depot-3254952.

Kenya Investment Authority. n.d. "Special Economic Zones." Accessed August 16, 2023. https://www.invest.go.ke/special-economic-zones/.

Kenya Vision 2030. n.d.a. "About the Vision." Accessed August 16, 2023. https:// vision2030.go.ke/about-vision-2030/.

Kenya Vision 2030. n.d.b. "Rural Water Supply Sub Programme." Accessed August 16, 2023. https://vision2030.go.ke/project/rural-water-supply-sub-programme/.

Kioko, Eric. 2016. "Turning Conflict into Coexistence: Cross-Cutting Ties and Institutions in the Agro-Pastoral Borderlands of Lake Naivasha Basin, Kenya". PhD diss., University of Cologne. Accessed September 14, 2023. http://kups.ub.uni-koeln.de /id/eprint/7064.

Kitimo, Anthony. 2020a. "Uganda Backs Out of Compulsory Use of Container Port in Naivasha." *The East African*, May 30, 2020. https://www.theeastafrican.co.ke/tea /business/uganda-backs-out-of-compulsory-use-of-container-port-in-naivasha -1442320.

Kitimo, Anthony. 2020b. "KPA Entices Importers with Extended Free Storage at Naivasha ICD." *The East African*, May 24, 2020. https://www.theeastafrican.co.ke/tea/business /kpa-entices-importers-with-extended-free-storage-at-naivasha-icd-1441924.

Kenya Railways Authority. 2020. "Environmental and Social Impact Assessement for Naivasha Inland Container Depot (ICD)." ESIA Study Report. Accessed September 14, 2023. https://www.nema.go.ke/images/Docs/EIA_1710-1719/EIA_1715%20Naivasha %20Inland%20Container%20Depot%20Report_1769-min.pdf.

Kubo, Benjamin M. 2003. "Environmental Management at Olkaria Geothermal Project, Kenya." Paper presented at the International Geothermal Conference, Reykjavík, September 2003. Accessed September 14, 2023. https://orkustofnun.is/gogn/unu-gtp -25-ann/UNU-GTP-IGC-2003-13.pdf.

Kuiper, Gerda. 2019. *Agro-industrial Labour in Kenya: Cut Flower Farms and Migrant Workers' Settlements*. Cham: Palgrave Macmillan.

Larkin, Brian. 2013. "The Politics and Poetics of Infrastructure." *Annual Review of Anthropology* 42: 327–43. https://doi.org/10.1146/annurev-anthro-092412-155522.

Mariita, Nicolas. 2007. "Exploration History of Olkaria Geothermal Field by Use of Geophysics." Paper presented at Short Course IV on Exploration for Geothermal Resources, Lake Naivasha, Kenya, November 2007. Accessed September 14, 2023. https://rafhladan.is/bitstream/handle/10802/8152/UNU-GTP-SC-05-20.pdf?sequence=1.

Mariita, Nicolas. 2002. "The Impact of Large-Scale Renewable Energy Development on the Poor: Environmental and Socio-Economic Impact of a Geothermal Power Plant on a Poor Rural Community in Kenya." *Energy Policy* 30 (11–12): 1119–28. https://doi.org/10.1016/S0301-4215(02)00063-0.

Matara, Eric. 2020. "Building of 2,400 Naivasha Affordable Houses Starts February." *Business Daily*, January 2, 2020. https://www.businessdailyafrica.com/bd/economy/building-of-2-400-naivasha-affordable-houses-starts-february-2275036.

Mbembe, Achille and Sarah Nuttall. 2004. "Writing the World from an African Metropolis." *Public Culture* 16 (3): 347–72. https://doi.org/10.1215/08992363-16-3-347.

Müller-Mahn, Detlef. 2019, "Envisioning African Futures: Development Corridors as Dreamscapes of Modernity." *Geoforum* 115:156–59. https://doi.org/10.1016/j.geoforum.2019.05.027.

Müller-Mahn, Detlef, Kennedy Mkutu, and Eric Kioko. 2021. "Megaprojects—Mega Failures? The Politics of Aspiration and the Transformation of Rural Kenya." *The European Journal of Development Research* 33: 1069–90. https://doi.org/10.1057/s41287-021-00397-x.

Murage, George. 2019. "Mad Scramble for Land as Construction Starts on Naivasha Industrial Park Inland Depot". *The Star*, November 15, 2019. https://www.the-star.co.ke/counties/rift-valley/2019-11-15-mad-scramble-for-land-as-construction-starts-on-naivasha-industrial-park-inland-depot/.

Murage, George. 2020. "World Bank Cautioned against Funding Malewa Dam Project: Flower Farmers and Environmentalists Say Project Will Kill Lake Naivasha." *The Star*, August 6, 2020. https://www.the-star.co.ke/counties/rift-valley/2020-08-06-world-bank-cautioned-against-funding-malewa-dam-project/.

Mwangi, Esther and Hannah Wangui. 2019. "Government to Provide Water at the Special Economic Zone." *KenyaNews*. May 2, 2019. https://www.kenyanews.go.ke/government-to-provide-water-at-the-special-economic-zone/.

Mwita, Martin. 2020. "Lack of Infrastructure Stalls Naivasha Dry Port." *The Star*, February 17, 2020. https://www.the-star.co.ke/business/kenya/2020-02-17-lack-of-infrastructure-stalls-naivasha-dry-port/.

Ngotho, Agatha and George Murage. 2020. "Construction of Sh35bn Malewa Dam to Proceed – CAS." *The Star*, September 21, 2020. https://www.the-star.co.ke/counties/rift-valley/2020-09-21-construction-of-sh35bn-malewa-dam-to-proceed-cas/.

Nyanchaga, Ezekiel N. 2016. *History of Water Supply and Governance in Kenya (1895–2005): Lessons and Futures*. Tampere: Tampere University Press.

Nweke-Eze, Chigozie, and Eric M. Kioko. 2021. "'But We Cannot Do It All': Investors' Sustainability Tensions and Strategic Selectivity in the Development of Geothermal Energy in Kenya." In *Sustainable Development in Africa*. World Sustainability Series, edited by Walter Leal Filho, Rudi Pretorius, and Luiza Olim de Sousa, 385–404. Cham: Springer. https://doi.org/10.1007/978-3-030-74693-3_21.

Odongo, Vincent Omondi, Christiaan van der Tol, Pieter R. van Oel, Frank M. Meins, Robert Becht, Japheth Onyando, and Zhongbo Su. 2015. "Characterisation of Hydroclimatological Trends and Variability in The Lake Naivasha Basin, Kenya." *Hydrological Processes* 29 (15): 3276–93. https://doi.org/10.1002/hyp.10443.

Ogutu, Joseph O., Bernard Kuloba, Hans-Peter Piepho, and Erustus Kanga. 2017. "Wildlife Population Dynamics in Human-Dominated Landscapes under Community-Based Conservation: The Example of Nakuru Wildlife Conservancy, Kenya." *PLoS ONE* 12 (1): 1–30. https://doi.org/10.1371/journal.pone.0169730.

Republic of Kenya. 2013. *Nakuru County: First County Integrated Development Plan (2013–2017)*. Kenya Vision 2030. Accessed September 14, 2023. https://www.devolution.go.ke/wp-content/uploads/2020/02/Nakuru-CIDP-2013-2017.pdf.

Rutten, Marcel. 1992. "Selling Wealth to Buy Poverty: The Process of the Individualization of Landownership Among the Maasai Pastoralists of Kajiado District, Kenya, 1890–1990." PhD diss., African Studies Centre, Leiden.

Savo Project Developers. n.d. "Oserian Two Lakes Industrial Park." Accessed August 3, 2021. https://www.savo.earth/oserian-two-lakes-industrial-park.

SEZA. 2020. "The Proposed Naivasha Special Economic Zone Master Plan on LR no. 8396/56 in Nakuru County Kenya." Nairobi, 2020. Accessed August 18, 2021. https://www.nema.go.ke/images/Docs/SEA%20Reports/SEA%20for%20Naivasha%20Special%20Economic%20zone.pdf.

Shiundu, Alphonce. 2019. "Railway to Nowhere." *Development and Cooperation (D+C)*, December 22, 2021. https://www.dandc.eu/en/article/why-kenyas-ambitious-cross-country-railway-controversial.

Thorn, J. P. R., L. Alpha, and B. Mwangi. 2019. "Report of Participatory Scenario Planning Workshop of Impacts of the Standard Gauge Railway on Land Use Change in Naivasha, Kenya." Nairobi: Development Corridors Partnership.

Wasserman, Gary. 1976. *Politics of Decolonization: Kenya Europeans and the Land Issue 1960–1965*. Cambridge: Cambridge University Press.

World Bank. 2017. "Special Economic Zones: An Operational Review of Their Impacts." World Bank, Washington, DC. Accessed September 14, 2023. http://hdl.handle.net/10986/29054.

World Bank. n.d. "Kenya Water Security and Climate Resilience Project." Accessed April 30, 2021. https://projects.worldbank.org/en/projects-operations/project-detail/P117635.

Index

www.ingramcontent.com/pod-product-compliance
Lightning Source LLC
Chambersburg PA
CBHW071413230926
51846CB00001B/70
9789004695412